Compact Stars in the QCD Phase Diagram V

AA004769

Journal of Physics: Conference Series Volume 861

L'Aquila, Italy
23-27 May 2016

ISBN: 978-1-5108-4269-4
ISSN: 1742-6588

Printed from e-media with permission by:

Curran Associates, Inc.
57 Morehouse Lane
Red Hook, NY 12571

Some format issues inherent in the e-media version may also appear in this print version.

Printed by Curran Associates, Inc. (2017)

For permission requests, please contact the Institute of Physics
at the address below.

Institute of Physics
Dirac House, Temple Back
Bristol BS1 6BE UK

Phone: 44 1 17 929 7481
Fax: 44 1 17 920 0979

techtracking@iop.org

Additional copies of this publication are available from:

Curran Associates, Inc.
57 Morehouse Lane
Red Hook, NY 12571 USA
Phone: 845-758-0400
Fax: 845-758-2633
Email: curran@proceedings.com
Web: www.proceedings.com

TABLE OF CONTENTS

IOP Conf. Series: Journal of Physics: Conf. Series **861** (2017) 011001 doi:10.1088/1742-6596/861/1/011001

Preface

The 5th international workshop Compact Stars in the QCD phase diagram (CSQCD V) was held, during May 23rd – 27th 2016, at the Gran Sasso Science Institute (GSSI) in L'Aquila and at the Gran Sasso National Laboratories (LNGS) of the Italian National Institute for Nuclear Physics (INFN), Assergi (L'Aquila).

This series of meetings started in 2001 in Copenhagen (Denmark), and then continued in 2009 in Beijing (China), 2012 in Guaruja (Brazil) and 2014 in Prerow (Germany).

The CSQCD workshops aim to bring together physicists working on strongly interacting dense matter with astrophysicists working on Compact Stars and related astrophysical phenomena as supernovae, gamma ray bursts and pulsars. A profound theoretical description of these astrophysical phenomena necessarily involves the equation of state and transport properties of QCD matter that are derived from nonperturbative quantum field theories, many-body physics of strongly interacting matter and phenomenological investigations.

The CSQCD V workshop has covered recent developments in the following main topics:

- Equation of state and QCD phase transitions
- QCD in astrophysics of compact stellar objects, supernovae and compact stars mergers
- Strangeness in Compact Stars
- Strange Stars
- Nonequilibrium and transport phenomena in dense matter

In addition, given that the LNGS is one of the world leading laboratories on neutrino physics, special attention has been given to neutrino emission from supernovae and their possible detection in underground detectors, as well as related topics in neutrino physics. A series of seminars by local researchers about ongoing underground experiments at the Gran Sasso National Laboratories have been presented on these topics.

The present volume of the *Journal of Physics: Conference Series* represents contributions from the participants of the CSQCD V workshop and of the Working Group 2 Meeting of COST Action MP1304, NewCompStar.

We deeply thank the authors for their high-level contributions and we are convinced that this volume of *Journal of Physics: Conference Series* will be of great interest for our community. Each submitted contribution has been peer reviewed and the Editors are very grateful to the referees for their careful support.

The Editors of the CSQCD V Proceedings

Massimo Mannarelli (LNGS, INFN)
Ignazio Bombaci (Università di Pisa)

L'Aquila, 16th of May 2017

IOP Conf. Series: Journal of Physics: Conf. Series **861** (2017) 011001 doi:10.1088/1742-6596/861/1/011001

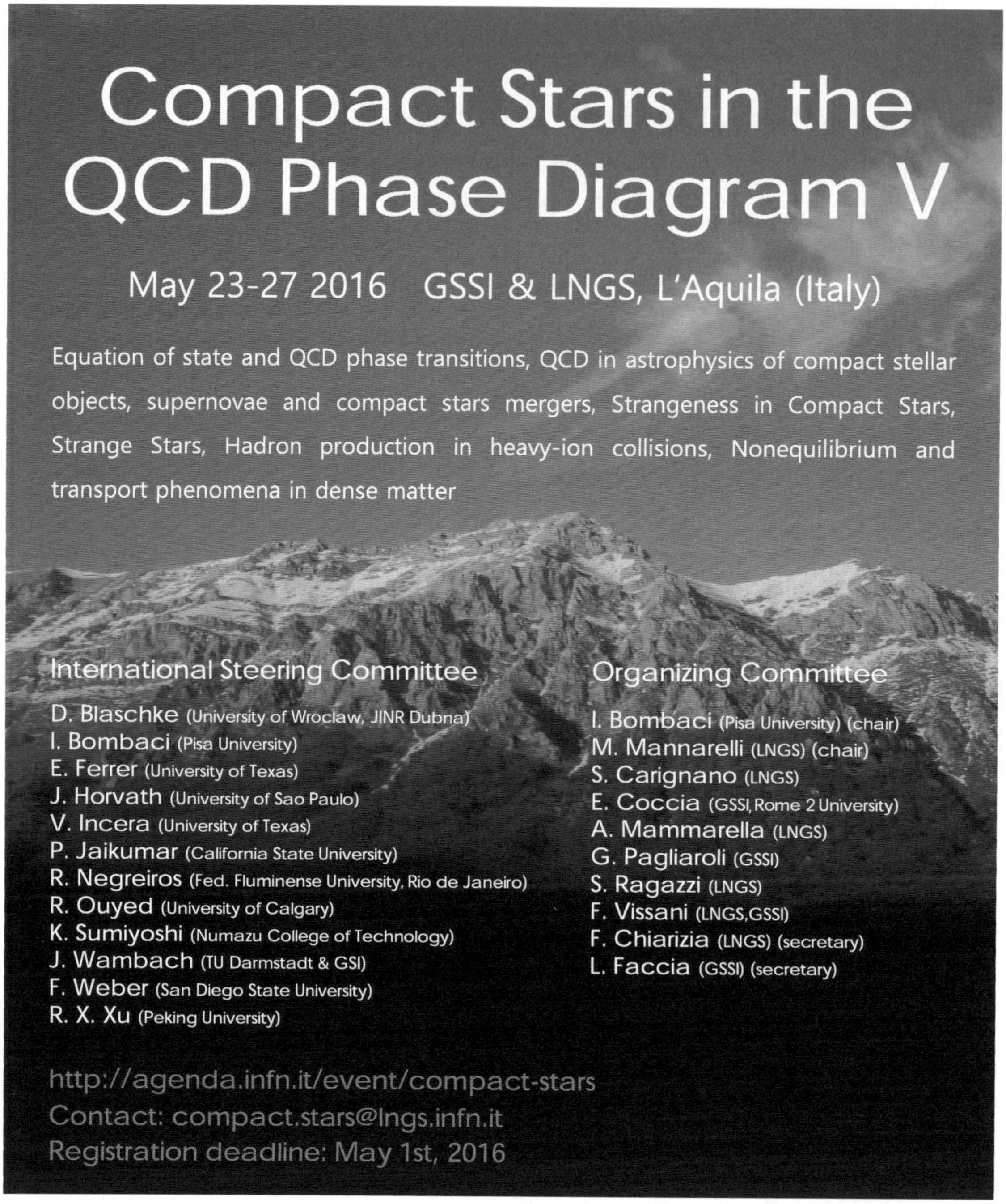

GRAN SASSO SCIENCE INSTITUTE
CENTER FOR ADVANCED STUDIES
Istituto Nazionale di Fisica Nucleare

INFN
Laboratori Nazionali del Gran Sasso

comp star

IOP Conf. Series: Journal of Physics: Conf. Series **861** (2017) 011002 doi:10.1088/1742-6596/861/1/011002

Peer review statement

All papers published in this volume of *Journal of Physics: Conference Series* have been peer reviewed through processes administered by the proceedings Editors. Reviews were conducted by expert referees to the professional and scientific standards expected of a proceedings journal published by IOP Publishing.

IOP Conf. Series: Journal of Physics: Conf. Series **861** (2017) 012001 doi:10.1088/1742-6596/861/1/012001

Quark degrees of freedom in nuclear matter

Marcello Baldo

Istituto Nazionale di Fisica Nucleare, Sez. Catania, via S. Sofia 64, 95123 Catania (I)

E-mail: baldo@ct.infn.it

Abstract. The microscopic theory of Nuclear Matter has been approached along the years by different many-body methods and different nucleon-nucleon (NN) interactions. The realistic NN interactions can be classified mainly into two categories. One relies on the meson-nucleon coupling scheme, within relativistic or non-relativistic framework. The so-called chiral interactions are based on the assumption that, once the pion exchange contribution has been explicitly isolated, it is possible to expand the NN interaction in a series of point interaction terms which respect the underlying QCD chiral symmetry. In both schemes three-body (TBF) or higher forces can be constructed, which can have different degrees of phenomenological character. The TBF are essential to get the Nuclear Matter saturation point close to the phenomenological one. However the QCD quark degrees of freedom are not explicitly introduced. It will be shown that a realistic NN interaction that is constructed from the quark degrees of freedom can produce the correct saturation point without the need of TBF. The corresponding Equation of State is compatible with all the phenomenological constraints, including the Neutron Star maximum mass limit. Taking this result literally, one can say that quarks have been revealed in Nuclear Matter. Another conclusion is that the effect of TBF is model dependent.

1. Introduction

The Equation of State of Nuclear Matter is of great relevance for numerous area of physics and the corresponding phenomena [1]. It is essential for the physics of heavy ion (HI) reactions at intermediate energies, supernovae (SN) and Neutron Stars (NS) and gravitational wave (GW) emission from binary mergers or NS oscillations. The physical conditions of nuclear matter is quite different in each one of these systems. In HI matter can be compressed at few times saturation density, it is close to be symmetric and the temperature is relatively low with respect to the Fermi energy. In SN matter is strongly asymmetric, it is present in a very wide density range, from vanishing small values to about twice saturation density, and the temperature can be as high as several tens of MeV. In NS the matter is mainly cold, very asymmetric and compressed up to several times saturation density. Finally in GW emission the asymmetry is large, the temperature is close to the one for SN and the density is always close to the maximum one obtainable without collapse to a black hole.

These phenomena can be analyzed in a more or less model dependent methods, but in any case they provide severe constraints on the EoS. Within a phenomenological approach, e.g. on the basis of effective mean fields models, these constraints do not fix tightly the EoS, which therefore cannot be accurately known. Alternatively one can try to develop a microscopic many-body theory of nuclear matter and check to what extent they are compatible with the phenomenological constraints. The comparison among different theoretical methods can provide an estimate of the theoretical uncertainty on the EoS and the comparison with

IOP Conf. Series: Journal of Physics: Conf. Series **861** (2017) 012001 doi:10.1088/1742-6596/861/1/012001

the phenomenological constraints can eventually rule out some of the microscopic EoS. We will show that most of the microscopically derived EoS are compatible with the phenomenological constraints developed up to now, which implies that the microscopic structure of nuclear matter, as predicted by the different theoretical models, is not well constrained. In particular the presence and strength of three-body force (TBF) turns out to be dependent on the nucleon-nucleon (NN) interaction models. Furthermore if the quark degrees of freedom are explicitly introduced in the NN interaction model, the role of TBF seems to be strongly reduced. This shows the relevance of introducing explicitly the effective quark degrees of freedom.

An analysis of the future experiments and of the possible observations shows how they are expected to select severely among the different EoS.

2. Overview of the microscopic methods

A microscopic many-body method is characterized mainly by two basic elements, the realistic bare interaction among nucleons and the many-body scheme followed in the calculation of the EoS. There are three basic classes of bare nucleonic interactions

- Interactions mediated by meson exchanges
- Chiral expansion approach
- Models that include explicitly the quark-gluon degrees of freedom

The many-body methods can be enumerated as follows

- The Bethe-Brueckner-Goldstone (BBG) diagrammatic method and the corresponding hole-line expansion
- The Coupled Cluster (CC) expansion
- The relativistic Dirac-Brueckner (DBHF) approach
- The self-consistent Green' s function (SCGF)
- The variational method
- The renormalization group (RG) method
- Different methods based on Monte-Carlo (MC) techniques

We will proceed along the list of the nucleonic interactions, and for each class the results with different many-body methods that have been employed will be discussed. The simplest constraint that has to be considered is the reproduction of the phenomenological saturation point, and we will see to what extent this condition is fulfilled. Later other constraints will be discussed.

The Argonne v18 [2] potential is a meson exchange potential that was widely used in the calculation of the EoS with different many-body methods. The meson degrees of freedom in this case are not explicitly introduced, but the form factor of the series of operators which define the potential are mainly suggested by the static limit of meson exchange processes. The parametrization of these form factors results in an excellent fit of the available NN phase shifts and deuteron data. The result of the BBG calculation of the EoS for symmetric matter using this interaction is reported in Fig. 1, taken from ref. [3]. Here the full lines correspond to the calculations including only two-body correlations, i.e. at the Brueckner-Hartree-Fock level. The two full lines correspond to two different choices of the effective single particle potential, the so-called gap choice and continuous choice. In the former the single particle potential is set to zero above the Fermi energy, while in the latter it is calculated for all relevant momenta. One can see that there is some discrepancy. However when the three-body correlations are included, corresponding to the three-body cluster contribution, the final EoS look quite close to each other (square and circle symbols). This indicates a convergence of the BBG expansion, since if all the diagrammatic contributions, corresponding to the hole-line expansion, would be added, the result must be independent on the single particle potential. The latter is just an

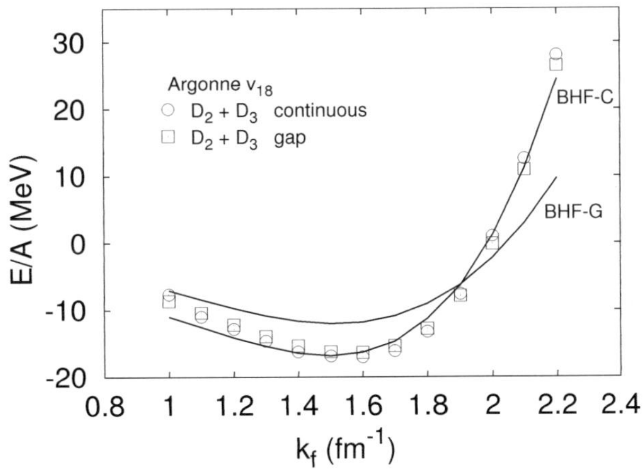

Figure 1. EoS for the NN interaction Av18 within the BBG expansion. The two full lines correspond to the calculations at the two hole-line level of approximation (Brueckner-Hartree-Fock) with the gap choice (BHF-G) and with continuous choice (BHF-C), respectively, for the auxiliary single particle potential. The symbols correspond to the EoS when the three hole-line contributions are included. Figure taken from ref. [3]

auxiliary quantity, useful to speed up the rate of convergence of the expansion. However one can see that the saturation point is not reproduced, since it can be nominally taken at energy per particle E/A = -16 MeV and k_F = 1.33 fm^{-1}, with some uncertainty, much smaller than the discrepancy. This result is systematic for all the meson exchange potential, independently of the many-body method employed. This calls for the introduction of three-body forces (TBF). This can be performed at phenomenological level or by evaluating again the processes involving three nucleons in an irreducible way [4]. In Fig. 2, taken from ref. [5], are reported the EoS calculated with phenomenological TBF within the BBG scheme (squares) and the variational method (circles). The TBF were adjusted to reproduce the saturation point, by varying mainly one parameter. The agreement is somehow misleading since the EoS from variational method has been modified with an additional correction by hand around saturation, which turns out to be of about 3 MeV at saturation density. The agreement at higher density is a genuine one. The effect of TBF is quite moderate around saturation but they are essential to get the correct saturation point.

Two-body forces based on meson exchange model can be used in a relativistic framework. One can try to formulate the BHF equations, in particular the one for the G-matrix, within a relativistic formalism. This is a difficult task, and in general one keeps the interaction as instantaneous (static limit) and a reduction to a three-dimensional formulation from a four-dimensional one. This is the Dirac-Brueckner scheme, where the main relativistic effect is due to the use of the spinor formalism. It turns out that the relativistic formulation produces a saturation effect, and one can get the correct saturation point by a suitable realistic NN interaction [6]. It has been shown [7] that the use of spinors has the effect of introducing a particular TBF, if they are projected into the bare particle states. In fact the presence of the single particle potential in the the so-called small components of the spinors produces a "rotation" of the spinors with respect to the free ones, which results in the appearance of anti-particle states. At non relativistic level this is equivalent to the effect of a TBF where a particle-antiparticle pair is created in the intermediate state [7], the so-called Z-diagram. However at the non-relativistic level other TBF can be present, and it is unclear if they would affect these results at relativistic level.

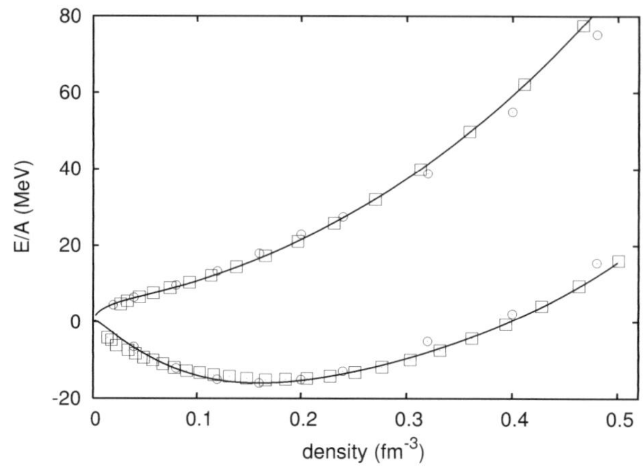

Figure 2. EoS for the NN interaction Av18 plus phenomenological three-body forces, for pure neutron matter (upper part) and symmetric matter (lower part). The squares correspond to the BHF calculation [5], the circles to the variational one of ref. [8]. The full lines correspond to a polynomial fit to the BHF calculation. Figure taken from ref. [5].

An approach more closely related to the underlying QCD structure of nucleons is the chiral scheme, originally devised by Weinberg [9]. In this case one includes in the interaction a series of operators which reflect the partially broken chiral symmetry of QCD. To this purpose a non-linear representation of the $SU(2) \times SU(2)$ chiral group is introduced and the form of the operators invariant with respect to the representation is then fixed. The strength parameters associated to each operator are then determined by fitting the NN phase shifts, the properties of deuteron and eventually of few-body nuclear systems. The method is then implemented in the framework of the Effective Field Theory (EFT), i.e. by ordering the terms according their dependence on the physical parameter Q/M, where M is the nucleon mass and Q a generic momentum that appears in the Feynman diagram for the considered process. This parameter is assumed to be small. Each term is dependent on a given power of this parameter and therefore this fixes its relevance. In this way a hierarchy of the different terms of the forces is established. In particular one finds that the three-body forces so introduced are of higher order than the simplest two-body forces. In practice this procedure amounts to expand the nuclear force in a series of point interactions with possible derivatives, except for the pion degrees of freedom that are treated separately and with the proper range. In comparison with the meson exchange model this is equivalent to treat the pion as an explicit degree of freedom, while the exchange of heavier meson is expanded in the transfer momenta. This is illustrated in Figs. (3,4), where some two-body and three-body forces respectively are depicted. The dashed lines indicate pion propagation, a full dot a derivative coupling. The assumption of a small Q/M parameter in principle restricts the applications of these forces to not too large momenta and therefore to a not too large density of nuclear matter. It turns out that the safe maximum density is around the saturation value. This power counting has been refined along the years and many applications can be found in the literature. We restrict the discussion on few calculations that illustrate the different approaches to the nuclear matter EoS.

One of the main characteristics common to these forces is the critical relevance of the TBF in the saturation mechanism. In fact, as illustrated in Fig. 5, taken from ref. [10], without TBF the EoS does not display an apparent saturation, and anyhow close to saturation density the TBF

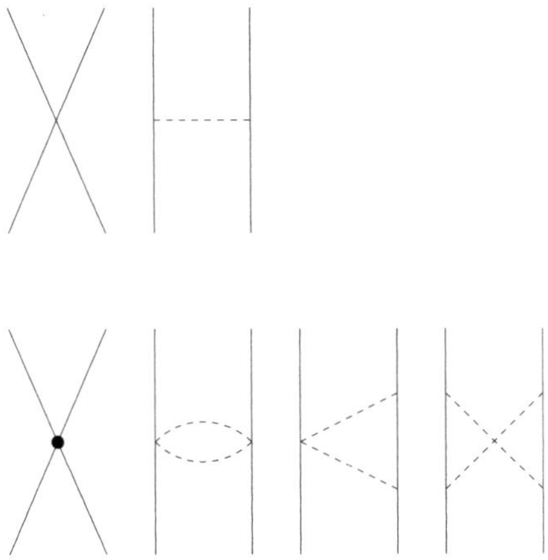

Figure 3. Chiral two-body forces. The full lines indicate nucleon propagation, the dashed lines indicate pion propagation, which results in a finite range interaction. A simple crossing between two nucleon lines indicates a point interaction, and a full dot correspond to a derivative coupling.

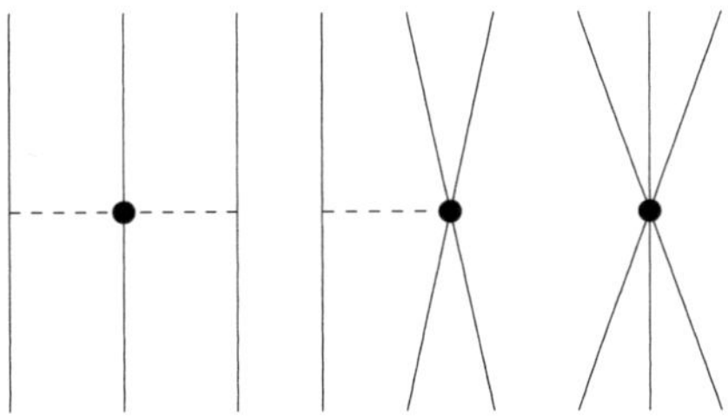

Figure 4. Chiral three-body forces. The meaning of the symbols is the same as in Fig. 3

contribution is quite large, of several MeV, in contrast to the case of meson exchange interactions, where the TBF contribution around saturation is of the order of 1 MeV. These calculations are perturbative in character, as indicated in the labels, but this feature holds true also in more refined calculations. Notice the relevance of the momentum cutoff, that is introduced in order to control the point interaction forces, that otherwise would produce a divergent contribution. In general the chiral two-body forces are evolved according the Renormalization Group (RG) method before they are employed in the many-body calculations, like in ref. [10]. The same procedure has been followed in the BHF calculations of ref. [11], where a similar relevance of the TBF was found. The most sophisticated many-body calculations with chiral forces is probably the ones of ref. [12], where the CC method was employed up to a selected set of three-body clusters. In this paper it was also found that it is difficult with the same chiral forces to fit both the binding of few nucleon systems (H, ^{3}He) and the saturation point. This feature is common to the meson exchange forces discussed above, for which the same difficulty was found. A similar conclusion was found in the BHF calculations of ref. [13], where it is suggested to fit simultaneously the few-body binding and the saturation point.

Another approach inspired by the QCD theory of strong interaction has been developed by a few groups [14, 15, 16, 17, 18, 19, 20]. In this approach the quark degree of freedom is explicitly introduced and the nucleon-nucleon interaction is constructed from gluon and meson exchange between quarks, the latter being confined inside the nucleons. One of these quark models (QMS) of the nucleon-nucleon interaction, named fss2 [20, 21], is able to reproduce closely the experimental phase shifts and the few-body binding energies [21, 22, 23, 24]. In particular triton binding energy is reproduced within 300 KeV. More recently it has been shown [25] that the fss2 interaction is able to reproduce also the correct nuclear matter saturation point without any additional parameter or need to introduce TBF. This is illustrated in Fig. 6, taken from ref. [25], where the EoS for symmetric matter is reported. The scheme of this figure follows

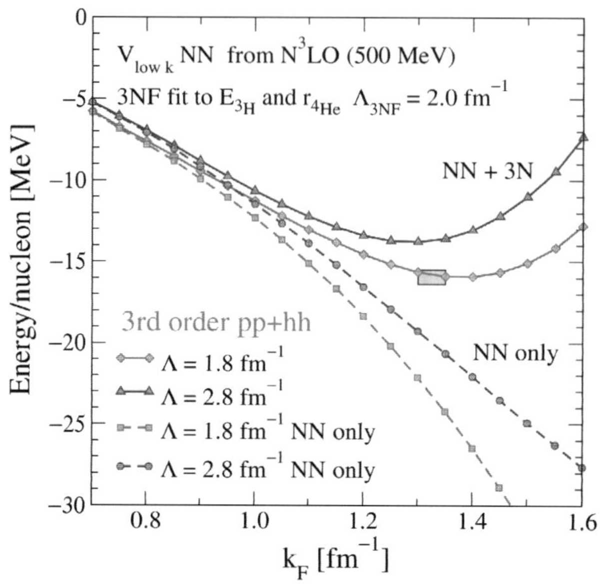

Figure 5. EoS of symmetric nuclear matter including third order perturbative corrections based on chiral interaction at N^3LO level. The cutoff for three-body forces is fixed at $\Lambda_{3N} = 2.0$ fm^{-1}, while the cut-off Λ of the (evolved) chiral two-body interaction is set equal to the two indicated values. Figure taken from ref. [10].

the one of Fig. 1. The open symbols correspond to the EoS calculated at the BHF level of approximation with the gap and the continuous choices, while the full symbols correspond to the EoS calculated by including the three hole-line contribution. One can sees that also in this case the final EoS is insensitive to the choice of the single particle potential. However in this case the three hole line contribution is minimal for the gap choice.

The main result of this calculation is that the saturation point is reproduced without the introduction of three-body forces.

Notice that this is the only two-body interaction that is able to reproduce with a fair accuracy both the binding of few nucleon systems and the saturation point of nuclear matter, without the need of three-body forces. The remaining small discrepancy for the few nucleon systems can be due to a (small) contribution of TBF or higher and relativistic effects.

The conclusion one can draw from this rapid review of results with different forces is that the relevance of TBF is model dependent and that the explicit introduction of the quark degrees of freedom reduce strongly the relevance of the TBF and allows to connect few-body systems to nuclear matter.

3. Phenomenological tests of the different EoS.

Several phenomenological constraints on the nuclear matter EoS have been devised along the years. They are discussed in few review papers [26, 1, 27]. Here we present a discussion on the comparison of different EoS with these phenomenological constraints with the purpose of possibly selecting the EoS which better respect the boundaries. We limit the analysis to the EoS that were derived from a microscopic many-body theory.

The first simple constraint is the saturation point, which is known to a certain accuracy. If we assume that the saturation density ρ_0 must be in the range $0.15 < \rho_0 < 0.17$ fm^{-3}, and

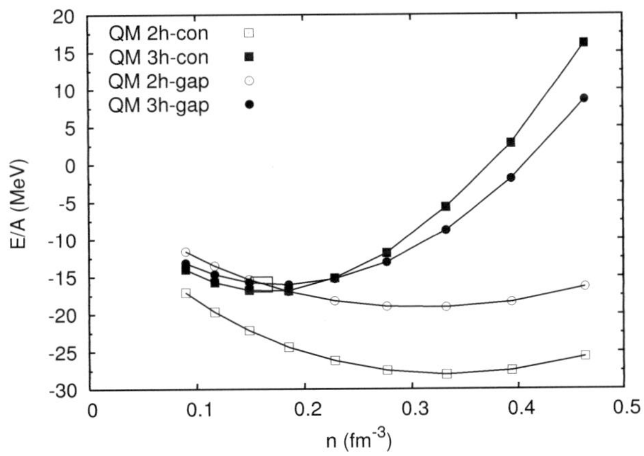

Figure 6. EoS of symmetric nuclear matter from the quark model interaction fss2. The open and full symbols correspond to the two hole line and three hole-line calculations, respectively. The circles and the squares indicate the calculation with the gap and continuous choices for the single particle potential, respectively. Figure taken from ref. [25]

the corresponding energy per particle E/A in the range $-17 < E/A < -15$MeV, then already several EoS must be ruled out. We are therefore going to mention only a set of EoS which satisfy this saturation point constraint.

A comparison among different EoS, obtained with different interactions and many-body methods, along the lines of Sec. 2, is reported in Fig. 7, taken from ref. [28], for both symmetric matter (SM) and pure neutron matter (PN).

The considered set of EoS includes variational calculations (APR) [8], BHF calculations with TBF, both phenomenological [5] and microscopically derived [4], and relativistic Dirac-Brueckner calculations [29]. The EoS from the QM interaction (FSS2), described in Sec. 2, appears to be among the softest ones in the set.

Once the energy per particle is known as a function of density, by differentiation one gets the pressure as a function of density. This allows to confront the EoS with the results of the analysis on the flow data in heavy ion reactions of ref. [30] and the ones on kaon production of ref. [31]. Such a comparison is shown in Fig. 8, taken from ref. [28], for the considered set of EoS. One can see that most of the EoS fall inside the constraint region, and the only marginal deviations occur at the highest density where the analysis is less reliable due to the onset of particle production. Again the EoS from the QM interaction looks a soft one, with respect to the others, at least at the higher density. In any case there is a fair agreement among the considered EoS and with the phenomenological boundary, a result which cannot be considered trivial or obvious. It is also clear that a refinement of the boundary could in principle put a more stringent constraint on the microscopic EoS, or even to rule out some of the EoS.

One of the main physical quantity that characterizes a given EoS is the symmetry energy as a function of the density. In ref. [32] the symmetry energy up to saturation density has been constrained from a study of the Isotopic Analog States (IAS) and the neutron skin size in neutron rich nuclei. Both of them are sensitive to the density dependence of the symmetry energy. These constraints are reported in Fig. 9. The largest box comes from the IAS analysis, while the smaller one is obtained by adding the analysis of the skin size obtained from pionic atoms and polarized hadron scattering. As a comparison the symmetry energy extracted from the EoS of the previous set is also reported. The symmetry energy from the effective interaction method

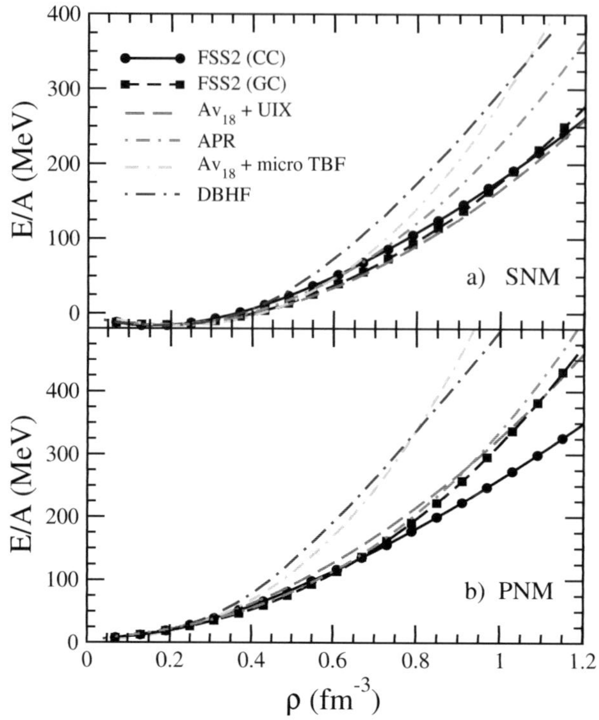

Figure 7. Comparison among the EoS obtained by different forces and different many-body methods. For the meaning of the labels see the text. Figure taken from ref. [28].

[33] is also reported. One can notice also in this case a fair agreement with the phenomenological constraints. Despite the shape of the symmetry energy as a function of density can be different for different EoS, the microscopic calculations fall approximately inside the allowed region. This means that the constraints cannot be used to extract a simple functional parametrization of the symmetry energy density dependence. In particular, if one assume a power law dependence, i.e. $E_{sym} \approx \rho^{\alpha}$, the exponential index α cannot be constrained within a meaningful accuracy. It means also that the constraints are not appreciably selective among the microscopic EoS. In principle the BHF calculation based on Av_{18} interaction plus phenomenological TBF, the BBG calculation based on the QM interaction and the EoS from chiral forces look well inside both boundary regions of the constraints.

The extension of these calculations beyond the saturation density reveals that the agreements among the different EoS is lost. This is illustrated in Fig. 10, where one can see that the different predictions are divergent. A phenomenological constraints in this density range is likely to be strongly selective on the EoS. Unfortunately there is no stringent enough constraints at above saturation density, but it can be expected that in the relatively near future such constraints will be possibly formulated, e.g. from the planned CBM experiment [34] at FAIR. The EoS calculated with the QM interaction displays a soft dependence of the symmetry energy at high density.

As a further phenomenological constraint we consider the largest Neutron Star (NS) mass observed up to now. This is a particularly relevant constraint since it is based on direct observations, without the use of numerical simulations. Although the analysis of the data needs models for the NS atmosphere, they can be considered reliable enough to draw firm conclusions.

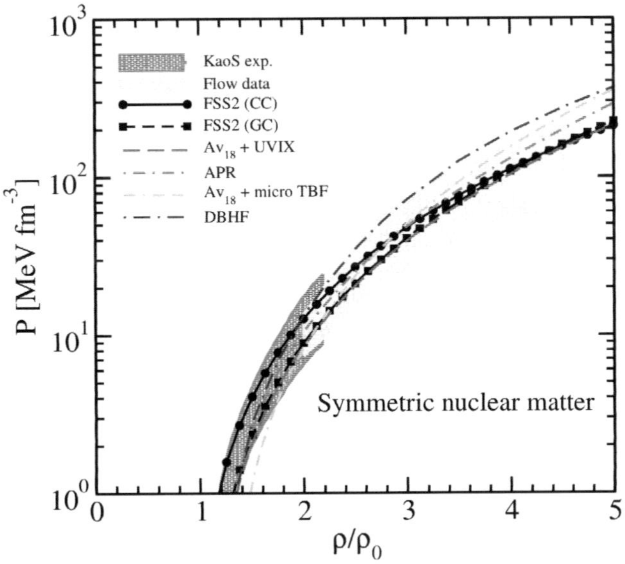

Figure 8. (Color online) The pressure of symmetric matter for several EoS. The larger (yellow) and smaller (violet) bands represent the phenomenological constraints from experimental data. See text for details. Figure taken from [28].

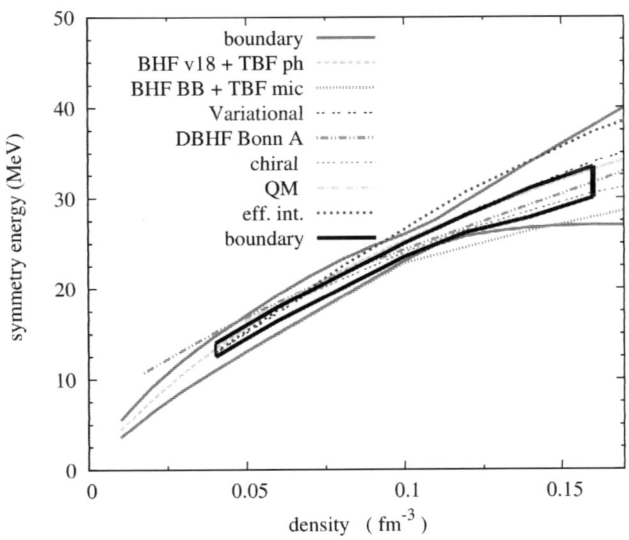

Figure 9. Symmetry energy as a function of density calculated with different microscopic many-body methods. The larger (red) box indicates the constraints from the analysis of the Isobaric Analog States of ref. [32]. The smaller (black) box is the boundary obtained in the same reference by adding the constraints from the neutron skin data. Figure taken from ref. [27]

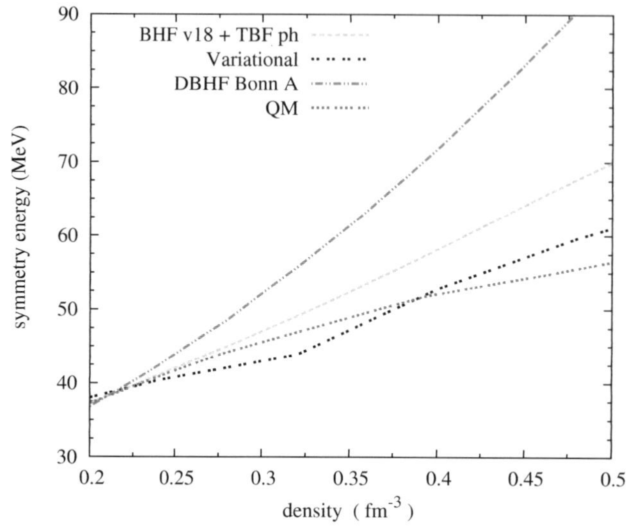

Figure 10. The symmetry energy above saturation density calculated with the different EoS included in Fig. 9, with exception of the EoS from the chiral approach. Figure taken from Ref. [27].

The largest mass is close to 2 solar masses [35], and in Fig. 11, taken from ref. [28] is reported the mass vs. radius and mass vs. central density relationship for NS calculated for the different EoS. These calculations employ the EoS for strongly asymmetric nuclear matter and are assuming that only nucleonic and leptonic (electrons and muons) components are present inside NS, and therefore they are sensitive also to the symmetry energy. In both panels the plots display a maximum, which is the maximum mass predicted by the corresponding EoS. This must be equal or larger than 2 solar masses if the EoS can be acceptable. The microscopic EoS which are included in this analysis seem to be compatible with observations, taking into account the uncertainty of the data [35]. For the EoS derived from QM the calculation has been performed both with the continuous choice (CC) and the gap choice (GC), and the difference between the two results can be taken as an estimate of the uncertainty of the theoretical prediction. Although the EoS in the CG and CC cases are quite close around saturation, at very high density, typical of the NS interior, they have a different incompressibility, with the EoS in the GC case slightly stiffer.

Microscopic calculations that include also the hyperon degrees of freedom, as reported in e.g. ref. [28], show that the NS matter EoS become softer, and the value of the NS maximum mass is substantially reduced, well below the observational limit of 2 solar masses. This poses a serious problem for the microscopic theory of NS interior. A way out of this puzzle was proposed in ref. [36], where it was proposed that an additional repulsion in all channels, both in the nucleonic and hyperonic sectors, can be provided by the process of multi-pomeron exchange. The pomeron has the quantun number of the vacuum, and as such the process acts universally in all channels. The problem was overcome also in the calculations of ref. [37], where the DBHF method was extended to the heperonic sector assuming SU(6) symmetry, which is mainly equivalent to the introduction of a three-body force which acts in analogous way in the nucleonic and hyperonic sectors. Both approaches show that the observational limit of the NS masses can be reached only if one assumes a TBF in the hyperonic sector at least as strong as in the nucleonic sector. These solutions of the puzzle rely on specific models for the strange sector interaction, and they could be justified only by more microscopic approaches.

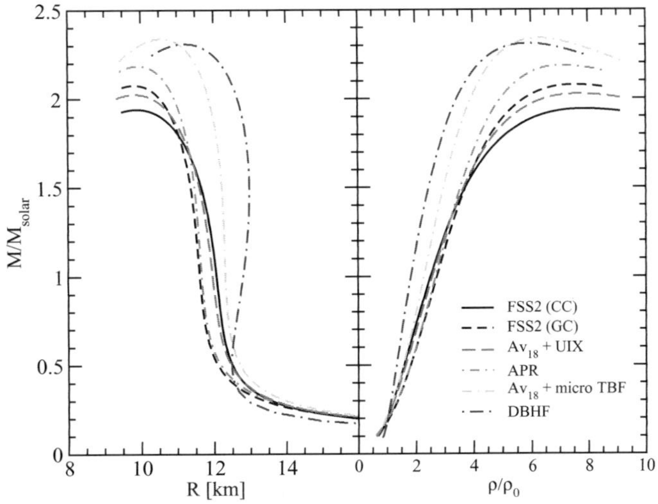

Figure 11. Neutron Star mass as a function of radius (left panel) and as a function of central density (right panel) for different microscopic EoS. Figure taken form [28].

Finally one should also consider the possibility of the appearance of quark matter in the NS core. Also in this case one usually finds that the EoS for NS matter becomes softer, and it is again quite uncertain if the observational limit can be reached. This is especially true if the simple bag model is used for the quark matter EoS. A discussion on this point can be found in ref. [38], where an alternative approach is presented based on the Nambu-Jona Lasinio model with the addition of a vector and di-quark interaction. With a suitable choice of the parameters it is indeed possible to overcome the observational limit. A microscopic approach to this problem based on QCD is still lacking. It would be challenging to try the extension of the QM interaction, devised for the nucleonic matter, to the region of deconfined quark matter.

4. Conclusions

It has been shown that a NN interaction based on quark degree of freedom [25, 28] can be performing at least at the same level as the best NN interaction based on meson exchange processes or on the chiral symmetry of QCD. In addition, without the introduction of TBF or parameters other than the ones introduced in the fitting of the NN phase shifts and deuteron properties, the interaction is able to reproduce the three-body properties, also in the continuum, and at the same time the saturation point of nuclear matter. The latter result can be obtained only by including in the many-body calculation the three-body correlations within the hole-line expansion of the BBG formalism. These results seem to indicate that the explicit introduction of the quark structure of the nucleons is of decisive relevance for the NN interaction, in particular the effect of the TBF processes, introduced in the more traditional interactions, are to a large extent automatically included. In the physics of NS one has to explore the possible appearance of hyperons and eventually deconfined quark matter in the star core. Furthermore it would be desirable to consider other interactions based on quark degrees of freedom in order to understand if the properties of the considered interaction is a common feature of this type of interactions, and eventually to pin down the main reasons of their high performances. All that is left to future works.

IOP Conf. Series: Journal of Physics: Conf. Series **861** (2017) 012001 doi:10.1088/1742-6596/861/1/012001

Acknowledgments

The author wishes to acknowledge the “NewCompStar” COST Action MP1304.

References

[1] For a brief review see Baldo M and Burgio G F 2012 *Rep. Prog. Phys.* **75** 026301
[2] Wiringa R B, Stoks V G J and R. Schiavilla 1995 *Phys. Rev. C* **51** 38
[3] Baldo M and Maieron C 2007 *J. Phys. G :Nucl. Part. Phys.* **34** R1
[4] Grangé P, Lejeune A, Martzolff M and Mathiot J-F 1989 *Phys. Rev. C* **40** 1040; Zuo W, Lejeune A, Lombardo U and Mathiot J-F 2002 *Nucl. Phys. A* **706** 418; 2002 *Eur. Phys. J. A* **14** 469
[5] Baldo M, Robledo L M, Schuck P and Viñas X 2013 *Phys. Rev. C* **87** 064305
[6] Machleidt R 1989 *Adv. Nucl. Phys.* **19** 189
[7] Brown G E, Weise W, Baym G and Speth J 1987 *Comm. Nucl. Part. Phys.* **17** 39
[8] Akmal A, Pandharipande V R and Ravenhall D G 1989 *Phys. Rev. C* **58** 1804
[9] Weinberg S 1990 *Phys. Lett. B* **251** 288; 1991 *Nucl. Phys. B* **363** 3; 1992 *Phys. Lett. B* **295** 114; 1968 *Phys. Rev.* **166** 1568
[10] Hebeler K, Bogner S K, Furnstahl R J, Nogga A and Schwenk A 2011 *Phys. Rev. C* **83** 031301
[11] Sammaruca F, Coraggio L, Holt JW, Itaco N, Machleidt R and Marcucci L E 2015 *Phys. Rev. C* **91** 054311
[12] Hagen G, Papenbrock T, Ekström A, Wendt K A, Baardsen G, Gandolfi S, Hjorth-Jensen M and Horowits CJ 2014 *Phys. Rev. C* **89** 014319
[13] Logoteta D, Bombaci I and Kievsky A 2016 *Phys. Lett. B* **758** 449
[14] Oka M and Yazaki K 1980 *Phys. Lett. B* **90** 41; 1981 *Prog. Theor. Phys.* **66** 556; 1981 *Prog. Theor. Phys.* **66** 572; in *Quarks and Nuclei*, edited by W. Weise (World Scientific, Singapore, 1984), p. 489.
[15] Wong C W 1986 *Phys. Rep.* **136**, 1
[16] Oka M, Shimizu K and Yazaki M 1987 *Nucl. Phys. A* **464** 700
[17] Shimizu K 1989 *Rep. Prog. Phys.* **52** 1
[18] Shimizu K, Takeuchi S and Buchmann A J 200 *Prog. Theor. Phys. Suppl.* **137** 43
[19] Valcarce A, Garcilazo H, Fernández F and González P 2005 *Rep. Prog. Phys.* **68** 965
[20] Fujiwara Y, Suzuki Y and Nakamoto C 2007 *Prog. Part. Nucl. Phys.* **58** 439
[21] Fujiwara Y, Fujita T, Kohno M, Nakamoto C and Suzuki Y 2001 *Phys. Rev. C* **65** 014002
[22] Fujiwara Y, Suzuki Y, Kohno M and Miyagawa K 2008 *Phys. Rev. C* **77** 027001
[23] Fujiwara Y and Fukukawa K 2013 *Few-Body Systems* **54** 2357
[24] Fujiwara Y 2014 *Few-Body Syst.* **55** 993
[25] Baldo M and Fukukawa K 2014 *Phys. Rev. Lett.* **113** 242501
[26] Dutra M, Laurenço O, Sá Martins J S, Delfino A, Stone J R and Stevenson P D 2012 *Phys. Rev. C* **85** 035201
[27] Baldo M and Burgio G F 2016 *Progress in Particle and Nuclear Physics* **91** 203
[28] Fukukawa K, Baldo M, Burgio G F, Lo Monaco L and Schulze H-J 2015 *Phys. Rev.* **92**, 065802 (2015)
[29] Gross-Boelting T, Fuchs C and Faessler A 1999 Ñucl. Phys. A **648** 105
[30] Danielewicz P, Lacey R and Lynch W 2002 *Science* **298** 1592
[31] Miskowiec D *et al.* 1994 *Phys. Rev. Lett.* **72** 3650
[32] Danielewicz P and Lee J 2014 *Nucl. Phys. A* **922**, 1
[33] Benhar O and Valli M 2007 *Phys. Rev. Lett.* **99** 232501
[34] *Compressed Baryonic Matter in Laboratory Experiments*, Friman B, Höhne C, Knoll J, Leupold S, Randrup J, Rapp R and Senger P (editors) 2011 *Lecture Notes in Physics Vol.* **814** Springer
[35] Antoniadis J *et al.* 2013 *Science* **340** 1233232
[36] Yamamoto Y, Furumoto T, Yasutake N and Rijken Th A 2014 *Phys. Rev. C* **90** 045805
[37] Katayama T and Saito K 2015 *Phys. Lett.* **B747** 43
[38] Koyo T, Powell P D, Song Y and Baym G 2015 *Phys. Rev. D* **91** 045003

IOP Conf. Series: Journal of Physics: Conf. Series **861** (2017) 012002 doi:10.1088/1742-6596/861/1/012002

Looking for extra dimensions in compact stars

Germán Lugones[1] and José D V Arbañil[2]

[1]Centro de Ciências Naturais e Humanas, Universidade Federal do ABC, Avenida dos Estados 5001, 09210-580 Santo André, São Paulo, Brazil.

[2]Departamento de Ciencias, Universidad Privada del Norte, Av. Alfredo Mendiola 6062 Urb. Los Olivos, Lima, Perú.

E-mail: [1]german.lugones@ufabc.edu.br

E-mail: [2]jose.arbanil@upn.pe

Abstract. The properties of spherically symmetric static compact stars are studied in the Randall-Sundrum II type braneworld model assuming that the spacetime outside the star is described by a Schwarzschild metric. The integration of the stellar structure equations employing the so called causal limit equation of state (EoS) shows that the equilibrium solutions can violate the general relativistic causal limit. An analysis of the properties of hadronic and strange quark stars using standard EoSs confirm the same result: there is a branch in the mass-radius diagram that shows the typical behaviour found within the frame of General Relativity and another branch of stars that are supported against collapse by the nonlocal effects of the bulk on the brane. Stars belonging to the new branch can violate the general relativistic causal limit, may have an arbitrarily large mass, and are stable under small radial perturbations. If they exist in Nature, these objects could be hidden among the population of black hole candidates. The future observation of compact stars with masses and radii falling above the causal limit of General Relativity but below the Schwarzschild limit maybe a promising astrophysical evidence for the existence of extra dimensions.

1. Introduction

Braneworld models represent the universe as a three-dimensional brane where elementary particles live embedded in a higher-dimensional spacetime called the bulk, only accessed by gravity [1]. Within this framework, astrophysical and cosmological models can be constructed where the gravitational effect of extra-dimensions can be explored. In Randall-Sundrum models, ultraviolet modifications to General Relativity are introduced such that significant deviations from Einstein's theory occur at very high energies, e.g. in the early universe, in gravitational collapse and in compact objects.

In the Randall-Sundrum model, our universe is a brane embedded in one extra dimension (the bulk) which is a portion of a 5D anti-de Sitter spacetime (AdS_5); i.e. the extra dimension is curved or warped rather than flat. At low energies, gravity has an exponentially suppressed tail into the extra dimension due to a negative bulk cosmological constant, $\Lambda_5 = -6/\ell^2$ where ℓ is the curvature radius of AdS_5. The brane gravitates with self-gravity in the form of a brane tension λ, where $\lambda = 3M_p^2/(4\pi\ell^2)$ and $M_p^2 = M_5^3\ell$. On the brane, the negative Λ_5 is counterbalanced by the positive brane tension λ.

The Einstein's field equation takes the conventional form but with an effective energy-momentum tensor $T^{\rm eff}_{\mu\nu}$, i.e., it reads:

$$G_{\mu\nu} = 8\pi G\, T^{\rm eff}_{\mu\nu}, \tag{1}$$

where $G_{\mu\nu}$ is the usual Einstein field tensor, and we consider $c = 1$.

The effective energy-momentum tensor has the form [2]

$$T^{\rm eff}_{\mu\nu} = T_{\mu\nu} + \frac{6}{\lambda} S_{\mu\nu} - \frac{1}{8\pi G}\,\mathcal{E}_{\mu\nu}\,, \tag{2}$$

where the first term contains the standard energy momentum tensor; e.g. for a perfect fluid we have $T_{\mu\nu} = \rho u_\mu u_\nu + p h_{\mu\nu}$, where p is the pressure of the fluid, ρ is its energy density, u^μ is the four-velocity and $h_{\mu\nu} = g_{\mu\nu} + u_\mu u_\nu$ is the projection orthogonal to u^μ. The second and third terms include modifications with respect to the standard 4D Einstein's field equation. The bulk correction includes a local term and a nonlocal one (second and third terms respectively) [1]. For a perfect fluid, the local contribution reads

$$S_{\mu\nu} = \frac{1}{12}\rho^2 u_\mu u_\nu + \frac{1}{12}\rho(\rho + 2p)h_{\mu\nu}. \tag{3}$$

The nonlocal contribution for static spherical symmetry reads:

$$\mathcal{E}_{\mu\nu} = -\frac{6}{8\pi G\lambda}\left[\mathcal{U} u_\mu u_\nu + \mathcal{P} r_\mu r_\nu + \frac{(\mathcal{U} - \mathcal{P})}{3} h_{\mu\nu}\right], \tag{4}$$

where $\mathcal{U}$ and $\mathcal{P}$ are respectively the nonlocal energy density and nonlocal pressure on the brane and r_μ is a unit radial vector. Notice that, as $\lambda \to \infty$, the bulk corrections vanish and General Relativity is recovered.

The braneworld generalization of the stellar structure equations for a static fluid distribution with spherical symmetry has been derived by Germani and Maartens [3], and read:

$$\frac{dm}{dr} = 4\pi r^2 \rho_{\rm eff}, \tag{5}$$

$$\frac{dp}{dr} = -(\rho + p)\frac{d\phi}{dr}, \tag{6}$$

$$\frac{d\phi}{dr} = \frac{Gm + 4\pi G r^3 \left(p_{\rm eff} + \frac{4\mathcal{P}}{(8\pi G)^2\lambda}\right)}{r(r - 2Gm)}, \tag{7}$$

$$\frac{d\mathcal{U}}{dr} = -(4\mathcal{U} + 2\mathcal{P})\frac{d\phi}{dr} - 2(4\pi G)^2(\rho + p)\frac{d\rho}{dr} -2\frac{d\mathcal{P}}{dr} - \frac{6}{r}\mathcal{P}, \tag{8}$$

where

$$\rho^{\rm eff} = \rho\left(1 + \frac{\rho}{2\lambda}\right) + \frac{6\mathcal{U}}{(8\pi G)^2\lambda}, \tag{9}$$

$$p^{\rm eff} = p + \frac{\rho}{2\lambda}(\rho + 2p) + \frac{2\mathcal{U}}{(8\pi G)^2\lambda}. \tag{10}$$

To solve Eqs. (5)$-$(8) we need an equation of state $\rho = \rho(p)$ and a relation of the form $\mathcal{P} = \mathcal{P}(\mathcal{U})$ relating the nonlocal components ("dark" equation of state).

IOP Conf. Series: Journal of Physics: Conf. Series **861** (2017) 012002 doi:10.1088/1742-6596/861/1/012002

Two of the boundary conditions of the stellar structure equations on the brane are the same as for the standard General Relativistic equations. Specifically, at the center of the star ($r = 0$) the enclosed mass is zero:

$$m(r = 0) = 0, \tag{11}$$

and at the surface of the object the pressure vanishes:

$$p(R) = 0. \tag{12}$$

The remaining boundary condition is determined by the Israel-Darmois matching condition $[G_{\mu\nu}r^{\nu}]_{\Sigma} = 0$ at the surface of the object Σ, where $[f]_{\Sigma} \equiv f(R^{+}) - f(R^{-})$.

In brane-world models, the Schwarzschild solution is no longer the unique asymptotically flat vacuum exterior (for a discussion on the validity of the Schwarzschild exterior see Ref. [4] and references therein). In general, the exterior carries an imprint of nonlocal bulk graviton stresses and knowledge of the 5D Weyl tensor is needed as a minimum condition for uniqueness. Here, for simplicity, we focus on a class of models with a Schwarzschild exterior ($\mathcal{U}^{+} = \mathcal{P}^{+} = 0$). We also assume $\mathcal{P}^{-} = 0$, which is consistent with the isotropy of the physical pressure in the star. As a consequence, the interior must have nonvanishing nonlocal Weyl stresses ($\mathcal{U}^{-} \neq 0$). Therefore, the boundary condition for $\mathcal{U}$ at $r = R$ reads:

$$(4\pi G)^2 \rho^2(R) + \mathcal{U}^{-}(R) = 0. \tag{13}$$

Due to the latter boundary condition at the stellar surface, the numerical integration of the structure equations is less straightforward than for the standard structure equations in General Relativity. In the present case, a shooting method is used in order to match iteratively the boundary condition for $\mathcal{U}^{-}$ at $r = R$ (for more details see Ref. [4]).

2. The causal limit in braneworld models

The equation of state of neutron star matter can be determined with confidence up to $\sim 2\rho_{sat}$, being $\rho_{sat} \approx 151\,\mathrm{MeV/fm^3}$ the nuclear saturation density. For larger densities, the determination of the EoS depends strongly on the knowledge of strong interactions in a regime that cannot be reached experimentally. As a consequence, there is a large amount of high-density EoSs in the literature that incorporate several aspects that may play a crucial role at the inner core of the star, such as three-body forces, bosonic condensates, hyperonic degrees of freedom and quark matter [5, 6].

An important aspect of neutron stars within the frame of General Relativity, is that there exists a maximum gravitational mass above which there are no stable stellar configurations. The maximum mass exists no matter what the EoS, but its determination depends on a deep comprehension of the EoS up to several times ρ_{sat}. However, using the so called *causal limit EoS*, it is possible to circumvent the uncertainties related to the properties of high-density matter and obtain upper bounds to the maximum allowed mass of a neutron star [5, 6]. The causal limit EoS can be constructed by using a detailed EoS at density ranges where they can be safely regarded as accurate and imposing generic constraints at densities exceeding some fiducial density, e.g., subluminal sound velocity and thermodynamic stability (see e.g. [5, 6, 7]). Here, we adopt the well established Baym, Pethick, and Sutherland (BPS) EoS [8] at densities below a fiducial density ρ_t, and a causal equation of state (i.e. sound velocity = speed of light) $p = \rho - a$ above ρ_t [6, 7]. Since both EoSs are matched at an energy density ρ_t and a pressure p_t, the constant a in the high density EoS is given by $a = \rho_t - p_t$, where ρ_t and p_t also fulfill the BPS EoS.

Varying the value of a and integrating the stellar structure equations we obtain the results shown in Fig. 1. Within General Relativity, the causal limit EoS allows obtaining a straight line in the mass-radius diagram that connects the maximum mass stars for different values of

IOP Conf. Series: Journal of Physics: Conf. Series **861** (2017) 012002 doi:10.1088/1742-6596/861/1/012002

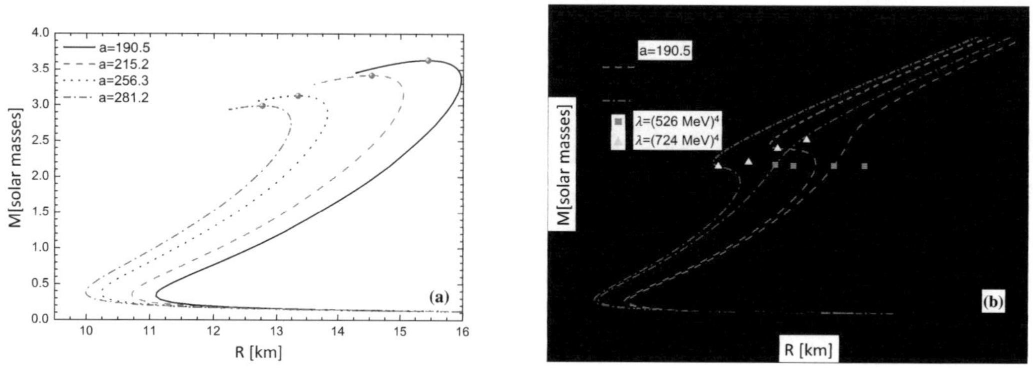

Figure 1. Mass-radius relationship for the causal limit EoS matched continuously with the BPS EoS in (a) General Relativity and (b) braneworld models. Both EoSs are matched at different fiducial densities that lead to different values of $a = \rho_f - p_f$. In the case of General Relativity, the dots over the curves indicate the maximum masses. For braneworld stars, the values of λ lead to $M_5 = (\frac{4}{3}\pi\lambda M_p^2)^{1/6} \sim 2000\,\mathrm{TeV}$; i.e. larger than 10 TeV, in compatibility with LHC.

a. Above such line, known as *causal limit*, stellar configurations are not allowed. The forbidden region is given by [6, 5, 9, 10]:

$$M > 0.345R, \tag{14}$$

or, equivalently:

$$\left(\frac{M}{M_\odot}\right) > 0.234\left(\frac{R}{\mathrm{km}}\right). \tag{15}$$

In the case of braneworld stars, we find that for small masses, the curves show the typical behavior found within the frame of General Relativity. Specifically, very small mass stars have very large radii, and as the mass increases above a few tenths of solar masses the radii fall within a range of few kilometers around ~ 10 km. Nevertheless, for large mass objects, local high-energy effects as well as nonlocal corrections lead to significant deviations with respect to General Relativity. At around $1.5 - 2\,M_\odot$ the $M(R)$ curves bend anticlockwise as in the general relativistic case. However, instead of reaching a maximum mass as in General Relativity, the curves bend once more (clockwise) for larger masses and thereafter they increase roughly linearly (see Fig. 1).

A striking feature of this behavior is that once the $M - R$ curves bend clockwise they may fall above the causal limit obtained within General Relativity (c.f. Eqs. (15) – (14)). It can also be checked that as the masses and radii increase, the curves tend asymptotically to the Schwarzschild limit $M = 2R$.

Since the $M - R$ curves for the causal EoS approach asymptotically the line $M = 2R$, but do not go beyond it, the Schwarzschild limit $M = 2R$ is a good representation of the causal limit in the braneworld model. In other words, the equilibrium solutions found in the braneworld can violate the limit of causality for General Relativity (Eqs. (15) – (14)) and, for sufficiently large mass, can occupy the region between the straight lines shown in Fig. 2.

3. Hadronic and strange stars in braneworld models

Now, we consider two representative equations of state for hadronic and strange quark stars. To describe hadronic matter, we use a relativistic mean-field model which is widely used in stellar

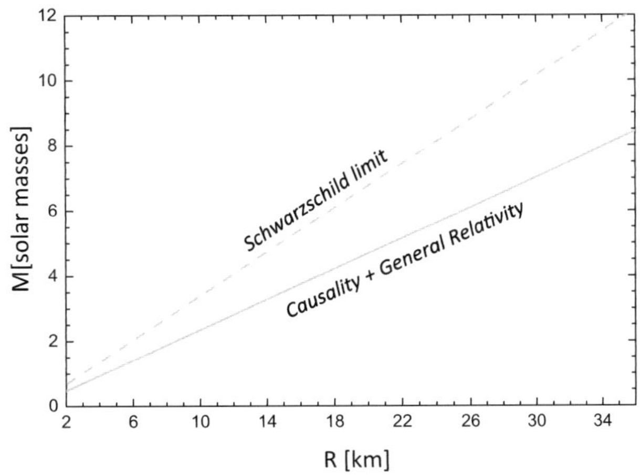

Figure 2. The causal limit for General Relativity and the Schwarzschild limit $M = 2R$. In the braneworld model, static stellar configurations fulfilling a causal EoS (sound velocity < speed of light) can occupy the region between both straight lines.

structure calculations within General Relativity. We adopt the following standard Lagrangian for matter composed by nucleons, hyperons and electrons [5, 6],

$$\begin{aligned}\mathcal{L}_H &= \sum_B \bar{\psi}_B[\gamma_\mu(i\partial^\mu - g_{\omega B}\omega^\mu - \frac{1}{2}g_{\rho B}\vec{\tau}.\vec{\rho}^\mu) \\ &- (m_B - g_{\sigma B}\sigma)]\psi_B + \frac{1}{2}(\partial_\mu\sigma\partial^\mu\sigma - m_\sigma^2\sigma^2) - \frac{1}{3}bm_n(g_\sigma\sigma)^3 \\ &-\frac{1}{4}c(g_\sigma\sigma)^4 - \frac{1}{4}\omega_{\mu\nu}\omega^{\mu\nu} + \frac{1}{2}m_\omega^2\omega_\mu\omega^\mu - \frac{1}{4}\vec{\rho}_{\mu\nu}.\vec{\rho}^{\mu\nu} \\ &+\frac{1}{2}m_\rho^2\vec{\rho}_\mu.\vec{\rho}^\mu + \sum_L \bar{\psi}_L[i\gamma_\mu\partial^\mu - m_L]\psi_L. \end{aligned} \tag{16}$$

Leptons L are treated as non-interacting and baryons B are coupled to the scalar meson σ, the isoscalar-vector meson ω_μ and the isovector-vector meson ρ_μ. The five constants in the model are fitted to the bulk properties of nuclear matter; here we use the parametrization GM1 [11]. The explicit form of the EoS obtained from the above Lagrangian as well as the coupling constants used for the GM1 parametrization can be found in [12, 13, 14, 15] and references therein. At low densities we use the Baym, Pethick and Sutherland model [8].

For quark matter we use the MIT bag model with zero strong coupling constant and massless quarks. The equation of state adopts the simple form $\rho = 3p + 4B$, where B is the bag constant. Witten [16] conjectured that, at zero pressure and temperature, three flavor quark matter may have an energy per baryon smaller that ordinary nuclei. This would make strange quark matter the true ground state of strongly interacting matter and would lead to the existence of strange quark stars, i.e. stellar objects completely composed by strange quark matter [17, 18]. Within the MIT bag model for massless quarks and zero strong coupling constant, the Witten hypothesis is verified if the bag constant is in the range $57\,\mathrm{MeV/fm^3} < B < 94\,\mathrm{MeV/fm^3}$. In this paper we consider $B = 60\,\mathrm{MeV/fm^3}$.

In Fig. 3 we show the mass-radius relationship for strange quark stars and for hadronic stars. We also include the causal limit in General Relativity and the Schwarzschild limit $M = 2R$. For small masses ($< 1.5 - 2M_\odot$), the curves show the typical behavior found within the frame

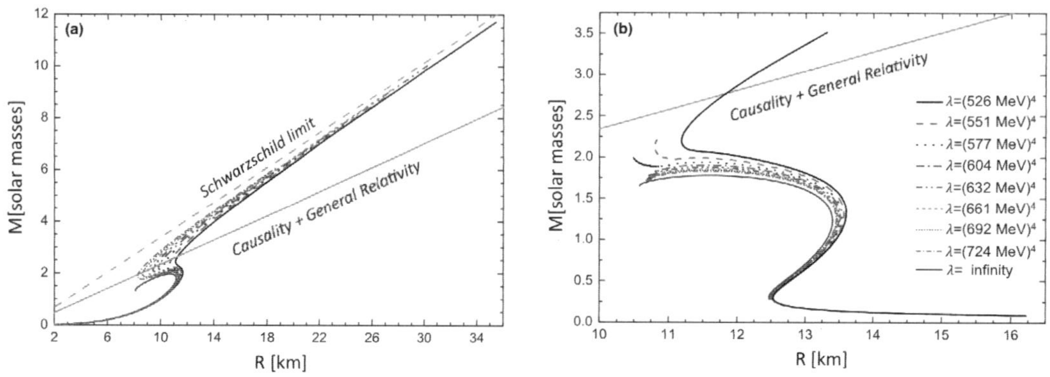

Figure 3. Mass-radius relationship for several values of the brane tension λ: (a) strange quark stars and (b) hadronic stars.

of General Relativity. Specifically, very small mass hadronic stars have very large radii, while strange stars follow roughly $M \propto R^3$. For large mass objects, local high-energy effects as well as nonlocal corrections lead to the same deviations with respect to General Relativity that were found in the previous section for the causal EoS. Instead of reaching a maximum mass as in general relativity, the $M(R)$ curves have a new branch that approaches asymptotically the Schwarzschild limit $M = 2R$. In some cases there is a local maximum in the $M(R)$ curves.

4. Stability of stellar configurations

In the previous sections we showed that in braneworld models there is a new branch of stellar configurations that is not present within General Relativity. Since only compact objects in stable equilibrium are acceptable from the astrophysical point of view, we will check here the stability of these models. A well known static criterion that is widely used in the literature states that a necessary condition for a model to be stable is that its mass M increases with growing central density, i.e.

$$\frac{dM}{d\rho_c} > 0. \tag{17}$$

The latter is a necessary but not sufficient condition.

Here, we employ a sufficient criterion which enables one to determine the precise number of unstable normal radial modes using the $M(R)$ curve [5, 19]. According to such criterion, at each critical point of the $M(R)$ curve (local maxima or minima) one and only one normal radial mode changes its stability, from stable to unstable or vice versa. There are no changes of stability associated with radial pulsations at other points of the $M(R)$ curves. Moreover, one mode becomes unstable (stable) if and only if the $M(R)$ curve bends counterclockwise (clockwise) at the critical point.

In our previous results we found two qualitatively different types of $M(R)$ curves. One type presents one local maximum and one local minimum in $M(R)$ (and in $M(\rho_c)$ as well). The other one has no critical points. These two types are represented separately in Fig. 4, where we show the $M(R)$ and $M(\rho_c)$ curves for strange quark stars for two different values of λ (for simplicity we do not show hadronic stars because the stability analysis is completely equivalent, as we shall see below).

In order to analyze the stellar stability using the above criterion, we assume that the low mass branch (up to $< 1.5 - 2M_{\odot}$) of the $M(R)$ curves is stable for all radial modes, as it is in the general relativistic case. For the curves with two critical points (panels (c) and (d) of Fig. 4),

 doi:10.1088/1742-6596/861/1/012002

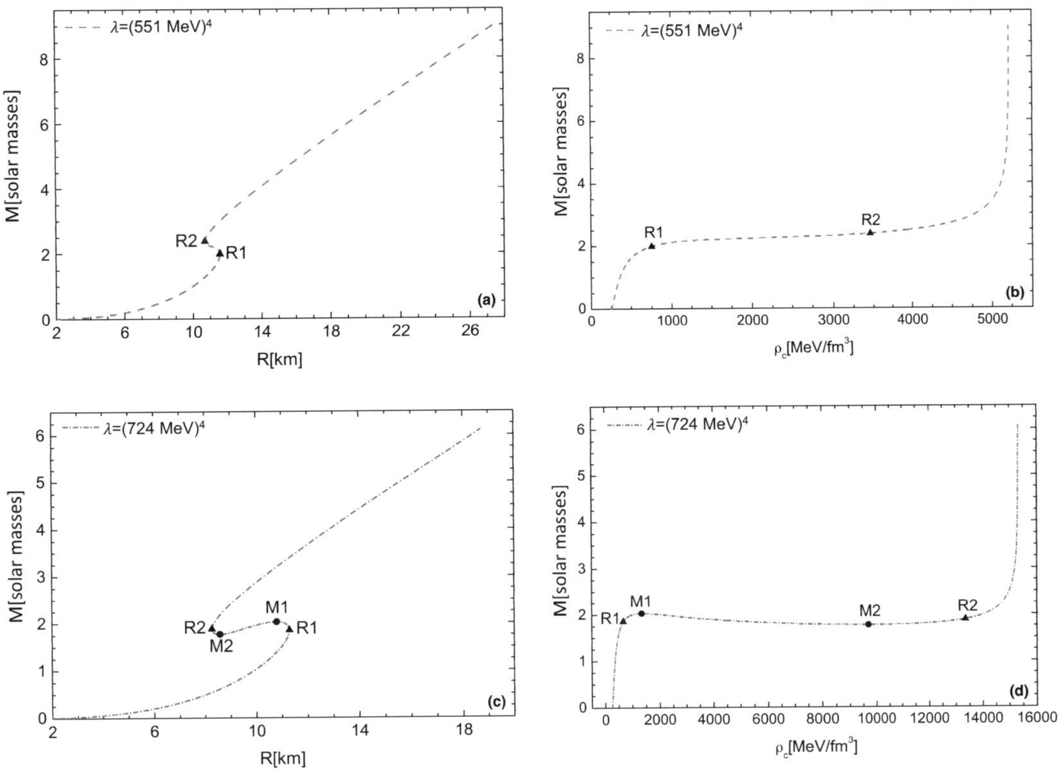

Figure 4. Analysis of stellar stability in braneworld models based on the critical points in the M versus R diagram. The curves represent strange quark stars but the stability analysis for hadronic stars is completely equivalent. Figures (a) and (b) correspond to a value of the brane tension, $\lambda = (551\,\mathrm{MeV})^4$, that results in no critical points. Figures (c) and (d) correspond to a different value of the brane tension, $\lambda = (724\,\mathrm{MeV})^4$, that results in a local maximum and a local minimum in both the $M(R)$ and $M(\rho_c)$ curves.

the $M(R)$ curve bends counterclockwise at the local maximum and the fundamental oscillation mode becomes unstable. However, at the local minimum the fundamental mode becomes stable again because the curve bends clockwise there. Beyond the local minimum there are no more critical points and all the radial modes remain stable. In the case without critical points (panels (a) and (b) of Fig. 4), the whole sequence remains stable for all radial modes provided that the low mass configurations are stable. Thus, we can conclude that the branches that approach asymptotically to the Schwarzschild limit are always stable under small radial perturbations, and therefore, stellar configurations of arbitrarily large mass would be allowed within braneworld models.

The previous analysis concerns the stability of non-rotating stars. However, if the star rotates, there is a handful of perturbations to be considered for a full stability analysis. For example, rapidly rotating neutron stars can be unstable to the gravitational-wave-driven Chandrasekhar, Friedman and Schutz (CFS) mechanism [20, 21]. These instabilities deserve further study within braneworld models.

5. Conclusions

In this work we have studied the structure of spherically symmetric static compact stars in the Randall-Sundrum II type braneworld model assuming that the spacetime outside the star is described by a Schwarzschild metric. We obtain the low mass branch of compact star configurations already known from general relativistic calculations, but we also find a new branch that approaches asymptotically to the Schwarzschild limit which is always stable under small radial perturbations. As a consequence, stellar configurations of arbitrarily large mass, supported against collapse by the nonlocal effects of the bulk on the brane, are allowed within braneworld models. It is worth emphasizing that black holes are still possible within the here studied braneworld models. Moreover, the stellar configurations that asymptotically approach to the Schwarzschild limit are expected to be stable under small perturbations, but not necessarily under large ones. Therefore, a very large mass braneworld compact star could collapse into a black hole if strongly perturbed in a catastrophic astrophysical event, e.g. in a binary stellar merging.

From the astrophysical point of view, the existence of more than 20 compact objects, whose masses are in the range $M = 5 - 30 M_{\odot}$, has been confirmed via dynamical observations. These objects are usually assumed to be stellar mass black holes but, strictly speaking, they are only black hole candidates. They are located in X-ray binary systems, where the X-rays are produced by gas that flows from the companion star on to the compact object via an accretion disk. The strongest argument for identifying them as black holes is that they are sufficiently massive and compact, and observations could not be matched with any object in stable equilibrium other than a black hole. However, this by itself does not prove that these objects are true black holes, defined as objects with event horizons [22]. At least some black hole candidates could, in principle, be exotic objects made of some kind of unusual matter that enables them to have a surface (no horizon), despite their extreme compactness. The hypothetic brane world stars discussed in this work fall in this category, and at least in principle, they could be hidden among the population of black hole candidates. While a number of tests have been devised to check whether black hole candidates have a surface, at present all evidence is still indirect.

As explained above, the very existence of the new branch of large mass objects can be tested through the observation of masses and radii of compact stars falling above the causal limit of General Relativity but below the Schwarzschild limit. If found, such objects could be an astrophysical manifestation of the existence of extra dimensions.

Acknowledgments

G L acknowledges the Brazilian agencies FAPESP and CNPq for financial support.

References

[1] Maartens R and Koyama K 2010 *Living Rev. Relativity* **13** 5
[2] Shiromizu T, Maeda K-I and Sasaki M 2000 *Phys. Rev. D* **62** 024012
[3] Germani C and Maartens R 2001 *Phys. Rev. D* **64** 124010
[4] Lugones G and Arbañil J D V 2017 *Phys. Rev. D* **95** 064022
[5] Haensel P, Potekhin A Y and Yakovlev D G 2007 *Neutron Stars 1: Equation of State and Structure* Springer Verlag, New York
[6] Glendenning N K 2000 *Compact Stars: Nuclear Physics, Particle Physics and General Relativity* 2nd ed. Springer Verlag, New York
[7] Rhoades C E and Ruffini R 1974 *Phys. Rev. Lett.* **32** 324
[8] Baym G, Pethick C and Sutherland P 1971 *Astrophys. J.* **299** 170
[9] Lattimer J M and Prakash M 2004 *Science* **304** 536
[10] Lattimer J M 2012 *Annu. Rev. Nucl. Part. Sci* **62** 485
[11] Glendenning N K and Moszkowski S A *Phys. Rev. Lett.* **67** 2414
[12] Lugones G, do Carmo T A S, Grunfeld A G and Scoccola N N 2010 *Phys. Rev. D* **81** 85012
[13] do Carmo T A S, Lugones G and Grunfeld A G 2013 *J. Phys. G: Nucl. Part. Phys.* **40** 035201

[14] Lugones G, Grunfeld A G, Scoccola N N and Villavicencio C 2009 *Phys. Rev. D* **80**
[15] Lugones G and Grunfeld A G *Phys. Rev. D* 2011 **84** 85003
[16] Witten E 1984 *Phys. Rev. D* **30** 272
[17] Lugones G *Eur. Phys. J. A* 2016 **52** 53
[18] Benvenuto O G and Lugones G 1998 *Int.J.Mod.Phys. D* **7** 29
[19] Harrison B K, Thorne K S, Wakano M and Wheeler J A 1965 *Gravitation Theory and Gravitational Collapse*, University of Chicago Press, Chicago
[20] Chandrasekhar S 1970 *Astrophys. J.* **161** 561
[21] Friedman J L and Schutz B F 1978 *Astrophys. J.* **222** 281
[22] Narayan R and McClintock J E 2013 arXiv:1312.6698 [astro-ph.HE]

IOP Conf. Series: Journal of Physics: Conf. Series **861** (2017) 012003 doi:10.1088/1742-6596/861/1/012003

Possible rotation-power nature of SGRs and AXPs

M. Malheiro[1], Jaziel G. Coelho[2], D. L. Cáceres[5], R. C. R. de Lima[3,5], R. V. Lobato[1,4,5], J. A. Rueda[4,5,6] and R. Ruffini[4,5,6]

[1]Departamento de Física, Instituto Tecnológico de Aeronáutica, ITA - DCTA, Vila das Acácias, São José dos Campos, 12228-900 SP, Brazil
[2]INPE - Instituto Nacional de Pesquisas Espaciais, Divisão de Astrofísica, Av. dos Astronautas 1758, São José dos Campos, 12227-010 SP, Brazil
[3]Universidade do Estado de Santa Catarina, Av. Madre Benvenuta, 2007 Itacorubi, 88.035–901, Florianópolis, Brazil
[4]Dipartimento di Fisica, Sapienza Universit di Roma, P.le Aldo Moro 5, I-00185 Rome, Italy
[5]ICRANet, P.zza della Repubblica 10, I65122 Pescara, Italy
[6]ICRANet-Rio, Centro Brasileiro de Pesquisas Físicas, Rua Dr. Xavier Sigaud 150, 2290–180 Rio de Janeiro, Brazil

E-mail: malheiro@ita.br

Abstract. We investigate the possibility of some Soft Gamma-ray Repeaters (SGRs) and Anomalous X-ray Pulsars (AXPs) could be described as rotation-powered neutron stars (NSs). The analysis was carried out by computing the structure properties of NSs, and then we focus on giving estimates for the surface magnetic field using both realistic structure parameters of NSs and a general relativistic model of a rotating magnetic dipole. We show that the use of realistic parameters of rotating neutron stars obtained from numerical integration of the self-consistent axisymmetric general relativistic equations of equilibrium leads to values of the magnetic field and radiation efficiency of SGRs/AXPs very different from estimates based on fiducial parameters. This analysis leads to a precise prediction of the range of NS masses, obtained here by making use of selected up-to-date nuclear equations of state (EOS). We show that 40% (nine) of the entire observed population of SGRs and AXPs can be described as canonical pulsars driven by the rotational energy of neutron stars, for which we give their possible range of masses. We also show that if the blackbody component in soft X-rays is due to the surface temperature of NSs, then 50% of the sources could be explained as ordinary rotation-powered pulsars. Besides, amongst these sources we find the four SGRs/AXPs with observed radio emission and six that are possibly associated with supernova remnants (including Swift J1834.9-0846 as the first magnetar to show a surrounding wind nebula), suggesting as well a natural explanation as ordinary pulsars.

1. Introduction

SGRs and AXPs constitute a class of pulsars with the following properties [1]: periods $P \sim (2$–$12)$ s, slowing down rates $\dot{P} \sim (10^{-15} - 10^{-10})$ s/s, persistent X-ray luminosity as large as 10^{35} erg s^{-1} and transient activity in the form of outbursts of energies around (10^{41}–10^{43}) erg. There is a current widely accepted view of SGR/AXPs as magnetars [2, 3]. Adopting NS's fiducial parameters (mass $M = 1.4M_{\odot}$, radius $R = 10$ km, and moment of inertia $I = 10^{45}$ g cm^2), one has that the rotational energy loss

$$\dot{E}_{\rm rot} = -4\pi^2 I \frac{\dot{P}}{P^3} = -3.95 \times 10^{46} \frac{\dot{P}}{P^3} \quad {\rm erg\ s^{-1}}\,. \tag{1}$$

When the above P and $\dot{P}$ are substituted into Eq. (1), we conclude that $\dot{E}_{\rm rot}$ is not enough to explain the observed X-ray luminosities from SGR/AXPs (i.e. $L_X > \dot{E}_{\rm rot}$), unlike the traditional rotation-powered pulsars. Thus, SGRs and AXPs need further scrutiny of their energy reservoirs. The magnetar model hypothesizes the release of magnetic energy via the decay of an ultrastrong NS surface magnetic field, larger than the critical field for vacuum polarization $B_c = m_e^2 c^3/(e^- h) = 4.4 \times 10^{13}$ G.

It is important to pointing out that there are alternative scenarios with NSs but with ordinary magnetic fields $B \sim 10^{12}$ G which already challenge the magnetar model (see e.g., [4, 5] and references therein). Other alternatives have been explored in terms of quark stars, i.e. accreting quark stars [6] or the quark-nova remnant model [7]. Alternatively to NSs, Malheiro et al. [8] have suggested that massive, fast rotating, highly magnetized WDs could explain as well the general properties of SGRs and AXPs. Despite the wide acceptance of the magnetar model, we have recalled and stressed that there are alternative scenarios not ruled out by the current data, motivating us to scrutinize further their predictions and evaluate their consistency [9]. Following this reasoning, the authors of [8] introduced a word of caution on labeling a source as a magnetar only because it shows a transient outburst activity. As we show below, this method of classification on which the current catalog of SGRs/AXPs has been constructed [10], is in clear contrast with the original idea underlying magnetars, which were introduced for the explanation of sources with $L_X > \dot{E}_{\rm rot}$.

Bearing this in mind, we examined in [8, 9, 11, 12] the possibility of some SGR/AXPs being rotation-powered pulsars (RPPs): we found at the time that, indeed, four sources (1E 1547.0–5408, SGR 1627–41, PSR J1622–4950, and XTE J1810–197) of the SGR/AXP catalog could be explained via the rotational energy loss of a NS. Once identified as RPPs, one is led to the theoretical prediction that some of the phenomena observed in ordinary pulsars could also occur in SGRs and AXPs. Indeed, the radio emission, a property common in pulsars but generally absent in SGRs and AXPs, is observed in three out of these four objects, as previously discussed in [9, 12]. It is interesting to notice that among these nine sources, we can also find six sources with possible associations with supernova remnants (SNRs). Very recently, Younes et al. [13] discovered a vast cloud of high-energy particles called a wind nebula around of Swift J1834.9-0846, for the first time. Wind nebula around other high-B sources have also been suggested. In this work, we have shown that this discovery is further evidence corroborating to the description of some SGRs/AXPs as rotation powered pulsars.

In addition, PSR J1846-0258, a source that although recognized as rotation-powered is often quoted as a magnetar-like or a transitional object due to its outburst event in June 2006, the energetics of its flaring activity is well explained from the gain of rotational energy during the accompanied glitch. Such a glitch-outburst connection is clearly not expected in a source not driven by rotational energy.

The aim of this work is to extend the analysis of [8, 9, 11, 12] on the possibility of SGRs/AXPs as rotation-powered NSs. In order to proceed with such an analysis, we first compute in section 2 the structure properties of NSs, and then in section 3 we focus on giving estimates for the surface magnetic field using both realistic structure parameters of NSs and a general relativistic model of a rotating magnetic dipole. We then show in section 4 that indeed nine of the currently confirmed twenty three SGR/AXPs can be ordinary rotation-powered pulsars. This analysis leads to a precise prediction of the range of NS masses, obtained here by making use of selected up-to-date nuclear equations of state (EOS), in both cases of local and global charge neutrality. We also show in section 5 an analysis of the glitch/outburst connection in the 9 aforementioned sources. Finally, in Section 6, we summarize the main conclusions and remarks.

2. Neutron star structure

In order to compute the rotational energy loss of a NS as a function of its structure parameters, e.g. mass and radius, we need to construct the equilibrium configurations of a uniformly rotating NS in the range of the observed periods. In the case of both static and rotating NSs, the Tolman-Oppenheimer-Volkoff (TOV) system of equations is superseded by the Einstein-Maxwell system of equations coupled to the general relativistic Thomas-Fermi equations of equilibrium, giving rise to what we have called the Einstein-Maxwell-Thomas-Fermi (EMTF) equations. In the TOV-like approach, the condition of local charge neutrality is applied to each point of the configuration, while in the EMTF equations the condition of global charge neutrality is imposed. The EMTF equations account for the weak, strong, gravitational and electromagnetic interactions within the framework of general relativity and relativistic nuclear mean field theory. In this work we shall use both global (EMTF) and local (TOV) charge neutrality to compare and contrast their results. We have recently discussed this point in [14], and for the rotational periods as the ones observed in SGR/AXPs ($P \sim 2$–12 s), the structure of the rotating NS can be accurately described by small departures from the spherically symmetric case [15, 16], for instance using Hartle's formalism [17]. Following this method we compute rotating configurations, accurate up to second-order in Ω, with the same central density as the seed static non-rotating configurations. The mass-radius relation for non-rotating configurations in the cases of global and local charge neutrality are shown in figure 1 of [14]. For the rotation periods of interest here, the mass-equatorial radius relation of the uniformly rotating NSs practically overlaps the one given by the static sequence (see figure 1 in [16]). Thus, we take here advantage of that result and consider hereafter, as masses and corresponding radii, the values of the non-rotating NSs. Although in general there is a dependence of all the structure parameters on the nuclear EOS, we use below, without loss of generality and for the sake of exemplification, only the GM1 EOS. Similar qualitatively and quantitatively results are obtained for the other EOSs.

3. Surface Magnetic Field

We have recently showed in [14, 18] that the range of P for SGRs and AXPs is similar to the one concerning the high-magnetic field pulsar class, we can directly apply the results of [16], taking into account only the most relevant of the above corrections for the present range of periods, namely the finite-size correction.

Figures 1 and 2 show our theoretical prediction for the surface magnetic fields of the SGR/AXPs as a function of the NS mass, using the general relativistic formula, for the GM1 parametrization and for the global and local charge neutrality cases, respectively (see [14], for details). We find that, both in the cases of global and local neutrality, some of the sources have inferred magnetic fields lower than the critical value, B_c, for some range of NS masses. Clearly this set of sources includes SGR 0418+5729, Swift J1822.3-1606 and 3XMM J185246.6+003317, which are already known to show this feature even using fiducial NS parameters and the classic magnetic dipole model (see e.g., [10]).

4. Efficiency of SGRs/AXPs

Another important quantity for the identification of the nature of SGRs and AXPs is the ratio between the observed luminosity and the rotational energy loss. We know that such a ratio is a function of the NS mass, hence the NS radius, via the moment of inertia. For SGR/AXPs the dominant emission is in X-rays, thus we analyze all the possible values of the ratio $L_X/\dot{E}_{\rm rot}$ in the entire parameter space of NSs. As we demonstrate below, some SGRs and AXPs allow a wide range of masses for which $L_X/\dot{E}_{\rm rot} < 1$, implying a possible rotation-powered nature for those sources. Figure 3 show the X-ray luminosity to rotational energy loss ratio as a function of the NS mass, for both global and local charge neutrality configurations. We can see from these

IOP Conf. Series: Journal of Physics: Conf. Series **861** (2017) 012003 doi:10.1088/1742-6596/861/1/012003

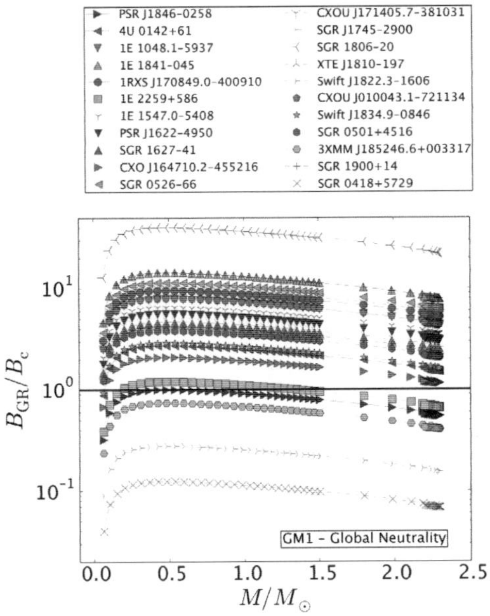

Figure 1. Magnetic field $B_{\rm GR}$ in units of critical field B_c, as function of the mass (in solar masses) in the global charge neutrality case.

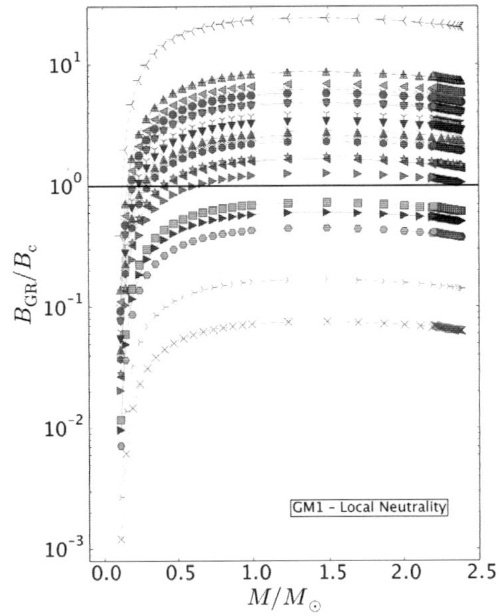

Figure 2. Magnetic field $B_{\rm GR}$ in units of critical field B_c, as function of the mass (in solar masses) in the local charge neutrality case.

figures that nine out of the twenty three SGR/AXPs could have masses in which $L_X < \dot{E}_{\rm rot}$, and therefore they could be explained as ordinary rotation-powered NSs. Such sources are: Swift J1834.9–0846, PSR J1846–0258, 1E 1547.0–5408, SGR J1745-2900, XTE J1810–197, PSR J1622–4950, SGR 1627–41, SGR 0501+4516, CXOU J171405.7381031 (see Table 1). In view of the proximity of some of the sources to the line $L_X/\dot{E}_{\rm rot} = 1$ (see e.g., SGR 1900+14, SGR 0418+5729, and Swift J1822.3–1606), and the currently poorly constrained determination of the distance to the sources, there is still the possibility of having additional sources as rotation powered NSs.

In this line, two sources are particularly interesting, namely SGR 1900+14 and SGR 1806-20, which appear very close (but above) to the limit of becoming rotation-powered NSs. The soft X-rays spectra of SGRs and AXPs are usually well fitted by a blackbody+power-law (BB+PL) spectral model. The BB temperature is usually of the order $k_BT \sim 0.5$ keV and surface radii of the emitting region are ~ 1 km. In the case of a NS, one could interpret the BB component as due to the surface temperature of the NS, while the PL component necessarily has a non-thermal nature and must be due to magnetospheric processes. Within this interpretation, the request that the rotational energy loss of a NS pays also for the contribution of the BB to the luminosity is unnecessarily rigorous since the surface BB luminosity can come from the thermal reservoir of the NS. Thus, the above ratio $L_X/\dot{E}_{\rm rot}$ becomes an upper limit to the actual efficiency for the conversion of rotational into electromagnetic energy. We have recently applied this interpretation to the above two sources (see [14], for more details). Figure 4 shows the ratio $L_X^{\rm PL}/\dot{E}_{\rm rot}$ as a function of the NS mass in the case of SGR 1900+14 and SGR 1806–20, adopting the GM1 EOS. It is clear from this analysis the importance of identifying the different contributions to the emission of the object. There is no doubt that the subtraction of the surface temperature contribution to the total flux in soft X-rays can be important for the correct identification of the nature of these sources: now there is a range of masses for which the luminosity to rotational

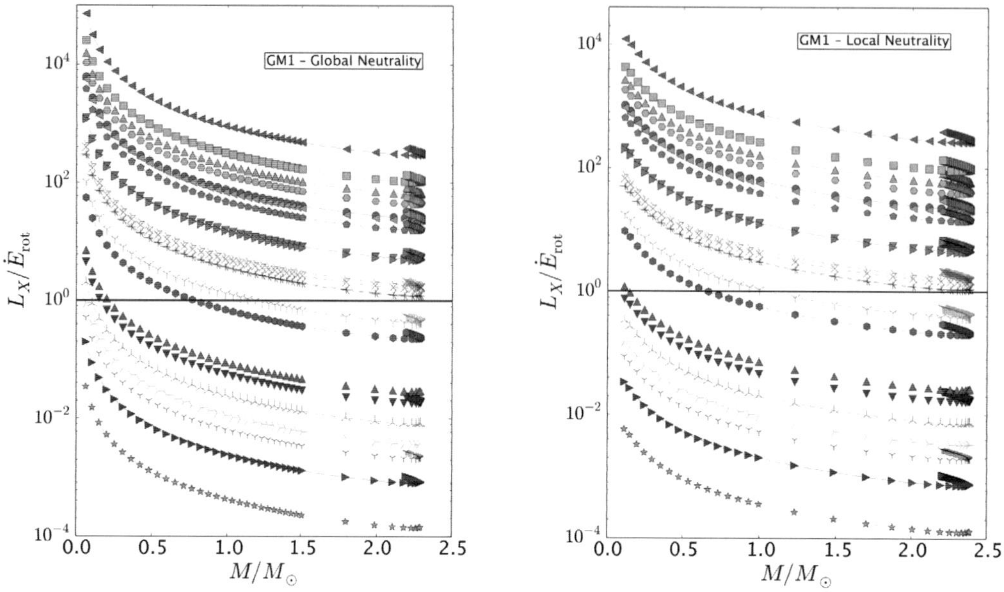

Figure 3. Ratio between the observed X-ray luminosity L_X and the loss of rotational energy $\dot{E}_{\rm rot}$ versus total mass of the rotating NS, in units of $M_\odot$, for the global (left panel) and local (right panel) neutrality cases.

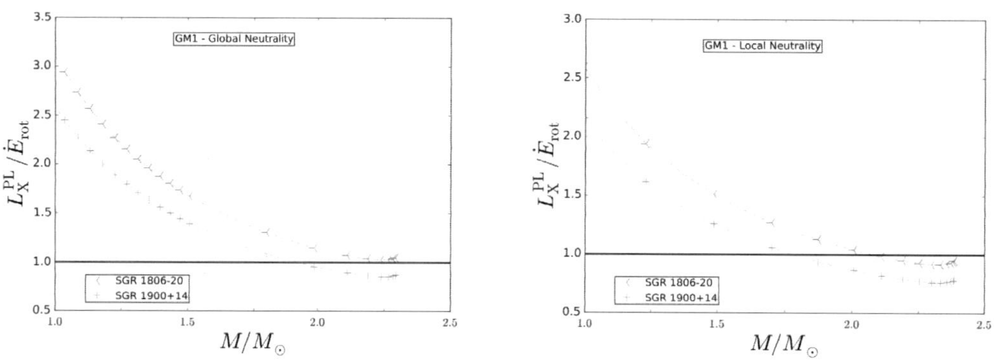

Figure 4. Ratio between the observed X-ray luminosity L_X and the loss of rotational energy $\dot{E}_{\rm rot}$ versus total mass of the rotating NS, in units of $M_\odot$, for the global and local neutrality cases.

energy loss ratio becomes smaller than unity. Again, it is worth recalling that there are still additional effects which could improve the above analysis: distances to the sources are not known accurately; the spectrum could be also fitted by a different spectral model (e.g. a double blackbody); the NS EOS is still unknown and so the moment of inertia and radius for a given mass might be different. These effects might lead to a different value of the total luminosity, contribution of the star's surface temperature to it, and rotational energy loss. Clearly the above analysis can be extended to all the other SGRs and AXPs, and in the case of the nine sources already identified with $L_X/\dot{E}_{\rm rot} < 1$, it will further diminish this ratio.

Table 1. The observational properties of nine sources explained as ordinary rotation-powered NSs. The parameters P,$\dot{P}$ and L_X have been taken from the McGill online catalog at www.physics.mcgill.ca/pulsar/magnetar/main.html. The surface magnetic field B_{NS} are calculated using fiducial neutron star model parameters.

Source	P (s)	$\dot{P}(\times 10^{-11}$ s/s)	$B_{NS}^{\rm fiducial}(\times 10^{14}$ G)	$L_X(\times 10^{33}$ erg/s)
SGR 0501+4516	5.8	0.59	1.9	0.81
1E 1547.0-5408	2.07	4.77	3.2	1.3
PSR J1622-4950	4.33	1.7	2.7	0.44
SGR 1627-41	2.59	1.9	2.2	3.6
CXOU J171405.7-381031	3.8	6.4	5.0	56
SGR J1745-2900	3.76	1.38	2.3	0.11
XTE J1810-197	5.54	0.77	2.1	0.043
Swift J1834.9-0846	2.48	0.79	1.4	0.0084
PSR J1846-0258	0.33	0.71	0.49	19

5. Glitches and bursts

It is natural to expect that in a rotation-powered star the energetics of its bursting activity of SGRs/AXPs could be explained by the gain of rotational energy during the glitch, $\Delta E_{\rm rot} = -2\pi^2 I \Delta P/P^3$ [8]. We have shown in the last section that nine of the twenty three SGR/AXPs can be explained as rotation-powered NSs. Thus, it is mandatory to scrutinize the outburst data of SGR/AXPs, to seek for associated glitches, and assess if the NS glitch rotational energy can explain the transient activity. It is important to start our analysis by recalling the case of PSR J1846–0258 ($L_X < \dot{E}_{\rm rot}$ when fiducial parameters to the NS are adopted), which showed an outburst activity in 2006 [19]. In view of the possible NS rotation-power nature of PSR J1846–0258, Malheiro et al. [8] assumed that the outburst with energy $(3.8\text{–}4.8) \times 10^{41}$ erg [20] was triggered by a glitch in the NS, and found that a glitch of $|\Delta P|/P \sim (1.73\text{–}2.2) \times 10^{-6}$ could explain the outburst of 2006. This theoretical result is in full agreement with the observational analysis of [21], who showed that indeed a major glitch with $|\Delta P|/P \sim (2\text{–}4.4) \times 10^{-6}$ occurred.

Following this reasoning, we proceed to theoretically predict the values of $|\Delta P|/P$ which could explain the energetics of the bursts of those SGR/AXPs with $L_X < \dot{E}_{\rm rot}$, obtained by assuming that $|\Delta E_{\rm rot}|$ equals the observed energy of the burst event, $E_{\rm burst}$, namely

$$\frac{|\Delta P|}{P} = \frac{E_{\rm burst} P^2}{2\pi^2 I}. \tag{2}$$

Figure 5 shows the value of $|\Delta P|/P$ obtained from equation (2), as a function of the NS mass. From this set of nine sources, the only glitch detected was the one of PSR J1846–0258 in June 2006 [21], whose observed value of $|\Delta P|/P$ is shown with the gray-shaded area in these plots. Indeed a minimum mass for the NS can be established for this source by requesting that 1) the entire moment of inertia is involved in the glitch and 2) that the theoretical value of $|\Delta P|/P$ coincides with the observed value: we obtain $M_{\rm min} = 0.72\ M_\odot$ and $M_{\rm min} = 0.61\ M_\odot$ for the global and local charge neutrality cases, respectively. On the other hand, if we substitute the moment of inertia I in equation (2) by $I_{\rm glitch} = \eta I$ where $\eta \leq 1$, being $I_{\rm glitch}$ the moment of inertia powering the glitch, then we can obtain a lower limit for the parameter η: we obtain $\eta = 0.20$ and $\eta = 0.18$ for the global and local charge neutrality cases, respectively. Tables 2 and 3 show the theoretically predicted value of $|\Delta P|/P$ assuming the mass of the NS is larger than 1 $M_\odot$ and $\eta = 1$, in the cases of global and local charge neutrality, respectively.

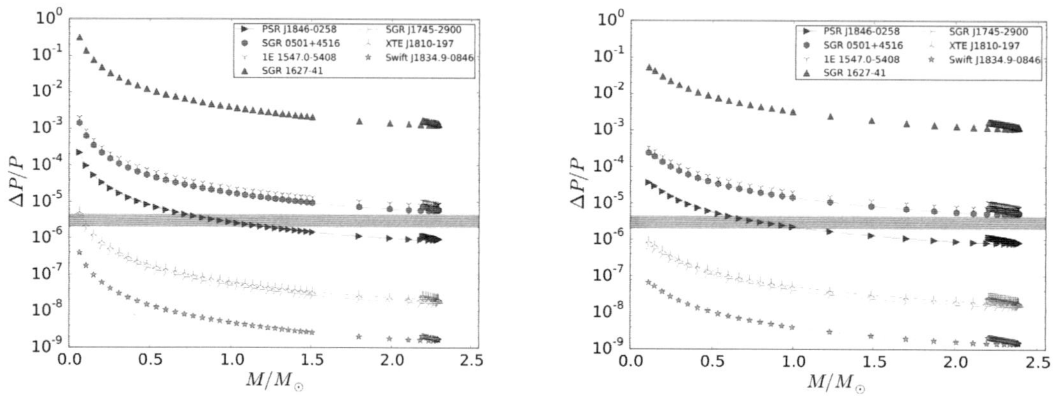

Figure 5. Inferred fractional change of rotation period during the glitch, $\Delta P/P$, obtained by equating the rotational energy gained during the glitch, $\Delta E_{\rm rot}$, to the energy of the burst, for globally neutral (left panel) and locally neutral (right panel) NSs. In this example the NS obeys the GM1 EOS. The gray-shaded area corresponds to the value of $|\Delta P|/P$ in the observed glitch of PSR J1846–0258 in June 2006 [21].

Table 2. Predicted values of $|\Delta P|/P$ assuming rotation-powered NSs - Global Charge Neutrality

Source name	Total isotropic burst energy [erg]	Predicted $\lvert\Delta P\rvert/P$ for $M > 1\,M_\odot$
PSR J1846–0258	4.8×10^{41}	$(8.8 \times 10^{-7} - 2.6 \times 10^{-6})$
1E 1547.0–5408	1.1×10^{41}	$(8.1 \times 10^{-6} - 2.4 \times 10^{-5})$
XTE J1810–197	4.0×10^{37}	$(2.1 \times 10^{-8} - 6.3 \times 10^{-8})$
SGR 1627–41	1.0×10^{41}	$(1.0 \times 10^{-5} - 3.8 \times 10^{-5})$
SGR 0501+4516	1.0×10^{40}	$(5.7 \times 10^{-6} - 1.7 \times 10^{-5})$
Swift J1834.9–0846	1.5×10^{37}	$(1.6 \times 10^{-9} - 4.8 \times 10^{-9})$
SGR 1745–2900	6.7×10^{37}	$(1.61 \times 10^{-8} - 4.9 \times 10^{-8})$

Table 3. Predicted values of $|\Delta P|/P$ assuming rotation-powered NSs - Local Charge Neutrality

Source name	Total isotropic burst energy [erg]	Predicted $\lvert\Delta P\rvert/P$ for $M > 1\,M_\odot$
PSR J1846–0258	4.8×10^{41}	$(7.9 \times 10^{-7} - 2.2 \times 10^{-6})$
1E 1547.0–5408	1.1×10^{41}	$(7.2 \times 10^{-6} - 2.0 \times 10^{-5})$
XTE J1810–197	4.0×10^{37}	$(1.9 \times 10^{-8} - 5.3 \times 10^{-8})$
SGR 1627–41	1.0×10^{41}	$(1.1 \times 10^{-5} - 3.2 \times 10^{-5})$
SGR 0501+4516	1.0×10^{40}	$(5.0 \times 10^{-6} - 1.4 \times 10^{-5})$
Swift J1834.9–0846	1.5×10^{37}	$(1.4 \times 10^{-9} - 3.9 \times 10^{-9})$
SGR 1745–2900	6.7×10^{37}	$(1.4 \times 10^{-8} - 4.1 \times 10^{-8})$

6. Conclusions

By extending the analysis of [8, 9, 11, 12] on the possibility of some SGRs and AXPs as rotation-powered NSs, we have explored in this work the consequences of a realistic model for NSs on the inference of the astrophysical observables of SGRs and AXPs. We showed that the X-ray luminosity of nine sources (Swift J1834.9–0846, PSR J1846–0258, 1E 1547.0–5408, SGR J1745–2900, XTE J1810–197, PSR J1622–4950, SGR 1627–41, SGR 0501+4516, CXOU J171405.7381031), which corresponds to the 40% of the entire SGR/AXP population, can be explained via the loss of rotational energy of NSs, thus falling into the family of ordinary rotation-powered pulsars. We have also given their associated lower mass limits. We have also shown that the interpretation of the blackbody component of the spectrum in soft X-rays as due to the surface temperature of the NS allow both SGR 1900+14 and SGR 1806-20 to be added to the above list, leading to $\approx$ 50% of the SGR/AXP population that could be explained via rotation-powered NSs. We also argued that the observational uncertainties in the determination of the distances and/or luminosities, as well as the uncertainties in the NS nuclear EOS, leave still room for a possible explanation in terms of spindown power for additional sources.

We have shown that both surface magnetic fields and the efficiencies of pulsars are overestimated, if computed with NS fiducial parameters. We estimated the surface magnetic fields of SGR/AXPs as a function of NS masses, from the self-consistent general relativistic model of the radiation of a rotating magnetic dipole in vacuum. We have explored for the nine sources with $L_X < \dot{E}_{\rm rot}$ the possibility that the energetics of their bursting activity, $E_{\rm burst}$, can be explained from the rotational energy gained in an associated glitch, $\Delta E_{\rm rot}$. We thus computed lower limits to the fractional change of rotation period of NSs caused by glitches, $|\Delta P|/P$, by requesting $\Delta E_{\rm rot} = E_{\rm burst}$. The fact that there exist physically plausible solutions for $|\Delta P|/P$ reinforces the possible rotation-powered nature for these sources (e.g., the case of PSR J1846–0258).

It is interesting to notice that four of the above nine sources, namely 1E 1547.0–5408, SGR J1745–2900, XTE J1810–197, and PSR J1622–4950, are actually the only ones with detected radio emission [10, 22, 23], as expected from ordinary rotation-powered pulsars, although it has interestingly different properties from that typical of radio pulsars. We shall perform a deeper analysis of the radio emission from SGRs and AXPs in a forthcoming publication. It is worth mentioning that, among these nine sources, we can also find six sources with possible associations with supernova remnants (SNRs): Swift J1834.9–0846 associated with SNR W41 (see [13] for the recent detection of a wind nebula around this source), PSR J1846–0258 (with SNR Kes75), 1E 1547.0–5408 with SNR G327.24–0.13, PSR J1622–4950 with SNR G333.9+0.0, SGR 1627–41 with SNR G337.0-0, CXOU J171405.7–381031 with SNR CTB37B [10]. All these findings support the description of some SGRs/AXPs as belonging to the class of RPPs. If these associations are observationally confirmed, then the NS was born from the core-collapse of a massive star which triggered the SN explosion.

Theoretical predictions and observations in additional bands of the electromagnetic spectrum such as the optical, high and ultra-high gamma rays and cosmic-rays are encouraged to discriminate amongst the different models and elucidate the actual nature of SGRs and AXPs.

References

[1] Mereghetti S 2008 *A&AR* **15** 225
[2] Duncan R C and Thompson C 1992 *ApJ* **392** L9
[3] Thompson C and Duncan R C 1995 *MNRAS* **275** 255
[4] Malov I F 2010 *Astronomy Reports* **54** 925
[5] Trümper J E, Dennerl K, Kylafis N D, Ertan Ü and Zezas A 2013 *ApJ* **764** 49
[6] Xu R X, Tao D J and Yang Y 2006 *MNRAS* **373** L85
[7] Ouyed R, Leahy D and Niebergal B 2011 *MNRAS* **415** 1590
[8] Malheiro M, Rueda J A and Ruffini R 2012 *PASJ* **64** 56
[9] Coelho J G and Malheiro M 2014 *PASJ* **66** 14
[10] Olausen S A and Kaspi V M 2014 *ApJSS* **212** 6
[11] Coelho J G and Malheiro M 2013 *arXiv:1307.8158*
[12] Coelho J G and Malheiro M 2015 in *Thirteenth Marcel Grossmann Meeting: On Recent Developments in*

Theoretical and Experimental General Relativity Astrophysics and Relativistic Field Theories ed. Rosquist K 2465–2467 *arXiv:1303.0863*
[13] Younes G, Kouveliotou C, Kargaltsev O *et al.* 2016 *ApJ* **824** 138
[14] Coelho J G *et al.* 2017 *A&A* **599** A87
[15] Belvedere R, Boshkayev K, Rueda J A and Ruffini R 2014 *Nuclear Physics A* **921** 33
[16] Belvedere R, Rueda J A and Ruffini R 2015 *ApJ* **799** 23
[17] Hartle J B 1967 *ApJ* **150** 1005
[18] de Lima R C R *et al.* 2015 *AIP Conference Proceedings* **1693** 030009
[19] Gavriil F P, Gonzalez M E, Gotthelf E V, *et al.* 2008 *Science* **319** 1802
[20] Kumar H S and Safi-Harb S 2008 *ApJ* **678** L43
[21] Kuiper L and Hermsen W 2009 *A&A* **501** 1031
[22] Lobato R V, Coelho J and Malheiro M 2015 *Journal of Physics Conference Series* **630** 012015
[23] Lobato R V, Malheiro M and Coelho J G 2016 *International Journal of Modern Physics D* **25** 1641025

IOP Conf. Series: Journal of Physics: Conf. Series **861** (2017) 012004 doi:10.1088/1742-6596/861/1/012004

Phase lags of quasi-periodic oscillations across source states in the low-mass X-ray binary 4U 1636–53

Marcio G B de Avellar

Instituto de Astronomia, Geofísica e de Ciências Atmosféricas, Universidade de São Paulo, Rua do Matão 1226, 05508-090, São Paulo, Brazil

E-mail: mgb.avellar@iag.usp.br

Abstract. The majority of attempts to explain the origin and phenomenology of the quasi-periodic oscillations (QPOs) detected in low-mass X-ray binaries invoke dynamical models, and it was just in recent years that renewed attention has been given on how radiative processes occurring in these extreme environments gives rise to the variability features observed in the X-ray light curves of these systems. The study of the dependence of the phase lags upon the energy and frequency of the QPOs is a step towards this end. The methodology we developed here allowed us to study for the first time these dependencies for all QPOs detected in the range of 1 to 1300 Hz in the low-mass X-ray binary 4U 1636–53 as the source changes its state during its cycle in the colour-colour diagram. Our results suggest that within the context of models of up-scattering Comptonization, the phase lags dependencies upon frequency and energy can be used to extract size scales and physical conditions of the medium that produces the lags.

1. Introduction

Low-mass X-ray binaries containing both, neutron stars (NS-) and black holes (BH-), show rapid (seconds to milliseconds) X-ray variability components in the power spectral density (PSD) of their X-ray light curves. In the case of NS-LMXBs besides a power-law component at very low frequencies, we identify quasi-periodic oscillations (QPOs) ranging from 0.01 to 1300 Hz (see [1] for a review). The frequencies of most of the QPOs correlate with each other in a given source and other of their properties like the fractional amplitude (rms) and the quality factor (Q), depend also on the spectral state of the source (see, for example, [2, 3]). In particular frequency-frequency correlations of some timing features in X-ray binaries containing black holes, neutron stars or white dwarfs (e.g. [4, 5, 6, 7]) suggests that the origin of these features does not depend on the nature of the central compact object, but are due to a physical component that is common to all these systems, likely the accretion disc or a corona.

Renewed attention has been given to the energy- and frequency-dependence of the time (or phase) lags of the X-ray signal (e.g. [8, 9, 10, 11, 12]), since it is believed that they possibly encode geometrical properties of the medium that produces the variability and the radiative processes occurring around the system. Time/phase lags are defined as Fourier-frequency-dependent measurements of the time (phase) delays between two concurrent and correlated signals, i.e. two light curves of the same source, in two different energy bands (see [8] for more details on the subject).

Compton up- and down-scattering are the most common mechanism to explain time/phase lags of X-ray photons in LMXBs. In this case, photons produced in the accretion disc of BH-

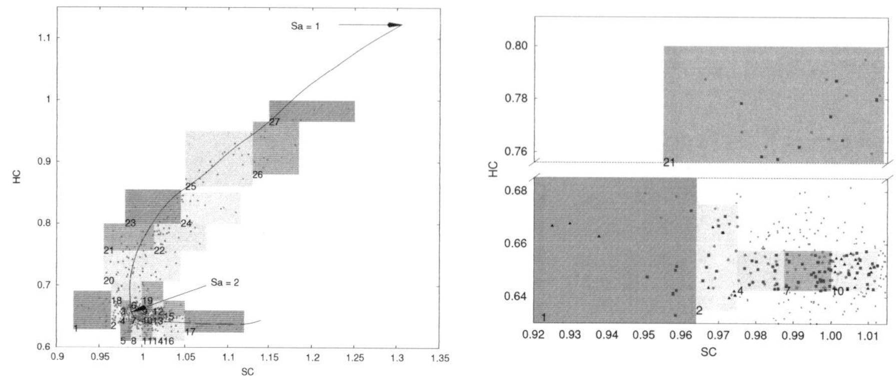

Figure 1. Left panel: the colour-colour diagram of 4U 1636–53 with the 27 originally defined boxes. Right panel: the boxes that were further subdivided; each one displays groups of observations (highlighted by different colour points) with similar PDSs grouped together creating the subdivisions. Each point corresponds to a single RXTE observation. We only include in this diagram the 511 observations with known kHz QPOs. For each box we fitted the QPOs in the average PSD. The dark line is the spline used to parametrise the position of the source on the CCD.

and NS-LMXBs, or even in the surface of the neutron star and/or its boundary layer in the case of NS-LMXBs, are scattered in a corona of hot electrons (see, for example, [13, 14, 15, 9, 16, 12] for a discussion about the models).

Here we study of the phase lags of all QPOs detected above 1 Hz in a large dataset of RXTE observations of the NS-LMXB 4U 1636–53 as a function of the position of the source on the colour-colour diagram (CCD). In §2 we describe our observations and methodology; in §3 we present the results for the frequency dependence of the phase lags (3.1) and for the energy dependence (3.2). We discuss our results and conclude in §4. For a more detailed discussion of the results presented here we kindly refer the reader to our paper [17].

2. Observations and data analysis

We defined in total seven narrow energy bands and two broad bands. The mean energies of the narrow bands are 4.2 keV, 6.0 keV, 8.0 keV, 10.2 keV, 12.7 keV, 16.3 keV and 18.9 keV and of the broad bands are 7.1 keV and 16.0 keV. These are the same energy bands defined in [10]. Additional details in [17].

To study the phase lags of all QPOs as a function of the position of the source in the CCD[1], we first defined boxes on it following two criteria: i) the boxes should be small to avoid changes in the physical conditions of the source, which would imply changes in the shape of the PSDs but with enough observations within each box in order to build the average spectra and ii) we then compared the shapes of the PSDs of each individual observation within each box both, visually and statistically[2]. If we identified, within a box, groups of observations with similar PDSs, then we further subdivided the box by grouping these similar observations. We ended up with 37 boxes as depicted in Figure 1.

The second step was to parametrise the position on the CCD with the quantity S_a as in

[1] The hard colour (HC) was defined as the 9.7-16.0 keV/6.0-9.7 keV count rate ratio and the soft colour (SC) as the 3.5-6.0 keV/2.0-3.5 keV count rate ratio. See [3].

[2] By statistically we mean that we analysed the average of two probabilities of similarity, KS ([19]) and MTT ([20]), and we say that two PDSs are similar if the average of the probabilities are greater than 65%.

[18, 21]. See Figure 1. Here we fixed $S_a = 1$ at (SC,HC) = (1.31,1.12) and $S_a = 2$ at (SC,HC) = (0.988,0.657). This is a phenomenological approach where we approximate the shape traced by the observations on the CCD with a spline and the quantity S_a along the spline is a measure of the position of the source on CCD.

The QPOs seen in the average PSD of each box were fitted with Lorentzian functions and the criteria we used to assess whether a QPO is significantly detected is if the ratio between the integral power and its uncertainty is larger than 3, but allowing to be $2.5 < \sigma < 3$ if the QPO lies in all the correlations mentioned above. In total, we detected eight different QPOs labelled L_{b2}, L_b, L_{LF}, L_h, L_{hHz}, $L_{hHz-harm}$, L_l, L_u as in [3].

For the energy-dependence of the lags, we calculated the phase lags of the photons of each narrow energy band relatively to the 10.2 keV band[3]. For the frequency-dependence of the lags, we calculated the phase lags of all photons in the broad band above 12 keV relatively to all photons in the broad band below 12 keV.

Finally, to calculate the lags of the QPOs we average the energy- and frequency-dependent lags over a frequency interval of one FWHM around the centroid frequency of the QPO, except in the cases where two QPOs are broad enough to overlap each other. In these cases we averaged over a fraction of the FWHM ranging from one half to one fifth. All phase lags are given in units of 2π rad.

3. Results

3.1. Frequency dependence of the phase lags

In Figure 2 we plot the phase lags of six out of the eight QPOs detected in the PSD as a function of the frequency of the QPO for photons with mean energy $\simeq 16.0$ keV relatively to photons with mean energy $\simeq 7.1$ keV. Uncertainties are given at a 68% confidence level. In Figure 3 we plot the phase lags of the QPO specified in each of the panels as a function of the quantity S_a.

In Table 1 we show constant fits of the phase lags as a function of S_a of all QPOs in order to compare them with each other. We see from Table 1 that the average phase lags are soft in the cases of L_{b2} and L_l. The phase lags of L_{b2} are significantly different from the phase lags of L_b, L_h and L_u, the phase lags of L_b are significantly different from the lags of L_h and L_l, the phase lags of L_h are significantly different from the lags of L_l, and the phase lags of L_l are significantly different from the lags of L_u. The phase lags of L_b, L_{hHz} and L_u are consistent with each other within 3σ. Also given in this Table are the average *time* lags as a function of S_a for each QPO, which we will use later.

The time lags of L_{b2} are inconsistent with the time lags of all other QPOs by more than 3σ. The time lags of L_b are consistent within 3σ with the time lags of all other QPOs but L_h. The time lags of L_h, in turn, are inconsistent within 3σ with the time lags of L_{hHz}, L_l and L_u. At last, the time lags of L_l are different from the time lags of L_u by more than 3σ, a result previously known [10].

3.2. Energy dependence of the phase lags

The results in Section 3.1 show that the phase lags do not depend strongly on the frequency of the QPO or S_a, except maybe for the lower kHz QPO (bottom left panel of Figure 2 and 3, respectively). Based on this, we averaged the phase lags over frequency to calculate the phase lags versus energy for each QPO under consideration (the details can be found in [17]). Notice for L_l this procedure is not completely correct since there is a dependence of the lags upon the frequency. The results of this procedure are shown in Figure 4. As with the frequency

[3] Being relative quantities, we chose the 10.2 keV band as the reference band that the lags of photons with other energies are calculated relatively to, because in this band both the variability and the source intensity are high, which reduces the uncertainties of the of time-lag measurements ([22]).

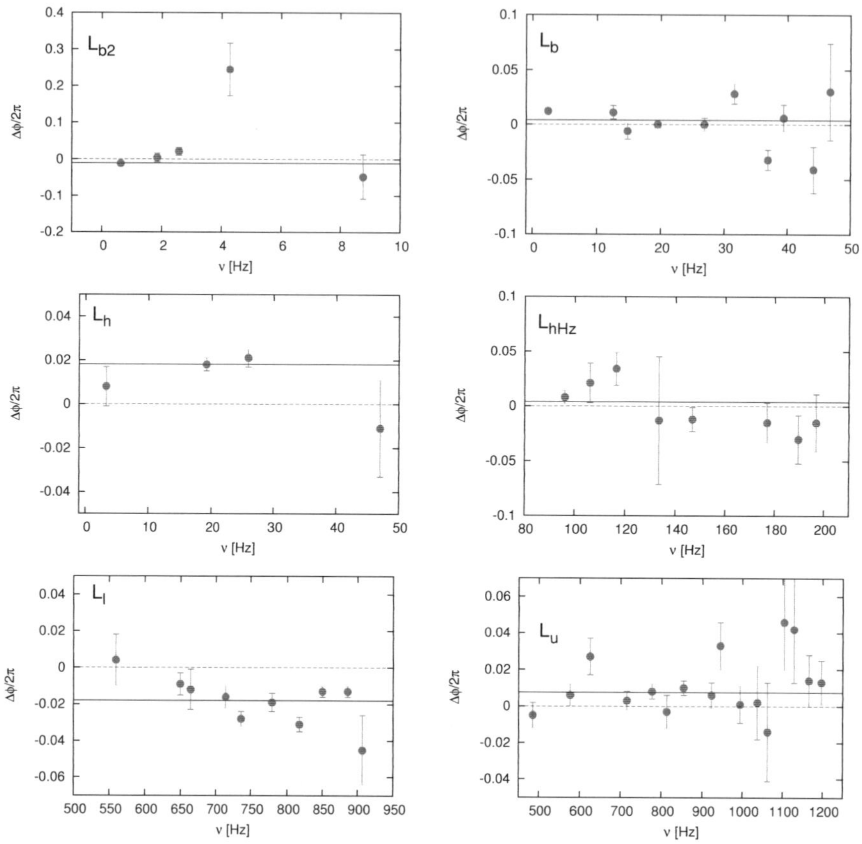

Figure 2. Phase lags, in units of 2π, of the different QPO components of 4U 1636–53 as a function of the frequency of the corresponding QPO. The lags represent the delay of photons with mean energy $\simeq$ 16 keV relative to the photons with mean energy $\simeq$ 7 keV. Left top panel: the QPO where the PDS shows a second break. Right top panel: the QPO at the break frequency. Left middle panel: the *hump* QPO. Right middle panel: the hecto-hertz (hHz) QPO. Left bottom panel: the lower kHz QPO. Right bottom panel: the upper kHz QPO. The blue lines are the best fit constant to the lags. The errors correspond to the 1σ confidence level.

Table 1. Weighted average phase lags and time lags as function of S_a (see Figure 3) for the L_{b2}, L_b, L_h, L_{hHz}, L_l and L_u QPOs of the NS-LMXB 4U 1636–53 of all photons in the broad band 12-20 keV relatively to all photons in the broad band 4-12 keV.

QPO	phase lag [2π rad]	χ^2	dof	time lag [msec]
L_{b2}	-0.010 ± 0.002	25.6	5	-8.7 ± 2.1
L_b	0.005 ± 0.002	40.7	11	0.05 ± 0.09
L_h	0.018 ± 0.002	4.1	5	0.8 ± 0.1
L_{hHz}	0.004 ± 0.004	13.5	10	0.008 ± 0.035
L_l	-0.017 ± 0.001	13.5	8	-0.020 ± 0.002
L_u	0.008 ± 0.002	19.7	14	0.010 ± 0.002

IOP Conf. Series: Journal of Physics: Conf. Series **861** (2017) 012004 doi:10.1088/1742-6596/861/1/012004

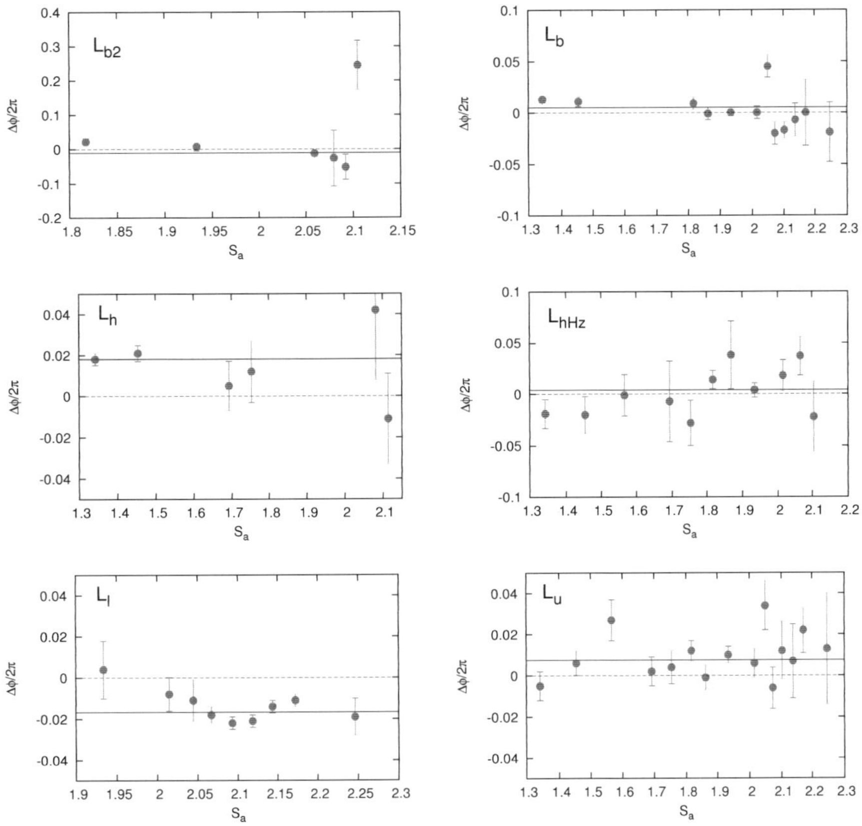

Figure 3. Phase lags, in units of 2π, of the QPOs in Figure 2 as a function of the parameter S_a which is representative of the position of the source in the CCD. The blue lines are the best fit constant to the lags.

dependence, the results for the energy dependence of the phase lags of the kHz QPOs are consistent with the ones obtained in [10].

It is interesting to note, based on the shapes of the lag vs E plots, that the trend with energy, being marginal or not, is an increasing trend for L_b, L_h, L_{hHz}, and L_u, but is decreasing for L_l and possibly for L_{b2}. In the case of L_{b2} the question arises if the phase lags of L_{b2} may be linearly correlated with the phase lags of L_l, although the F-test probability of 0.026 indicates the contrary. The same question arises about a possible anti-correlation between the phase lags of L_h and L_l (F-test probability 0.005).

Comparing the slopes of the lag vs E linear fits for each QPO, we found that we can separate the components into two groups having slopes consistent with each other: On the one hand L_{b2} and L_l, and on the other hand L_b, L_h, L_{hHz}, and L_u. To further investigate the slopes we also calculated the Pearson Product Moment Correlation (PPMC) coefficient for all 15 possible combinations of lag vs lag. This correlation coefficient is a statistic that calculates the actual relationship between two variables.

We have not found any correlation between the lags except in the case of the phase lags of L_h in relation to the phase lags of L_{hHz}. Even so, we still can separate the lags in two groups as mentioned above.

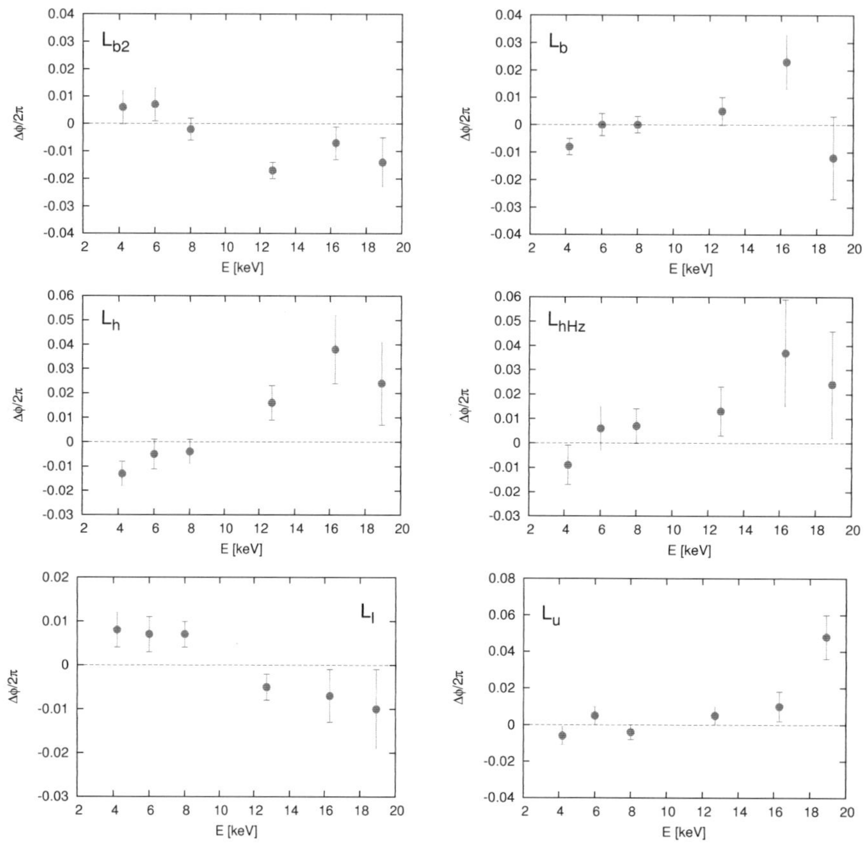

Figure 4. Phase lags, in units of 2π, of the different QPO components of 4U 1636–53 as a function of energy. The lags represent the delay of photons in the bands whose average energies are given by the values in the x-axis relative to the photons in the band whose average energy is 10.2 keV. The panels show the same QPOs as in Figure 2.

4. Discussion

4.1. Frequency dependence

We see from Figure 3 that the phase lags are practically independent of S_a, except for L_l (for which we see a complex behaviour) and maybe for L_{hHz}. A scenario where the accretion flow has two components ([23, 24, 25]) could explain the fluctuations of the phase lags of L_l to softest values and the hard lags of the L_u that would "keep its constancy" through compensations in the properties of the coronal component.

We derived upper limits to the size of the medium in which the time lags are produced by light travel time arguments in the context of models for the lags that involve reflection off the disc or Comptonization. From the values in Table 1 we can roughly estimate the scale size, a, of the medium where the lags of each QPO are produced with the expression $a \sim c\Delta t \frac{k_b T_e}{m_e c^2} \frac{4\tau}{ln(E_2/E_1)}$ (see, e.g. [13]). Here, $\tau \equiv 5$ is the optical depth and $k_b T_e \equiv 5$ keV is the plasma temperature (typical values in modelling the spectra of 4U 1636–53, e.g. [26]), Δt are the time lags, $E_2 = 16.0$ keV and $E_1 = 7.1$ keV are the energies of the photons in the broad energy bands we chose. The electron density comes from $n_e = \tau/(a\sigma_T)$. See Table 2.

The values in Table 2 suggest that at least two components, one for L_b, L_{hHz}, L_l and L_u with density $\sim 10^{19-20}$ cm^{-3} and another for L_{b2} and L_h with density $\sim 10^{17-18}$ cm^{-3}, are needed

IOP Conf. Series: Journal of Physics: Conf. Series **861** (2017) 012004 doi:10.1088/1742-6596/861/1/012004

Table 2. Estimates of the scale size and density of the medium where L_{b2}, L_b, L_h, L_{hHz}, L_l and L_u QPOs of the NS-LMXB 4U 1636–53 are produced. Here, $c\Delta t$ gives the light travel time corresponding to the measured time lags (see Table 1), a is the size scale and n_e is the electronic density of the medium.

QPO	$c\Delta t$ [km]	a [km]	n_e [10^{20} cm^{-3}]
L_{b2}	2610 ± 630	628.6 ± 151.7	0.0012 ± 0.0003
L_b	15 ± 27	3.6 ± 6.5	0.21 ± 0.37
L_h	240 ± 30	57.8 ± 7.2	0.013 ± 0.002
L_{hHz}	2.4 ± 10.5	0.58 ± 2.53	1.3 ± 5.7
L_l	6.3 ± 0.6	1.52 ± 0.14	0.50 ± 0.05
L_u	3.0 ± 0.6	0.72 ± 0.14	1.04 ± 0.21

for producing the measured phase lags. Notice the scale sizes for the first set are quite small (0.5 to 3.5 km), while for the latter are much bigger, 60 km and 630 km.

On one hand, the phase lags of L_{b2} and L_{hHz} could have the same mechanism of the lags seen in AGNs, given the similarity of their lag spectrum (see [27]). On the other hand, it is tempting to compare our results on L_l and L_u (bottom panels of Figure 2) with those phase lags measured by [28]. It is possible that the same mechanism that produces the soft lags of the QPO at 35 Hz and the hard lags of the QPO at 67 Hz also applies to the lower and upper kHz QPOs of 4U 1636–36, since they also have opposites signs and are inconsistent with each other (see [28] for additional discussion about this).

Based on our results and on the findings of [29], for example, we suggest that the time lags of the upper kHz QPO encode the properties of the medium at the magnetospheric radius (where the upper kHz QPO would be produced, at 16 to 21 km in our estimations[4], and where the scale size for the lags is $\sim$ 0.72 km), while the time lags of the lower kHz QPOs encode the properties of the medium at the boundary layer and nearby (where the lower kHz QPO would be produced, near the surface of the neutron star, where the scale size for the lags is $\sim$ 1.52 km). This idea would explain why the time lags of these two QPOs are inconsistent with each other and of opposite sign.

4.2. Energy dependence

The trends with energy show resemblance if we group together L_b, L_h, L_{hHz} and L_u on one hand and L_{b2} and L_l on the other. The lags of the first group become harder and those of the second group become softer as the energy increases. These trends may be a hint of a general mechanism behind the production of the lags of the QPOs of each group.

Recently, [12] developed a new model for the energy dependence of the time lags based on a thermal Comptonizing plasma. They were able to obtain soft lags (of the same magnitude as the ones reported here for 4U 1636–53 and the ones reported by [11] for 4U 1608–52) only when there is a variation in the heating rate of the corona and a significant fraction, $\eta = 0.4$, of the photons impinges back into the photon source. They constrained the size of the Comptonizing region to $L = 1.0$ km. The same mechanism could apply to L_{b2} since it also shows soft lags. However, the hard lags found in the other QPOs are beyond this model. This model also explains satisfactorily the increase of the fractional rms with energy, although cannot explain the flattening at energies above 10 keV seen in the data.

[4] We assumed the neutron star mass in the range $M \sim 1.6 - 1.9$ M$_{\odot}$, fixed the radius of the neutron star in 10 km, fixed the magnetic field at $B \sim 10^8$ G and assumed that the source accretes at a rate $0.03 - 0.06$ L$_{\rm edd}$.

IOP Conf. Series: Journal of Physics: Conf. Series **861** (2017) 012004 doi:10.1088/1742-6596/861/1/012004

More recently [30] performed an extensive study of the spectral-timing properties of the lower and upper kHz QPOs in the source 4U 1728–34 and they found that the QPO spectrum is compatible with the one of a Comptonized black body with temperature higher than the temperature of the continuum, which implies that a more compact inner region of the boundary layer is producing the QPO. They also found that the lag-energy spectrum of the lower kHz and the upper kHz QPO are systematically different in the same way we have found in [10] and that they also depend weakly on frequency. In a scenario where the lags of the upper kHz QPO are dominated by reverberation, we see the upper kHz QPO signal as a response of the boundary layer to the variations in the accretion rate. Evidence that broader and lower frequency noise components are also generated by the same kind of mechanism ([31, 32]) would then be corroborated by our results based on the general shapes of the lag-energy spectrum, except for L_{b2} which is similar to the lower kHz QPO L_l. Then, the more coherent signal of the lower kHz QPO may have its origin in a more compact region of the boundary layer itself due to internal oscillations in the heating rate of the boundary layer (as in [9, 12], for example). Our results for the lag energy spectrum of the QPOs seem to point also in that direction.

If extended to include all the other QPOs, these models provide an opportunity to study the dynamic and physical conditions of the Comptonising corona in neutron-star low-mass X-ray binaries.

References

[1] van der Klis M 2006 *Compact stellar X-ray sources* (Cambridge Astrophysics Ser. vol 39) ed Lewin W and van der Klis M (Cambridge: Cambridge University Press) p 39
[2] van Straaten S, van der Klis M, di Salvo T and Belloni T 2002 *Astrophys. J.* **568** 912
[3] Altamirano D, van der Klis M, Méndez M, Jonker P, Klein-Wolt M and Lewin W H G 2008 *Astrophys. J.* **685** 436
[4] Wijnands R A D and van der Klis M 1999 *Astrophys. J.* **514** 939
[5] Psaltis D, Belloni T and van der Klis M 1999 *Astrophys. J.* **520** 262
[6] Mauche C W 2002 *Astrophys. J.* **580** 423
[7] Warner B and Woudt P A 2002 *Mon. Not. R. Astron. Soc.* **335** 84
[8] Nowak M A, Vaughan B A, Wilms J, Dove J B and Begelman M C 1999 *Astrophys. J.* **510** 874
[9] Lee H C, Misra R and Taam R E 2001 *Astrophys. J.* **549** L229
[10] de Avellar M G B, Méndez M, Sanna A and Horvath J E 2013 *Mon. Not. R. Astron. Soc.* **433** 3453
[11] Barret D 2013 *Astrophys. J.* **770** 9
[12] Kumar N and Misra R 2014 *Mon. Not. R. Astron. Soc.* **445** 2818
[13] Sunyaev R A and Titarchuk L G 1980 *Astron. & Astrophys.* **86** 121
[14] Payne D G 1980 *Astrophys. J.* **237** 951
[15] Lee H C and Miller G S 1998 *Mon. Not. R. Astron. Soc.* **299** 479
[16] Falanga M and Titarchuk L 2007 *Astrophys. J.* **661** 1084
[17] de Avellar B, Méndez M, Altamirano D, Sanna A and Zhang G 2016 *Mon. Not. R. Astron. Soc.* **461** 79
[18] Zhang G, Méndez M, Belloni T M and Homan J 2013 *Mon. Not. R. Astron. Soc.* **436** 2276
[19] Kolmogorov A 1933 *Giornale dell' Instituto Italiano degli Attuari* **4** 83
[20] Melo I, Tomášik B, Torrieri G, Vogel S, Bleicher M, Koróny S and Gintner M 2009 *Phys. Rev. C* **80** 024904
[21] Méndez M, van der Klis M, Ford E C, Wijnands R and van Paradijs J 1999 *Astrophys. J.* **511** L49
[22] Vaughan B A and Nowak M A 1997 *Astrophys. J.* **474** L43
[23] Esin A A, McClintock J E and Narayan R 1997 *Astrophys. J.* **489** 865
[24] Meyer-Hofmeister E, Liu B F and Meyer F 2005 *Astron. & Astrophys.* **432** 181
[25] Done C, Gierliński M and Kubota A 2007 *Astron. & Astrophys.* **Rv15** 1
[26] Lyu M, Méndez M, Zhang G and Keek L 2015 *Mon. Not. R. Astron. Soc.* **454** 541
[27] Zoghbi A, Fabian A C, Uttley P, Miniutti G, Gallo L C, Reynolds C S, Miller J M and Ponti G 2010 *Mon. Not. R. Astron. Soc.* **401** 2419
[28] Méndez M, Altamirano D, Belloni T and Sanna A 2013 *Mon. Not. R. Astron. Soc.* **435** 2132
[29] Bult P and van der Klis M 2015 *Astrophys. J. Lett.* **798** L29
[30] Peille P, Barret D and Uttley P 2015 *Astrophys. J.* **811** 109
[31] Uttley P 2004 *Mon. Not. R. Astron. Soc.* **347** L61
[32] Uttley P, Wilkinson T, Cassatella P, Wilms J, Pottschmidt K, Hanke M and Böck M 2011 *Mon. Not. R. Astron. Soc.* **414** L60

IOP Conf. Series: Journal of Physics: Conf. Series **861** (2017) 012005 doi:10.1088/1742-6596/861/1/012005

Ultra-high energy cosmic rays from white dwarf pulsars and the Hillas criterion

Ronaldo V. Lobato[1], **Jaziel G. Coelho**[2] **and M. Malheiro**[3]

[1] Departamento de Física, Instituto Tecnológico de Aeronáutica, ITA - DCTA, Vila das Acácias, São José dos Campos, 12228-900 SP, Brazil
Dipartimento di Fisica, Sapienza Università di Roma, P.le Aldo Moro 5, I-00185 Rome, Italy
ICRANet, P.zza della Repubblica 10, I-65122 Pescara, Italy
[2] Divisão de Astrofísica, Instituto Nacional de Pesquisas Espaciais - DAS/INPE/MCTI, São José dos Campos, 12227-010, São Paulo, Brazil
[3] Departamento de Física, Instituto Tecnológico de Aeronáutica, ITA - DCTA, Vila das Acácias, São José dos Campos, 12228-900 SP, Brazil

E-mail: [1]ronaldo.lobato@icranet.org

Abstract. The origins of ultra-high-energy cosmic rays ($E \gtrsim 10^{19}$ eV) are a mystery and still under debate in astroparticle physics. In recent years some efforts were made to understand their nature. In this contribution we consider the possibility of Some Soft Gamma Repeaters (SGRs) and Anomalous X-ray Pulsars (AXPs) beeing white dwarf pulsars, and show that these sources can achieve large electromagnetic potentials on their surface that accelerate particle almost at the speed of light, with energies $E \sim 10^{20-21}$ eV. The sources SGRs/AXPs considered as highly magnetized white dwarfs are well described in the Hillas diagram, lying close to the AR Sorpii and AE Aquarii which are understood as white dwarf pulsars.

1. Introduction

In our previous work we have raised some issues concerning the origin of ultra-high-cosmic rays (UHECR) [1]. We pointed out that some soft Gamma Repeaters (SGRs) and Anomalous X-ray pulsars (AXPs) could be considered white dwarf pulsars. The idea of SGRs/AXPs as possible sources of the highest energy cosmic rays was investigated by Arons [2], that have considered these sources as neutron stars with ultrastrong magnetic field ($B \sim 10^{15}$ G) on their surface, the so-called magnetars. Rotation and strong magnetic fields can induce large potential differences [3, 4] on the star's surface, accelerating particles with high energy and fulfilling the magnetosphere with pairs of electron and positrons, which turns into a cascade of photons. More recently Kashiyama et al. [5] have done an important work in this perspective, suggesting that the excess of cosmic ray electron and positrons observed in PAMELA is produced in magnetized White Dwarfs (WDs), by the same mechanism as the one of neutron star pulsars.

In fact, highly magnetized white dwarfs are able to have a large electric field on their surface. Recently, a great number of massive and magnetic white dwarfs (10^6-10^9 G) have been observed [6], and in 2016 a white dwarf pulsar known as AR Scorpii has been discovered with some minutes of period, showing radiation in a broad range of frequencies [7, 8]. Thus, these fast and magnetized WDs can be in principle a potential source for ultra-high-energy-cosmic rays, and in this contribution we will investigate this possibility.

2. The sources

SGRs/AXPs are typically identified with very slow rotating pulsars $P \sim (2-12)$ s, with intense magnetic fields $\sim 10^{15}$ G [9]. Their spin-down rates $\dot{P} \sim (10^{-13} - 10^{-10})$ s/s are larger than the ones of normal pulsars $\dot{P} \sim (10^{-15} - 10^{-14})$ s/s [10]. The magnetar's X-ray luminosity L_X is explained by the decay of their enormous magnetic field. These sources do not can be understood as rotation powered neutron stars, because their rotational energy is much smaller than X-ray luminosity. Currently there are 23 SGRs/AXPs[1] that are classified as magnetars and 6 until now only candidates [11].

2.1. SGRs/AXPs as white dwarf pulsars

It has been proposed by Malheiro, Rueda and Ruffini an alternative model considering SGRs/AXPs as white dwarf pulsars [12]. As discussed in [13], the process of release energy for dipole radiation in a white dwarf can be explained in terms of a canonical spin-powered pulsars model, since in certain aspects they are similar (see e.g., [14, 15, 16]).
For example, if we consider a star with $M = 1.4M_\odot$ and $R = 10^6$ cm, the magnetic field at poles is given by,

$$B_{\rm p}^{\rm NS} = 3.2 \times 10^{19} (P\dot{P})^{1/2} \rm G. \tag{1}$$

In the case of a white dwarf with $M = 1.4M_\odot$ and $R = 3 \times 10^8$ cm, there is a new scale for the magnetic field at poles,

$$B_{\rm p}^{\rm WD} = 4.21 \times 10^{14} (P\dot{P})^{1/2} \rm G. \tag{2}$$

Following the last consideration there are new values for the mass density, moment of inertia, dipole moment and rotation energy [16, 17]. This description is supported by the observational growth of fast, massive and highly magnetized WDs [6, 18, 19, 20].

2.2. White dwarf pulsars

This description was raised by Ostriker [21, 22]. Later, the seminal works of Usov [23, 24] have discussed the generation of gamma rays by magnetic white dwarfs, showing that WD can produce pairs of electron-positron $e^{\pm}$. Zhang and Gil [25] interpreted the transient radio source, GCRT J1745-3009, as a white dwarf pulsar with a period of 77.13 minutes. In this situation the production of $e^{\pm}$ due curvature radiation is below of the pair-production threshold, still if the star is assumed with a field $B_{\rm p}^{\rm WD} = 10^9$ G. We highlight the recent discovery, AR Scorpii (AR Sco's), as a pulsating white dwarf in a binary system confirming the hypothesis of white dwarf pulsar [7]. AR Sco's has a range of mass $0.81M_\odot < M1 < 1.29M_\odot$ and pulses with a period of 1.97 minutes. These pulses reflect the spin of a WD, slowing down on a 10^7 yr timescale. The AR Sco's broadband spectrum is characteristic of synchrotron radiation, requiring relativistic electrons, possibly originate from near the white dwarf and accelerated to almost the speed of light. These particles produce radiation from X-ray to radio wavelengths, typically of neutron star pulsars. This rapidly rotating magnetized WD would simulate the neutron star pulsars as pointed out by Geng et al. [26]. Another specific example is AE Aquarii, the first white dwarf pulsar, with a short period $P = 33$ s and spinning down at a rate $\dot{P} = 5.64 \times 10^{-14}$ s/s. The rapid braking of the white dwarf and the nature of hard X-ray pulses detected with SUZAKU space telescope [27] can be explained in terms of spin-powered pulsar mechanism [28]. Although AE Aquarii is a binary system with orbital period 9.88 hr, pretty likely the matters' accretion is inhibited by the fast rotation of the white dwarf.

[1] `http://www.physics.mcgill.ca/~pulsar/magnetar/main.html`

IOP Conf. Series: Journal of Physics: Conf. Series **861** (2017) 012005 doi:10.1088/1742-6596/861/1/012005

3. General constrains from geometry and radiation to UHECR

There are some constrains that UHECR particles should satisfy, as shown by [29]:

- geometry - the accelerated particle should need be kept inside the source while being accelerated;
- power - the source should posses the required amount of energy to give it to accelerated particles;
- radiation losses - the energy lost by a particle for radiation in the accelerating field should not exceed the energy gain;
- interaction losses - the energy lost by a particle in interactions with other particles should not exceed the energy gain;
- emissivity - the total number (density) and power of sources should be able to provide the observed UHECR flux;
- accompanying radiation of photons, neutrinos and low-energy cosmic rays should not exceed the observed fluxes, both for a given source and for the diffuse background.

4. Hillas criterion

If a particle escapes from the region where it was being accelerated, it will not be unable to gain more energy. The one in question establish a limit on maximum energy $E_{\rm max}$ acquired by a particle passing in a medium with magnetic field B,

$$E_{\rm max} = ZqBR_s, \tag{3}$$

where q is the electric charge of the particle, B is the magnetic field, R_s is the size of the accelerator and Z the atomic number of the particle (for the case of iron, $Z = 26$). This equation considers the Larmour radius of the particle, $R_{\rm L} = E_{\rm max}/(ZqB) \leqslant R_s$. This is a general geometrical criterion known as the *Hillas criterion* for all types of cosmic ray sources [30]. Neglecting energy losses, i.e., the accelerator is 100% efficient, we see that only the parameters R_s and B can describe the source, showing a relationship between the sources' magnetic field strength and its size.

5. Particle acceleration

As we discussed above, there are a great number of magnetic white dwarfs and according Usov [24] if the surface temperature is less than 10^4 K (see [31], for the thermal emission in fast rotating, highly magnetized white dwarfs), a scale height of its atmosphere is essentially smaller than the radius of white dwarf R and the WD will have a strong electric field E in its magnetosphere,

$$E_{\parallel} = \vec{E}\cdot\vec{B}/|\vec{B}| \neq 0. \tag{4}$$

This parallel electric field determines a charge distribution, known as Goldreich-Julian charge density [32],

$$\rho_{\rm c} = \frac{1}{4\pi}\vec{\nabla}\cdot\vec{E} = -\frac{1}{2\pi c}\vec{\Omega}\cdot\vec{B}. \tag{5}$$

Particles in that charge distribution co-rotate with the star until the distance where the linear velocity reaches the speed of light. This region is commonly known as light cylinder, with a radius $R_l = c/\Omega$. Near the surface of the star, the intense electric field tear away particles. Ultra relativistic particles flowing out, move along the open magnetic field lines producing curvature radiation,

$$E_\gamma = \frac{3}{2}\frac{\hbar\gamma^3 c}{r_c}, \tag{6}$$

IOP Conf. Series: Journal of Physics: Conf. Series **861** (2017) 012005 doi:10.1088/1742-6596/861/1/012005

where

$$r_c \sim (Rc/\Omega)^{1/2} \tag{7}$$

is the curvature radius of the field lines. The perpendicular energy of the particle is rapidly dissipated by synchrotron radiation. This curvature radiation interact with the magnetic field and produce secondary $e^{\pm}$ pairs, $\gamma + B \to e^- + e^+$, leading to a pair creation avalanche. The value of the Lorentz γ factor of the $e^{\pm}$ pairs produced, is given by the potential difference in the polar cap region,

$$\gamma = \frac{e\Delta V}{mc^2}. \tag{8}$$

Neutron star pulsars are able to produce electrons with Lorentz factors $\gamma \gtrsim 10^7$. There is a gap h above the polar region, and the potential difference becomes [33]

$$\Delta V = \frac{B_{\rm p}\Omega h^2}{2c}, \tag{9}$$

where h is given by

$$h \approx \left(\frac{R^3\Omega}{c}\right)^{1/2}. \tag{10}$$

6. Discussions

Using the dipole formula given in Ref. [34] to calculate the magnetic field, we have shown that the magnetic fields for SGRs/AXPs as highly magnetized white dwarf are at the order of [17, 35, 36, 37, 38],

$$B_{\rm p}^{\rm WD} \sim 10^9{\rm G}. \tag{11}$$

This allows us to estimate the charge density in the magnetosphere, i.e., the Goldreich-Julian density given by equation 5,

$$\rho_c^{\rm WD} \sim 10^{-2}{\rm cm}^{-3}, \tag{12}$$

this value is 10^5 smaller than the ones when SGRs/AXPs are considered as neutron stars, which have a magnetic field $B \sim 10^{14-15}$ G. Thus, in the magnetosphere of SGRs/AXPs being white dwarfs, there are less charged particles than the neutron stars ones.

From equation 9, we can calculate the maximum potential difference achieved on the surface of SGRs/AXPs as white dwarfs, using $\Omega \sim 1$ Hz, $R \sim 3 \times 10^3$ km,

$$\Delta V \sim 10^{16}{\rm V}. \tag{13}$$

This difference is achieved in a length of $h \sim 10^3$ cm, and allows the particles accelerated to reach ultra relativistic energies, larger than that of neutron star pulsars. Therefore, white dwarfs can accelerate electrons with Lorentz factors at least $\gamma \sim 10^{10}$, one thousand times larger than those accelerated by neutron star pulsars, being capable to produce curvature photons with an energy up to $\sim 10^{21}$ eV.

Figure 1 show the Hillas plot for a maximum energy of $E_{\rm max} = 10^{20-21}$ eV for protons and $E_{\rm max} = 10^{20-21}$ eV for iron. The sources above the green and black lines are able to accelerate atoms of iron up to 10^{20} eV and 10^{21} eV, respectively. Similarly, the sources above the blue and orange lines are able to accelerate protons up to 10^{20} eV and 10^{21} eV, respectively. This figure shows that SGRs/AXPs described as WDs and the two known white dwarf pulsars are all on the line of the Hillas plot obtained by Hillas criterion, and consistent with all the others cosmic ray sources known in the universe. Moreover, SGRs/AXPs as magnetars (neutron star pulsars) are out and much above this line, which is not the case for ordinary neutron star pulsars as

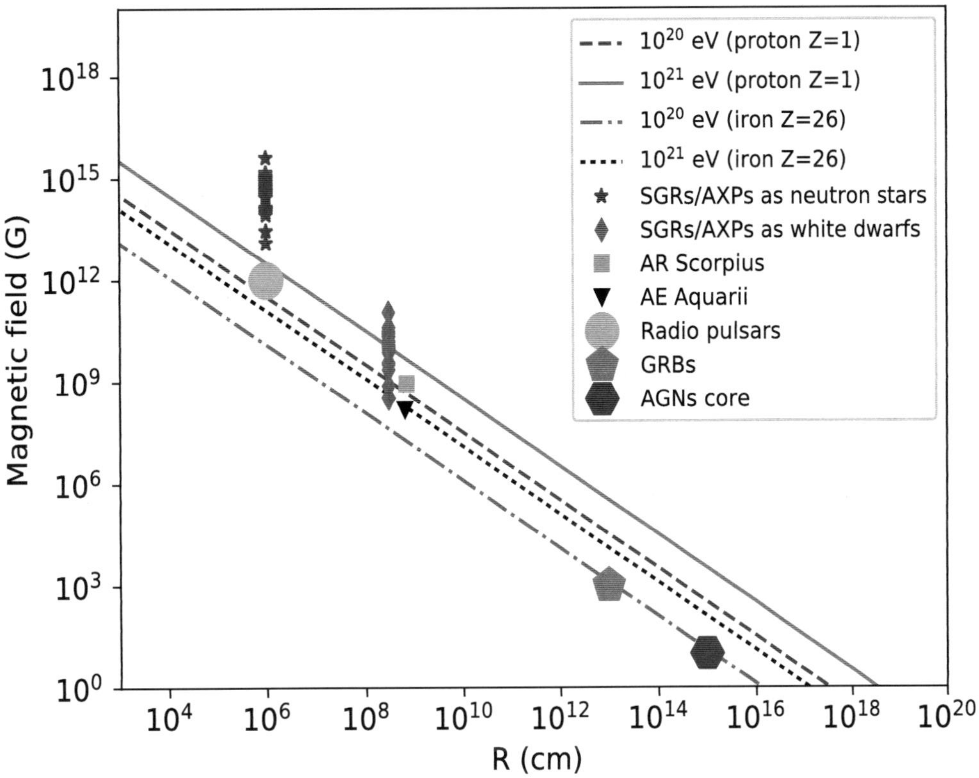

Figure 1. Hillas plot showing the magnetic field strength versus size of SGRs/AXPs as NSs (magnetic field $\sim 10^{13-15}$ G and radius of 10^6 cm) and WDs (magnetic field $\sim 10^{8-10}$ G and radius of 3×10^8 cm). The blue-stars and magenta-diamonds describe the SGRs/AXPs as neutron stars and white dwarfs, respectively. Here, we are considering these cosmic ray sources for a maximum energy of $E_{\rm max} = 10^{20-21}$ eV for protons (blue and orange lines) and $E_{\rm max} = 10^{20-21}$ eV for iron (green and black lines). The cyan-square and black-triangles represent AR Scorpius with magnetic field of $\approx 9\times10^8$ G and radius of 7×10^8 cm, and AE Aquarii with magnetic field of $\approx 1.5\times10^8$ G and radius of 6.5×10^8 cm, respectively. As we see these two sources are able to accelerate ultra-high energy cosmic rays. It's worth mentioning other possible sources of cosmic rays pointed in the literature: the yellow-circle, red-pentagon and blue-hexagon are the known radio pulsars, gamma rays-bursts (GRBs) and active Galaxy nuclei (AGNs), respectively.

we also show in Figure 1. Thus, it is quite important to obtain precision measurements of the radius and surface magnetic field of SGRs/AXPs with the new telescopes.

The interaction losses are minimum in the stars' magnetosphere, considering the photon is emitted in a cone with an angular aperture $1/\gamma$ and there is a low charged particle density. Among all SGRs/AXPs, CXOU J010043.1 is the most distant source, located $\sim$ 62.4 kpc, thus all sources are within the GZK limit, which is a theoretical upper limit on the energy of cosmic rays coming from outside of our Galaxy ($\approx$ 10 Mpc for protons with energy of 10^{19} eV) [39, 40]. Particles accelerated in these sources could be a fraction of ultra-high cosmic ray observed in our planet. All these findings support the plausibility of ultra-high energy cosmic rays from SGRs/AXPs (fast and magnetic massive white dwarfs), in line with important and recent astronomical observations of white dwarf pulsars. We encourage future observational campaigns to verify the radius and the magnetic field of these sources.

Acknowledgments

RVL would like to thank CNPq (Conselho Nacional de Desenvolvimento Científico e Tecnológico) process 141157/2015-1 and CAPES (Coordenação de Aperfeiçoamento de Pessoal de Nível Superior) CAPES/PDSE/88881.134089/2016-01 for financial support. JGC acknowledge the support of FAPESP through the project 2013/15088-0. MM thanks FAPESP thematic project 13/26258-4, CNPq and CAPES.

References

[1] Lobato R V and Malheiro M 2016 *Journal of Physics: Conference Series* **706** 052032 ISSN 1742-6588
[2] Arons J 2003 *The Astrophysical Journal* **589** 871 ISSN 0004-637X
[3] Aharonian F, Atoyan A and Voelk H 1995 *Astronomy and Astrophysics* **294** L41L44 ISSN 0004-6361
[4] Hooper D, Blasi P and Serpico P D 2009 *Journal of Cosmology and Astroparticle Physics* **2009** 025–025 ISSN 1475-7516
[5] Kashiyama K, Ioka K and Kawanaka N 2011 *Physical Review D - Particles, Fields, Gravitation and Cosmology* **83** 023002 ISSN 15507998
[6] Kepler S O, Pelisoli I, Jordan S, Kleinman S J, Koester D, K B, P V, Castanheira B G, Nitta A, Costa J E S, Winget D E, Kanaan A and Fraga L 2013 *Monthly Notices of the Royal Astronomical Society* **429** 2934–2944 ISSN 0035-8711, 1365-2966
[7] Marsh T R, Gänsicke B T, Hümmerich S, Hambsch F J, Bernhard K, Lloyd C, Breedt E, Stanway E R, Steeghs D T, Parsons S G, Toloza O, Schreiber M R, Jonker P G, van Roestel J, Kupfer T, Pala A F, Dhillon V S, Hardy L K, Littlefair S P, Aungwerojwit A, Arjyotha S, Koester D, Bochinski J J, Haswell C A, Frank P and Wheatley P J 2016 *Nature* 1–15 ISSN 0028-0836
[8] Buckley D A H, Meintjes P J, Potter S B, Marsh T R and Gänsicke B T 2017 *Nature Astronomy* **1** 0029 ISSN 2397-3366
[9] Duncan R C and Thompson C 1992 *The Astrophysical Journal* **392** L9 ISSN 0004-637X
[10] Turolla R, Zane S and Watts A 2015 *Reports on Progress in Physics* **78** 116901 ISSN 0034-4885
[11] Olausen S A and Kaspi V M 2014 *The Astrophysical Journal Supplement Series* **212** 6 ISSN 0067-0049, 1538-4365
[12] Malheiro M, Rueda J A and Ruffini R 2012 *Publications of the Astronomical Society of Japan* **64** 56 ISSN 0004-6264, 2053-051X
[13] Usov V 1988 *Pisma v Astronomicheskii Zhurnal* **14** 606609
[14] Coelho J G and Malheiro M 2013 *arXiv:1307.8158 [astro-ph, physics:nucl-th]*
[15] Malheiro M and Coelho J G 2013 *arXiv:1307.5074 [astro-ph, physics:gr-qc, physics:nucl-th]*
[16] Coelho J G and Malheiro M 2014 *Publications of the Astronomical Society of Japan* **66** 14–14 ISSN 0004-6264, 2053-051X
[17] Coelho J G and Malheiro M 2013 vol 1520 (AIP Publishing) pp 258–263
[18] Boshkayev K, Rueda J A, Ruffini R and Siutsou I 2013 *The Astrophysical Journal* **762** 117 ISSN 0004-637X
[19] Boshkayev K, Rueda J A, Ruffini R and Siutsou I 2014 *Journal of the Korean Physical Society* **65** 855–860 ISSN 0374-4884, 1976-8524
[20] Das U and Mukhopadhyay B 2013 *Physical Review Letters* **110** 071102
[21] Ostriker J 1968 *Nature* **217** 1227–1228 ISSN 00280836
[22] Gunn J E and Ostriker J P 1969 *Nature* **221** 454–456
[23] Usov V V 1988 *Soviet Astronomy Letters, Vol.14, NO. 4, P. 258, 1988* **14** 258
[24] Usov V V 1993 *The Astrophysical Journal* **410** 761 ISSN 0004-637X
[25] Zhang B and Gil J 2005 *The Astrophysical Journal* **631** L143–L146 ISSN 0004-637X
[26] Geng J J, Zhang B and Huang Y F 2016 *The Astrophysical Journal* **831** L10 ISSN 2041-8213
[27] Terada Y, Ishida M, Mukai K, Dotani T, Makishima K, Naik S, Hayashi T, Okada S, Nakamura R and Enoto T 2008 *Advances in Space Research* **41** 512–517 ISSN 02731177
[28] Ikhsanov N R 1998 *Astronomy and Astrophysics* **338** 521–526 ISSN 0004-6361
[29] Ptitsyna K V and Troitsky S V 2010 *Physics-Uspekhi* **53** 691–701 ISSN 1063-7869
[30] Hillas A M 1984 *Annual Review of Astronomy and Astrophysics* **22** 425–444

[31] Caceres D L, de Carvalho S M, Coelho J G, de Lima R C R and Rueda J A 2017 *Monthly Notices of the Royal Astronomical Society* **465** 4434–4440 ISSN 0035-8711
[32] Goldreich P and Julian W H 1969 *The Astrophysical Journal* **157** 869 ISSN 0004-637X
[33] Ruderman M A and Sutherland P G 1975 *The Astrophysical Journal* **196** 51–72 ISSN 0004-637X
[34] Ferrari A and Ruffini R 1969 *The Astrophysical Journal Letters* **158** L71 ISSN 0004-637X
[35] Coelho J G and Malheiro M 2012 *International Journal of Modern Physics: Conference Series* **18** 96–100 ISSN 2010-1945
[36] Lobato R V, Malheiro M and Coelho J G 2016 *International Journal of Modern Physics D* **25** 1641025 ISSN 0218-2718
[37] Lobato R V, Coelho J and Malheiro M 2015 *Journal of Physics: Conference Series* **630** 012015 ISSN 1742-6588
[38] Lobato R V, Coelho J G and Malheiro M 2015 *The Second César Lattes Meeting: Supernovae, Neutron Stars and Black Holes* (Rio de Janeiro - Niterói - João Pessoa - Recife - Fortaleza: AIP Publishing) p 030003
[39] Zatsepin G T and Kuz'min V A 1966 Upper Limit of the Spectrum of Cosmic Rays
[40] Greisen K 1966 *Physical Review Letters* **16** 748–750 ISSN 00319007

IOP Conf. Series: Journal of Physics: Conf. Series **861** (2017) 012006 doi:10.1088/1742-6596/861/1/012006

Rotational stellar structures based on the Lagrangian variational principle

Nobutoshi Yasutake[1], **Kotaro Fujisawa**[2], **Shoichi Yamada**[2]

[1] Physics Department, Chiba Institute of Technology, Shibazono 2-1-1, Narashino, Chiba, 275-0023, Japan

[2] Advanced Research Institute for Science and Engineering, Waseda University,Okubo 3-4-1, Shinjuku, Tokyo 169-8555, Japan

E-mail: [1]nobutoshi.yasutake@p.chibakoudai.jp

Abstract. A new method for multi-dimensional stellar structures is proposed in this study. As for stellar evolution calculations, the Heney method is the defacto standard now, but basically assumed to be spherical symmetric. It is one of the difficulties for deformed stellar-evolution calculations to trace the potentially complex movements of each fluid element. On the other hand, our new method is very suitable to follow such movements, since it is based on the Lagrange coordinate. This scheme is also based on the variational principle, which is adopted to the studies for the pasta structures inside of neutron stars. Our scheme could be a major break through for evolution calculations of any types of deformed stars: proto-planets, proto-stars, and proto-neutron stars, etc.

1. Introduction

Stellar evolution is a mature topic in astrophysics. After the seminal works [1-4], the Heney method [5] is the defacto standard now for stellar evolution calculations. Although this method is proposed originally for spherical stars without rotation and magnetic field, there are the some subspecific techniques taking into account the effects of rotation and/or magnetic field [6]. Such subspecific techniques are, however, based on the spherical symmetric assumption, and no one succeeds to propose a new scheme for fully-multi-dimensional stellar-evolution calculations.

One of the difficulties of fully-multi-dimensional stellar-evolution calculations is to obtain hydrostatic equilibria of deformed stars considering "*baroclinicity*", in which pressure depends not only on density but also temperature and chemical composition. It is necessary when we adopt a realistic equation of state (EoS) in stellar evolution calculations. Note that we can neglect the baroclinicity for neutron stars, because thermal effects on pressure are negligible, and chemical composition is derived uniquely from the beta equilibrium condition and neutrino free condition. Such cases are known as "*barotropic*" ones, in which pressure depends only on density. On the other hand, the baroclinicity appears in proto-neutron stars.

Even if some equilibria are obtained, it is highly non-trivial whether we can make a sequence of them that represents stellar evolutions appropriately, because the fluid elements composing rotating stars change their positions non-radially in complicated ways. In this study, we have introduced the Lagrange scheme to trace the moving of fluid elements in conservative ways. Otherwise, in a Eulerian scheme, it would be difficult to describe such displacements. This

IOP Conf. Series: Journal of Physics: Conf. Series **861** (2017) 012006 doi:10.1088/1742-6596/861/1/012006

difficulty appears in both barotropic cases (neutron stars) and baroclinic cases (proto-neutron stars).

It is not a new attempt to adopt the Lagrange scheme into multi-dimensional calculation, indeed already Freedman & Shutz have pointed out that trivial Lagrange displacements do not provide unique solution because of the gauge degree of freedom [7]. In this study, we solve this problem by using the Monte Carlo method.

2. Formalism

Let us introduce an idea of the Lagrangian variational principle. The details are shown in our papers [8, 9].

First we prepare nodes, which are connected with each other by edges making triangulations. Each node has mass, angular momentum, entropy, and compositions, which are conserved during the process to obtain hydro-static equilibria. We do not take into account energy equation, and just focus on the balance equation assuming pure hydrogen matter for simplicity, since this study is the preliminary step. Note that there is no problem to take account realistic composition in our scheme. It however changes just the pressure (EoS) slightly in our current formalism. If we considered nuclear burnings, advection of matter, and convections, etc., the composition would have a critical role. It remains as a future task on us.

Second, the finite volume is calculated by the triangulations around each node, then density, pressure, temperature, internal energy, etc. are derived. Finally, we obtain the most optimal structures changing the positions of all nodes to find the minimal energy. We adopt the Monte Carlo method in the iterative process: we change the positions of nodes gradually, one by one, at random. Note that periodical smoothing is necessary to avoid a stagnancy, in which the system is sticked with local minimum energy.

The idea of this work is quite simple and similar to the studies of *the pasta-structures*, which are the deformed nuclear-structures in neutron stars. In Table 1, we show the duality between hydrostatic equilibria and the pasta structures.

Table 1. Duality between deformed stellar structures and "*pasta*" structures in the crust of NSs, or in the quark-hadron phase transition in the core of NSs.

	Deformed stellar structures	Pasta structures
Basic equations	· Hydro-static equilibrium · Poisson equation (gravitation) · Equation of state	· Chemical equilibrium · Poisson equation (Coulomb interaction) · Equation of state
Method	Variational principle in Lagrange coordinate	Variational principle in Euler coordinate
Origin of deformation	Balance between the repulsive force of rotation, pressure and the attractive force by gravitation	Balance between the repulsive force of Coulomb interaction, pressure and the attractive force by surface tension
Reference	[8][9]	[10]

3. Results

Let us show some examples of obtained hydrostatic equilibria with baroclinic cases in this section.

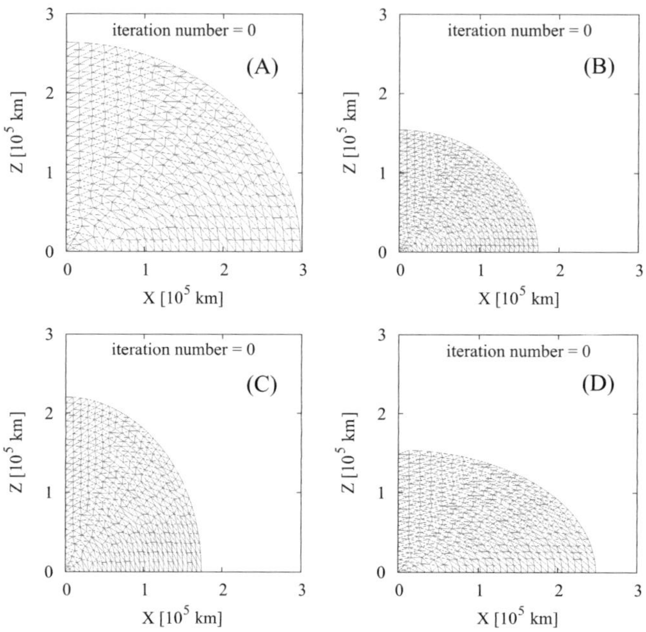

Figure 1. Initial models. Details are described in the text.

As described in the last section, first we need to prepare some profiles of mass, angular momentum, and entropy. Here, we introduce four types of initial models to check whether the resultant solutions are free or not from the local energy minima, since our method is based on the variational principle. These initial models are shown in Fig. 1. We extend 30 % radially an solution, which is given by Fujisawa method [11], and name it as initial model (A). For other three models, we shrink the solution by 30 % in three different ways: (B) radially, (C) horizontally, and (D) vertically. The rotation profile of the solution is given on the equatorial axis as following,

$$\Omega^2 = \frac{\Omega_0^2}{(\varpi^2/R_e^2) + d^2)}, \tag{1}$$

where the arbitrial parameter Ω_0, R_e and d are set as $\Omega_0 = 1.9 \times 10^{-3}$ rad s^{-1}, $R_e = 2.57 \times 10^5$ km, and $d = 0.90$ respectively. This value of Ω is a few hundred times the solar angular velocity. Here, ϖ is distance from the rotational axis. The rotation profile of the solution in other area is obtained self-consistently to satisfy the entropy distribution given as following,

$$K = K_0\left\{1 + e_1\left(1 + e_2 P_2(\cos\theta)\right)\frac{r^2}{{R_e}^2}\right\}, \tag{2}$$

where r and K is the distance from center and position-dependent value in the polytrope-like EoS, $p = K\rho^\gamma$ (p and ρ are pressure and density, and γ is fixed as $\gamma = 5/3$ in this study). The Legendre polynomials are denoted by P_n with θ being colattitude. The dimensionless parameter e_1 specifies radial non-uniformity whereas e_2 gives colatitudinal dependency, set as $e_1 = 0.45$ and $e_2 = 0.80$ here. The values of K_0 is chosen to be 5.49×10^{13} in cgs unit, respectively. These models are highly artificial admittedly, but it is known that such entropy distribution gives the rotation profile of the shellular type, which has an onion-skin-like angular velocity profile, known as the standard type based on the Zahn's conjecture [12].

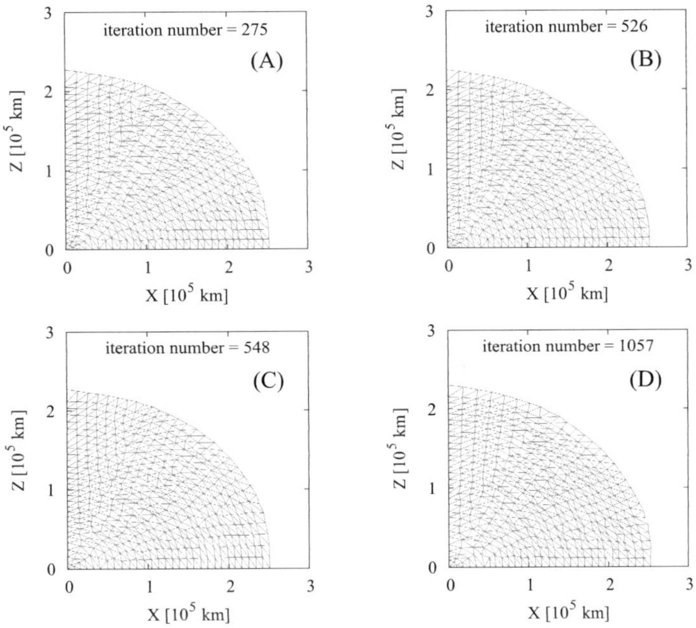

Figure 2. Final results corresponding to the initial models shown in Fig. 1.

The resultant structures for models (A)-(D) are shown in Fig. 2. Clearly, they look similar, though the initial models are extremely different from each other. This means that these solutions are free from the local energy minima. We have also checked that the results are consistent with the Bjerknes-Rosseland rules, and do not break the Solberg-Hoiland criterion.

4. Discussion & Summary

We study hydro-static equilibria for baroclinic cases using the Lagrangian variational principle and the Monte Carlo method. The solutions are free from the local energy minima. Our method could be applicable to stellar evolution calculations, since we can trace the displacements of fluid elements in conservative ways, although we do not take into account convection for these models.

Possible applications of our scheme will be extended to compact stars, proto-stars and proto-planets to mention a few. Magnetic fields and/or general relativity should be considered in due course.

Acknowledgments

We are grateful to K. Suzuki, T. Yamasaki, M. Okamoto, and T. Maruyama for fruitful discussions. This work was partially supported by JSPS KAKENHI Grant Numbers 25105510, 24244036 and Grant-in-Aid for Scientific Research on Innovative Areas, No.24103006.

References

[1] Burbidge K M, Burbidge G R, Fowler W A, Hoyle F 1957 *Review of Modern Physics* **29** 4
[2] Cameron A G W 1957 *Chalk River Report* CRL-41
[3] Hayashi C 1961 *PASJ* **13** 442
[4] Hayashi C 1961 *PASJ* **13** 450
[5] Henyey L G, Forbes J E and Gould N L 1964 *ApJ* **139** 306
[6] Woosley S E, Heger A, and Weaver T A 2002, *Reviews of Modern Physics* **74** 1015 and references therein
[7] Friedman J L, and Schutz B F 1978 *ApJ* **221** 937

[8] Yasutake N, Fujisawa K, Yamada S 2015 *MNRAS letter* **446** L56
[9] Yasutake N, Fujisawa K, Yamada S 2016 *MNRAS* **463** 3705
[10] Yasutake N, Łastowiecki, Benić S, Blaschke D, Maruyama T, Tatsumi T 2014 *Phys. Rev. C* **89** 065803 and references therein
[11] Fujisawa K 2015 *MNRAS* **454** 3060
[12] Zahn J-P 1992 *Astro. Astrophys.* **265** 115

IOP Conf. Series: Journal of Physics: Conf. Series **861** (2017) 012007 doi:10.1088/1742-6596/861/1/012007

Properties of relativistically rotating quark stars

Enping Zhou

Department of Astronomy, Peking University, 100871, Beijing, P.R. China

E-mail: zhouenpingz715@sina.com

Abstract. In this work, quasi-equilibrium models of rapidly rotating triaxially deformed quark stars are computed in general relativistic gravity, assuming a conformally flat spatial geometry (Isenberg-Wilson-Mathews formulation) and a polynomial equation of state. Especially, since we are using a full 3-D numerical relativity initial data code, we are able to consider the triaxially deformed rotating quark stars at very high spins. Such triaxially deformed stars are possible gravitational radiation sources detectable by ground based gravitational wave observatories. Additionally, the bifurcation from axisymmetric rotating sequence to triaxially rotating sequence hints a more realistic spin up limit for rotating compact stars compared with the mass-shedding limit. With future observations such as sub-millisecond pulsars, we could possibly distinguish between equation of states of compact stars, thus better understanding strong interaction in the low energy regime.

1. Introduction

The direct detection of gravitational waves (GWs) [1, 2] has opened a brand new GW astronomy window for us to explore the Universe. In addition to binary black hole and binary compact star mergers, rapidly rotating compact stars have also been considered as important candidate of GW sources that could be detected by ground based GW observatories [3, 4, 5, 6, 7]. Therefore, it will be interesting and essential to study the configuration of rapidly rotating compact stars.

The equilibrium models of self-gravitating, uniformly rotating incompressible fluid stars has been systematically studied in a Newtonian scheme long ago [8]. Depending on the rotational kinetic energy, the configuration could be axisymmetric Maclaurin ellipsoids as well as nonaxisymmetric ellipsoids, such as Jacobian (triaxial) ellipsoid. For the GW astronomy of rotating compact stars, however, general relativity is required to replace Newtonian gravity. The configuration of relativistic rotating stars is also studied in [9, 10].

By assuming more realistic equation of states (EoS) , such as piecewise polytropic EoS, scientists have conducted more realistic calculations for rotating neutron stars. It has been shown that a rotating neutron star will spontaneously break its axial symmetry if the rotational kinetic energy to gravitational binding energy ratio ($T/|W|$) exceeds a critical value [11]. This high $T/|W|$ ratio is possible to be reached for a newly born rotating compact star during a core collapse supernova or for a neutron star which is spun up by accretion. Quasi-equilibrium configurations of triaxially rotating neutron stars have also been created and studied in a full general relativity scheme [12].

Nevertheless, it's worth noting that the EoS of compact stars is not fully clear yet since it's related to the non-perturbative quantum chromodynamics. Other EoS model such as strange quark star has been favored after Bodmer and Witten have suggested that strange quark matter composed of de-confined u,d and s quarks could be absolutely stable [13, 14]. There are also some observational hints of the existence of quark stars (for example, see [15]). Unlike neutron stars which are bound by gravity, quark stars are self-bound by strong interaction. Consequently, rotating quark stars can reach a much larger $T/|W|$ ratio compared with neutron stars and the triaxial instability could play a more important role for quark stars [16, 17, 18]. In particular, the triaxial bar mode (Jacobi-like) instability for MIT bag models has also been investigated in a relativistic scheme [19]. In this work, we compute initial data for both axisymmetric and triaxial rotating quark star, as well as sequences with various quark star EoSs (MIT bag model and LaiXu09 model [20]) and different compactnesses.

2. Configuration of triaxially rotating quark stars

As an illustration of the calculated triaxial solution, we have shown the surface plot for the C=0.2 case of MIT bag model, see Fig.1.

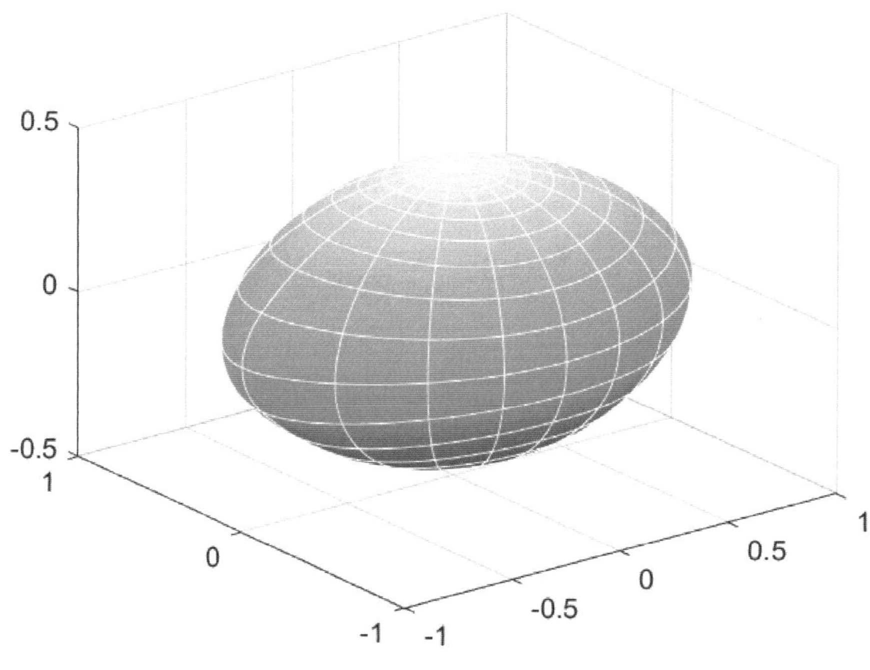

Figure 1. Illustration of the solution with the largest triaxial deformation in our simulation. The colors in this plot illustrates the z-axis value of the surface of the star while the white circles are the equal-phi/theta contours. The fact that the equal-phi circles are not parallel with the color patterns highlights the triaxial deformation, since in an axisymmetric case equal-phi contours should be identical with equal-z contours.

In Fig.2 and 3, the relation between the T/W ratio, $M\Omega$ and the ellipticity of the star has been plotted for 3 different compactness ($C = 0.1, 0.15$ and 0.2) for both the LaiXu09 model and the MIT bag model.

Unlike in the Newtonian case of an incompressible star, for which the bifurcation to triaxial deformation happens at $T/W \approx 0.1375$ for any compactness, it is found that in a full relativistic case, for quark stars the T/W ratio at bifurcation point strongly depends on the compactness. Similar results have been obtained for triaxially rotating neutron stars ([12]). The largest T/W ratio for the onset of secular instability is found to be roughly 0.17 for rotating quark stars in the configurations that we considered, and it will be even larger for a higher compactness.

It's worth noting that compared with the rotating neutron stars calculated by [12], rotating quark stars have much longer triaxial sequences. Especially, at the $C = 0.2$, the triaxial sequence terminates for neutron stars with $n = 0.5$ EoS, while for quark stars, the bifurcation point happens much earlier than the mass shedding limit is reached and there is still reasonably long triaxial sequence which are stable. This justifies the discussion mentioned above, that triaxial instability plays a more important role for quark stars since these stars can reach a much larger T/W ratio.

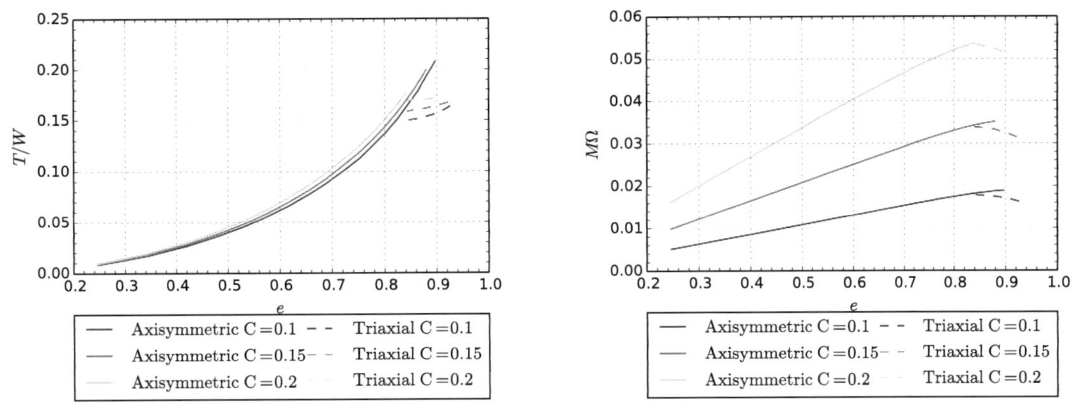

Figure 2. Plots for $T/|W|$ (left panel) and ΩM (right panel) versus eccentricity $e := \sqrt{1 - (\bar{R}_z/\bar{R}_x)^2}$ (in proper length) for MIT bag model sequences. Solid curves are axisymmetric solution sequences, and dashed curves are triaxial solution sequences. In each panel, from the top to the bottom the curves correspond to $M/R = 0.2$ (green), 0.15 (red) and 0.1 (blue) respectively.

3. Discussion

Triaxially rotating compact stars are important sources for ground based gravitational wave observatories. Our calculation has shown that for rotating quark stars with different EoSs, the bifurcation point to triaxial sequence happens at roughly 1 millisecond of the spin period. Therefore, the corresponding frequency of the gravitational waves from such sources will be at 2 kHz roughly, which is still detectable by GW observatories such as AdvLIGO or VIRGO. It has been shown that with the largest triaxial deformation solution in our calculation (which is also the one that can be seen in Fig.1), the GW amplitude can be as high as of the order of 10^{-23} at a distance of 30 Mpc, which is sufficient for detections.

Another astrophysical implication of the existence of the triaxial sequences is on the spin-up limit for rotating compact stars. A rough estimation is that pulsars can spin up to the mass shedding limit, which is obtained by requiring axisymmetry. This usually gives a spin period of 0.7 ms for a compact star with mass of $1.4\,M_\odot$ and radius of 10 km. However, as shown by both the calculation in this work and in [12], when triaxial deformations are taken into account, the spin period of a compact star actually decreases as it gains angular momentum (by accretion, for example) in the triaxial sequence. In another word, the 'spin-up' mechanism is actually

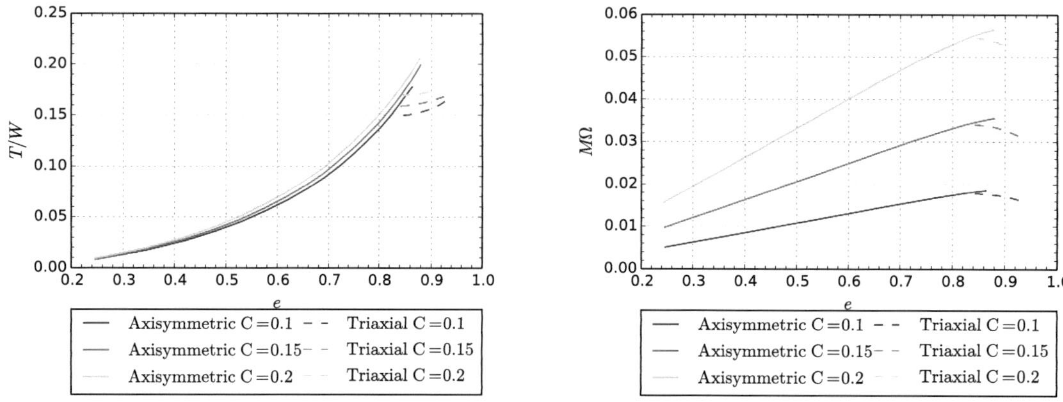

Figure 3. Same as Fig. 2 but for LaiXu09 EoS sequences. Dashed curves and solid curves from the top to the bottom in each panel correspond to $M/R = 0.2$ (green), 0.15 (red) and 0.1 (blue) respectively.

spinning down the pulsar when the bifurcation point is reached. Therefore, it's reasonable to believe that no pulsar could spin up faster than the period at the bifurcation point in a realistic astrophysical scheme. Therefore, in the future, as we detect more and more fast spinning pulsars when projects such as SKA and FAST are completed, we can also try to rule out EoSs with the observation of sub-millisecond pulsars, as the period of the bifurcation point also depends on the EoSs.

References

[1] Abbott B P, Abbott R, Abbott T D *et al* 2016 *Phys. Rev. Lett.* **116** 061102
[2] Abbott B P, Abbott R, Abbott T D *et al* 2016 *Phys. Rev. Lett.* **116** 241103
[3] Abramovici A A *et al* 1992 *Science* **256** 325
[4] Accadia T *et al* 2011 *Class. Quantum Grav.* **28** 114002
[5] Aso Y, Michimura Y, Somiya K, Ando M, Miyakwa O, Sekiguchi T, Tatsumi D and Yamamoto H 2013 *Phys. Rev. D* **88** 043007
[6] Kuroda K and LCGT Collaboration 2010 *Class. Quantum Grav.* **27** 084004
[7] Punturo M *et al* 2010 *Class. Quantum Grav.* **27** 084007
[8] Chandrasekhar S 1969 *Ellipsoidal Figures of Equilibrium* (New Haven, CT: Yale University Press)
[9] Meinel R, Ansorg M, Kleinwchter A, Neugebauer G and Petroff D 2008 *Relativistic Figures of Equilibrium* (Cambridge: Cambridge University Press)
[10] Friedman J L and Stergioulas N 2013 *Rotating Relativistic Stars* (Cambridge: Cambridge University Press)
[11] Bonazzola S, Frieben J and Gourgoulhon E 1996 *ApJ* **460** 379
[12] Huang X, Markakis C, Sugiyama N and Uryū K 2008 *Phys. Rev. D* **78** 124023
[13] Bodmer A R 1971 *Phys. Rev. D* **4** 1601
[14] Witten E 1984 *Phys. Rev. D* **30** 272
[15] Dai Z G, Wang S Q, Wang J S, Wang L J and Yu Y W 2016 *ApJ* **817** 132
[16] Gondek-Rosiska D, Haensel P, Zdunik J L and Gourgoulhon E 2000 *Astron. Soc. Pac.: Conf. Series* **202** 661
[17] Gondek-Rosiska D, Bulik T, Zdunik J L, Gourgoulhon E, Ray S, Dey J and Dey M 2000 *A&A* **363** 1005
[18] Gondek-Rosiska D, Stergioulas N, Bulik T, Kluniak W and Gourgoulhon E 2001 *A&A* **380** 190
[19] Gondek-Rosiska D, Gourgoulhon E and Haensel P 2003 *A&A* **412** 777
[20] Lai X Y and Xu R X 2009 *MNRAS* **398** L31-L35

IOP Conf. Series: Journal of Physics: Conf. Series **861** (2017) 012008 doi:10.1088/1742-6596/861/1/012008

Simulating Hadronic-to-Quark-Matter with Burn-UD: Recent work and astrophysical applications

Luis Welbanks[1], **Amir Ouyed, Nico Koning and Rachid Ouyed**

Department of Physics & Astronomy, University of Calgary, 2500 University Drive NW, Calgary, AB T2N 1N4, Canada

E-mail: [1]`lcwelban@ucalgary.ca`

Abstract. We present the new developments in Burn-UD, our in-house hydrodynamic combustion code used to model the phase transition of hadronic-to-quark matter. Our two new modules add neutrino transport and the time evolution of a (u, d, s) quark star (QS). Preliminary simulations show that the inclusion of neutrino transport points towards new hydrodynamic instabilities that increase the burning speed. A higher burning speed could elicit the deflagration to detonation of a neutron star (NS) into a QS. We propose that a Quark-Nova (QN: the explosive transition of a NS to a QS) could help us explain the most energetic astronomical events to this day: superluminous supernovae (SLSNe). Our models consider a QN occurring in a massive binary, experiencing two common envelope stages and a QN occurring after the supernova explosion of a Wolf-Rayet (WO) star. Both models have been successful in explaining the double humped light curves of over half a dozen SLSNe. We also introduce SiRop our r-process simulation code and propose that a QN site has the hot temperatures and neutron densities required to make it an ideal site for the r-process.

1. Introduction

The hypothesis that quark matter (matter constituted of up, down, and strange quarks) is the most stable state of matter in the universe [1, 2], makes possible the existence of compact objects made up entirely of deconfined quark matter called (u,d,s) quark stars (QS). The core of a neutron star (NS) can reach a high enough density initiating a QCD phase transition that would allow for the formation of a QS.

The transition between hadronic matter and quark matter is an active question of research. This conversion can be thought of as a combustion, where hadronic matter acts as the fuel and quark matter as the ashes. With the conversion comes a front, and its velocity determines the type of combustion. The particular conditions of the burning and their consequences are subject of constant discussion amongst scientist and there is yet a consensus to be found.

The combustion of hadronic-to-quark matter can be characterized as either a detonation or a deflagration based on the speed of the combustion front with respect to the speed of sound. If the combustion front is faster than the speed of sound, the transition is a detonation; otherwise it is a deflagration. The transition as a weak deflagration has been studied by Olinto et al. [3], Heiselberg et al. [4], Olesen et al. [5], amongst others, as a result of a conversion induced by the diffusion of seed quark matter. Making these assumptions, Olesen et al. [5] found the time

IOP Conf. Series: Journal of Physics: Conf. Series **861** (2017) 012008 doi:10.1088/1742-6596/861/1/012008

scale of the conversion of a NS to take place between 0.1 seconds to a few minutes. Drago et al. [6] found a fast deflagration by solving the relativistic, hydrodynamic jump conditions using a realistic equation of state (EoS) for nuclear matter.

Other studies suggest that the presence of an initial strange quark seed may not be necessary. Instead, a shock could compress hadronic matter to high densities which would trigger deconfinement into two flavored quark matter (u, d). The two-flavored quark matter can then decay into three-flavored quark matter (u, d, s) in a heat releasing process. This two-step process for hadronic-to-quark matter transition results in a detonation but depends on an initial shock that would push the hadronic matter to the required densities. Bhattacharyya et al. [7] found that the conversion from hadronic matter to two-flavored quark matter happens in the order of milliseconds whereas the conversion from two-flavored quark matter to three-flavored quark matter is in the order of hundreds of seconds. Benvenuto et al. [8, 9] found that the deconfinement of hadronic matter can lead to a fast combustion due to a shock induced detonation, making the conversion of a NS on the scale of microseconds.

While authors like Lugones et al. [10] suggest that the combustion happens as a detonation, others like Drago et al. [6] suggest that the conversion process can only happen as a deflagration. The scientific community has no consensus for the way in which the conversion happens and the result is really dependant on the assumptions made and the physics considered when building the model. A different approach to studying the combustion is to perform hydrodynamical oriented studies by solving numerically the equations driving the combustion front. Niebergal et al. [11] studied the micro-physics of the combustion front by solving the reaction-diffusion-advection equations in spherical symmetry and considering neutrino reactions. Their result gave a burning front faster by almost six orders of magnitude compared to previous estimates. Other efforts to solve the hydrodynamic equations were performed by Herzog et al. [12] by simulating the conversion of a NS into a QS. This approach found that turbulence can enhance the burning speed of the front. Manrique et al. [13] assumed an initial shock when solving numerically the special relativistic hydrodynamic equations of combustion and confirmed numerically that the assumption of an initial shock leads to a detonation.

To fully understand the phase transition of hadronic-to-quark matter we believe that solving numerically the hydrodynamic equations of the burning front is the best approach. By following this micro-physical approach, we are capable of looking at the effects induced by neutrinos and electrons from first principles. To do so, we created Burn-UD a code that simulates the transition of hadronic-to-quark matter using micro-physics, implementing different EoS and different initial conditions (see [14] for a review). In this paper, we present the progress and changes we have done to Burn-UD. Section 2 explores the advances and new functionalities implemented in the code and the results we have obtained using our code. Section 3 presents some of the astrophysical applications of our results and lastly we present our conclusions in Section 4.

2. Developments in the code BURN-UD

Burn-UD is a hydrodynamic combustion code used to model the hadronic-to-quark matter transition. It is written in java and will be publicly available on all major platforms (Windows, Linux and OSx). Its distinct graphical user interface makes it a compelling tool for scientist to explore the micro-physics, hydrodynamical effects and EoS of the burning front. Ouyed et al. [14] offers an introduction to the software and its interface. Here, we present the two major updates of the software: a neutrino transport module and a QS evolution module.

2.1. Neutrino module

Electrons and neutrinos play a role in quark matter (u,d,s) through the following two reactions:

IOP Conf. Series: Journal of Physics: Conf. Series **861** (2017) 012008 doi:10.1088/1742-6596/861/1/012008

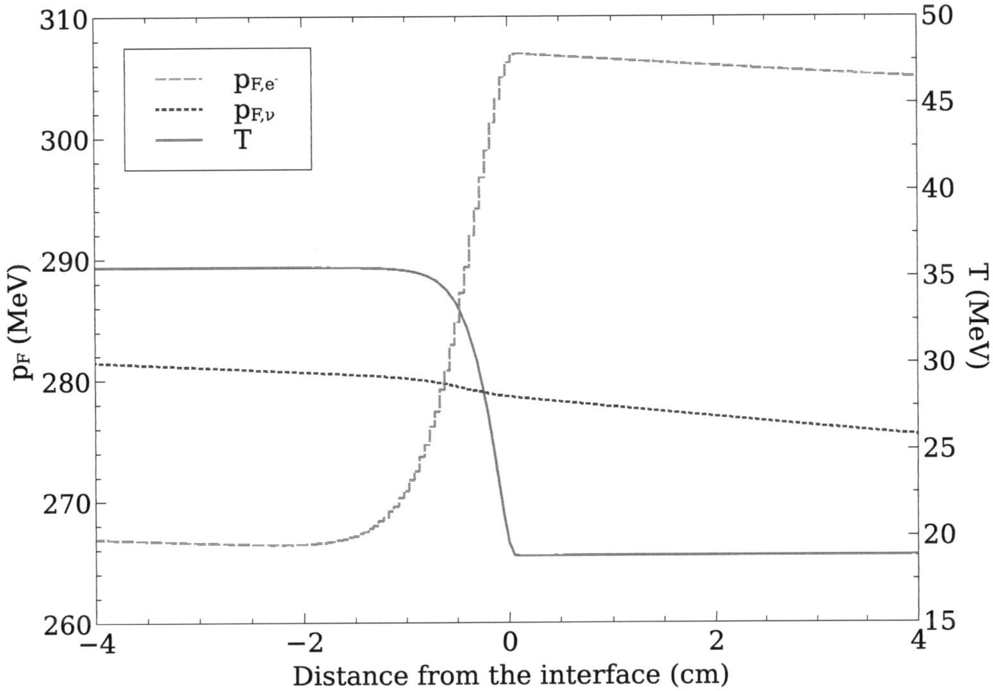

Figure 1. Electron fermi momentum ($p_{F_{e^-}}$), electron-neutrino fermi momentum ($p_{F_{\nu_e}}$), and temperature (T) as functions of distance from the interface. Burning raises the temperature and causes a strong electron pressure gradient.

$$u + e \leftrightarrow s + \nu_e \tag{1}$$

$$u + e \leftrightarrow d + \nu_e. \tag{2}$$

The above two reactions lead to a neutrino distribution function that moves across the burning interface. The neutrino transport module implements neutrino transport in a flux-limited, diffusion scheme. It uses the mean free paths of quark matter and couples the transport consistently to the momentum and entropy evolution equations:

$$\frac{\partial n_{\nu_e}}{\partial t} + \nabla \cdot (\mathbf{v}\nu_e) + \nabla \cdot (\nabla D_{\nu_e} n_{\nu_e}) = \Gamma_1 + \Gamma_2 \tag{3}$$

$$\frac{\partial Q}{\partial t} = \frac{\partial U_{\nu_e}}{\partial t}. \tag{4}$$

There, ν_e is the electron neutrino number density, $\mathbf{v}$ is the fluid velocity, D_{ν_e} is the diffusion coefficient, Q is the heat density, and U_{ν_e} is the internal energy density. Γ_1 and Γ_2 are the reaction rates due to equation (1) and equation (2) respectively.

We find that the addition of neutrino transport gives rise to various hydrodynamic effects, related to the pressure gradients induced by the lepton physics, such as electron capture and neutrino transport.

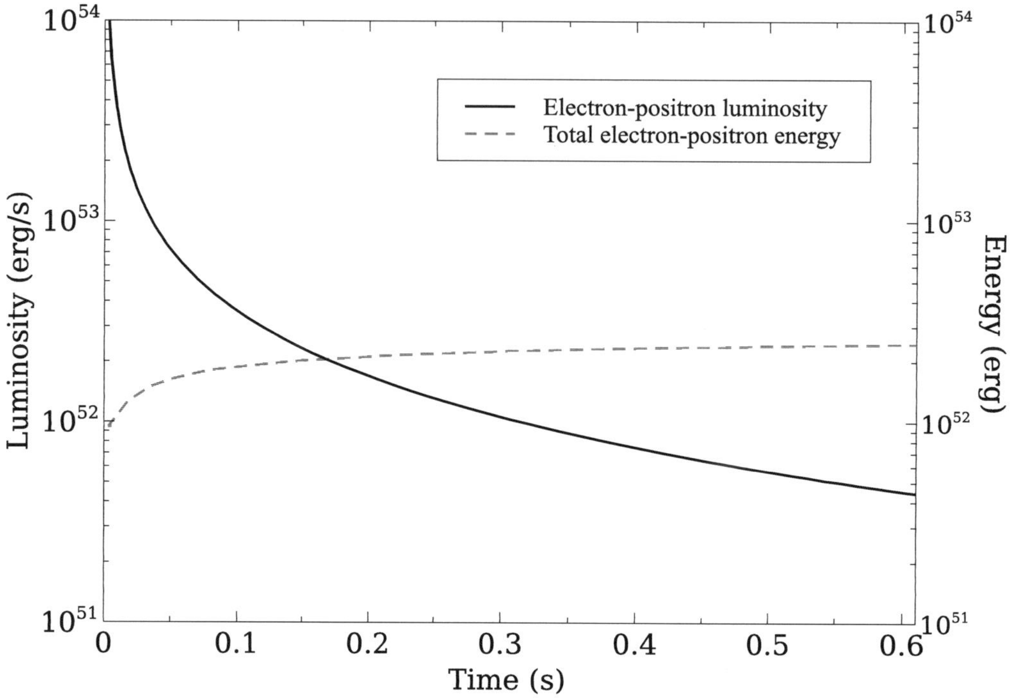

Figure 2. The luminosity of electron-positron pairs deposited through neutrino-antineutrino annihilation. The total energy deposited (the integrated luminosity) is shown. This is a few times $> 10^{52}$ ergs of kinetic energy.

2.2. Quark star evolution module

We also added a preliminary QS evolution module where the number densities and chemical potentials of quarks, electrons, and neutrinos are evolved at each time step for a whole compact star. The evolution is done by enforcing beta-equilibrium at each time-step while simultaneously solving the lepton and energy transport equations:

$$\frac{\partial E_{\nu_e}}{\partial t} + \frac{\partial}{\partial r}\frac{1}{r^2}\left(r^2\frac{\partial}{\partial r}D_{\nu_e}E_{\nu_e}\right) = S_1 \tag{5}$$

$$\frac{\partial E_{\nu_\mu}}{\partial t} + \frac{\partial}{\partial r}\frac{1}{r^2}\left(r^2\frac{\partial}{\partial r}D_{\nu_\mu}E_{\nu_\mu}\right) = S_1 \tag{6}$$

$$\frac{\partial n_{\nu_e}}{\partial t} + \frac{\partial}{\partial r}\frac{1}{r^2}\left(r^2\frac{\partial}{\partial r}D_{\nu_e}n_{\nu_e}\right) = S_3 \tag{7}$$

where E_i is the energy density of particle i and S_i are source terms.

It also calculates the e^+e^- luminosity caused by the process $\nu\overline{\nu} \to e^+e^-$ above the surface of the QS. This module will incorporate general relativistic effects in the transport equations in its upcoming versions.

2.3. Results with new updates.

2.3.1. Electron-gradient induced instability We find that the coupling of reactions (1), (2), and neutrino transport to the hydrodynamic equations gives rise to important hydrodynamic effects.

IOP Conf. Series: Journal of Physics: Conf. Series **861** (2017) 012008 doi:10.1088/1742-6596/861/1/012008

One of them is an electron-gradient induced instability. As equation (1) and equation (2) lead to electron capture behind the interface, and leptons deplete in the upstream region due to neutrino diffusion, a strong electron gradient develops that leads to a pressure gradient (see Fig. 1) that induces fluid velocities parallel to the upstream direction. In the case of trapped neutrinos, the front momentarily halts until the momentum and heat deposited by neutrinos revives it. However, in the case untrapped neutrinos, the electron pressure gradient is unopposed, halting the burning front in the time-scale of the simulation.

We speculate that in a multidimensional situation, uneven slowing down and halting of the interface due to the electron pressure gradient could wrinkle the interface, leading to turbulence. Niebergal et al. [11] discovered a similar instability (deleptonization instability) due to neutrino cooling as heat leaks from the upstream region due to escaping neutrinos and the thermal pressure behind the front decreases, which could lead to halting. Both the electron-gradient induced instability and deleptonization instability would exacerbate each other because both are caused by the depletion of leptons from the upstream region.

2.3.2. Mass ejection due to $\nu\bar{\nu}$ annihilation We used the QS evolution module (see section 2.2) and imposed an initial thermal energy distribution due to the release of binding energy through a hadron-quark phase transition. This is calculated by imposing hadronic EoS in the compact star, and then switching to an MIT Bag Model EoS, while conserving energy density. We find an intense e^-e^+ pair luminosity burst of $> 10^{53}$ erg/s for the duration of 0.5 s (see dashed curve in Fig. 2) due to $\nu\bar{\nu}$ annihilation near the surface of the QS, which suggests relativistic mass ejection of the hadronic matter overlaying on the surface of the QS. We find total integrated e^-e^+ pair luminosity of about 5×10^{52} ergs. This suggests a strong candidate for an engine of some of the most energetic events in astrophysics, such as superluminous supernovae, *hypernovae*, or gamma ray bursts.

The physical causes of this e^-e^+ pair luminosity burst are the extreme surface temperatures T $\sim$ 20 MeV of a nascent QS and its compactness. Large temperatures lead to large thermal ν and $\bar{\nu}$ number densities which would annihilate near the neutrino-sphere. This filters the neutrino energy by transforming it into electron-positron energy, the latter which is strongly coupled with matter. Most of the e^-e^+ pair luminosity is deposited as kinetic energy in the overlaying hadronic matter. Furthermore, the relative insensitivity of the quark matter EoS to temperature implies that the radius of QS is relatively small ($\sim$ 10 km) even at high temperatures which leads to general relativistic effects that enhance $\nu\bar{\nu}$ annihilation [15]. These results are studied and discussed in details in [16, 17].

3. Astrophysical applications

A complementary approach to these simulations is to look for astronomical signatures that help us probe the nature of this process. Particularly, we believe that by looking closely at the most luminous events in the sky, our understanding of stellar evolution, and the production of elements we can complement the state of the art simulations and theoretical work performed.

The Quark-Nova (QN) model assumes the explosive transition of a NS into a QS [18]. The explosion can eject $\sim 10^{-3}$ $M_\odot$ of the outer layers of the NS at relativistic speeds with an average Lorentz factor of $\Gamma_{QN} \sim 10$ [11, 19]. The ejecta could be visible by interacting with surrounding material or by the decay of unstable r-process isotopes. Ouyed et al. [20, 21] offer an in-depth review to the QN model and their observational signatures in astronomical data. In this section we present the latest updates of the QN model in superluminous supernova events (SLSNe).

3.1. Superluminous Supernova Events (SLSNe)

SLSNe are explosions seen in space with extreme luminosities that extend hundreds of days. Although they seem to resemble regular supernovae (SNe), their extreme luminosity and spectral

IOP Conf. Series: Journal of Physics: Conf. Series **861** (2017) 012008 doi:10.1088/1742-6596/861/1/012008

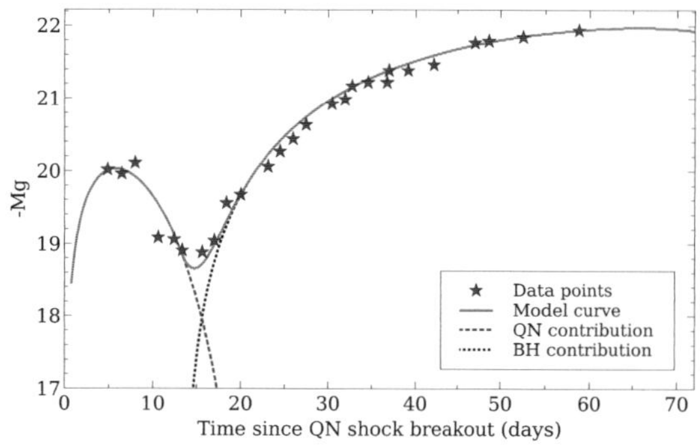

Figure 3. QN model fit to the g-band light curve of LSQ14bdq. The stars are the observations from Nicholl et al. [35], the black line is the BH accretion and the blue line is the QNe. The first model was employed because the SLSNe is not characterized by a strong oxygen absorption line.

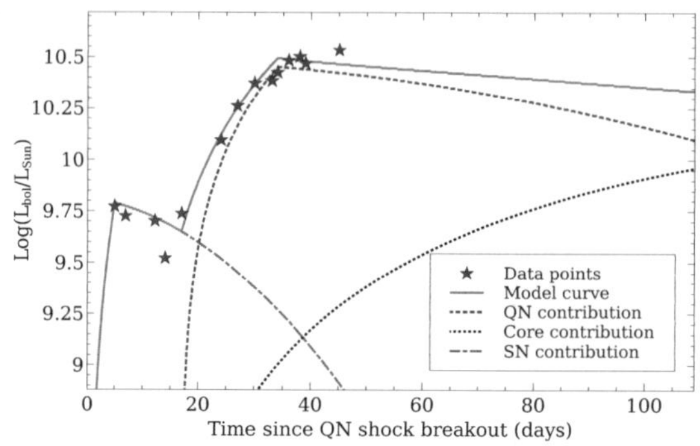

Figure 4. QN model fit to the bolometric light curve of SN2006oz. The stars are the observations from Leloudas et al. [36]. The green curve is the light curve of a WO type star SN. The blue curve is the envelope's contribution after being shocked by the QN while the black curve is the core's contribution after being heated by the QN. This event has an oxygen absorption line in its spectrum making it an ideal candidate for this model.

features point towards a yet unknown mechanism that has puzzled astrophysicists since their discovery a little over a decade ago. For some, the decay of ^{56}Ni can be the mechanism behind some of the events [22, 23, 24]. For others, the core collapse of massive stars, the thermonuclear explosion of white dwarfs or a combination of both on a binary arrangement that interacts with a circumstellar material could explain SLSNe [25, 26, 27, 28]. Another alternative is powering by a black hole (BH) experiencing accretion from surrounding material [29] or by a highly magnetic NS rotating with a millisecond period (magnetar) releasing rotational energy into a dense shell [30, 31].

When thinking about SLSNe, it is evident that a regular SN cannot provide the energy required to power them, so the idea of an even more energetic explosion seems appealing. The QN offers the energy required to explain these high luminosity events (see section 2). Currently we have two compelling scenarios that could explain SLSNe. Ouyed and his collaborators have published a series of papers on SLSNe and they have used the model to fit more than half a dozen SLSNe [21, 32, 33, 34]. Here we present an overview of the two models we have developed for double humped SLSNe, their achievements and limitations.

For SLSNe-I (hydrogen-poor), a QN occurring in a massive binary undergoing two common envelopes (CE) phases can account for the double-humped light curve. Two massive stars in a binary configuration can undergo their regular stellar evolution leading to the more massive going SN. The resulting NS can accrete matter until it reaches deconfinement densities. At this point the NS will go QN and the first hump in the light curve is explained. Later stages of the system allow for the formation of a BH accreting matter over a subtended period of time effectively powering a long lasting second hump in the light curve. Ouyed et al. [32] offer an in-depth explanation and step-by-step of the model.

The QN binary model's strength resides in the ejection of the first CE (H rich) and in being able to explain light curves that show multiple rebrightening epochs (i.e. bumps in the light curve). By ejecting the first CE, this model can explain the lack of H in the observed spectrum of SLSN-I or its later appearance. The rebrightening at different epochs naturally arises from the interactions between the QN and CO core and BH and CO disc; these elements behave as mass reservoirs that can be exploited and used at later times. This model has successfully explained the light curves of half a dozen SLSNe. However, one of its limitations is explaining events with strong oxygen absorption on the early stages of their light curve (see ASSASN-15lh [37] and its late UV rebrightening [38]).

Our second model is the basis for SLSNe with early strong oxygen absorption. The presence of a QN after the supernova explosion of a stripped oxygen sequence Wolf-Rayet star can explain the double hump nature of these events and the large presence of oxygen in the early spectra. First, the WO type star goes SN and the ejecta expands homologously. After the SNe, a two-component configuration is created by the extended low density envelope and the dense core. Later, a QN going off in this configuration will deposit energy in the extended envelope of the SN resulting in the first hump of the light curve. Lastly, the heated core, initially energized by the QN shock, acts as a *hot plate* that re-brightens the light curve after it is revealed by the receding photosphere of the shocked envelope giving result to the second hump in the light curve. This model has been successful in explaining the light curve, photospheric radius and effective temperature of ASASSN-15lh [39]. Furthermore, we believe that this model could yield similar SLSNe with different spectra if it were to happen inside other types of WR stars (e.g. WN and WC).

Both models have been successful in reproducing the light-curves of different SLSNe. Figure 3 and figure 4 are two of the multiple fits we have performed for SLSNe (more fits are available at http://quarknova.ca/LCGallery.html). The full understanding of hadronic-to-quark matter will allow us to make more robust models and eventually determine if one scenario is more likely to happen than the other. Furthermore, fitting more observations can help us delimit the energetics of this transition.

3.2. Stellar nucleosynthesis

The origin of elements heavier than iron is thought to be slow and rapid neutron capture processes, s-process and r-process respectively [40, 41]. While the s-process occurs at low neutron densities, the r-process is the opposite operating at much higher neutron densities. Simulating the conditions of r-process sites and matching our simulations to chemical abundances provides knowledge about element formation and allows us to look for astronomical signatures of proton rich, hot nucleosynthesis environments.

We have developed SiRop (Simulation of Rapid neutron capture) a code for the simulation of r-process under the same principles of Burn-UD, a friendly user interface, high optimization for running simulations and available for the public in all platforms. SiRop allows the user to study the chemical abundances of different astronomical sites (see [42] for a full review of SiRop and its manual). We have used SiRop (and r-Java, a previous version of the code) to simulate the conditions of a QN site (see [20, 43, 44]). Our findings show that the high temperature and proton rich ejecta of a QN make it an ideal site for the r-process (see Fig. 5). By continuing our study of the QN and the r-process we are looking at the elemental abundances of stars and matching our simulations to astronomical observations. Not only are we understanding the role of r-process sites but we are also finding signatures that help us establish constraints on the energetics of the transition of hadronic-to-quark matter and the QN events.

4. Conclusion

We believe that the study of hadronic-to-quark matter transition, its causes and consequences is vital for both theoretical physics and astronomy. The use of computational tools to perform high level hydrodynamical simulations of the burning front of a NS to a QS is important to

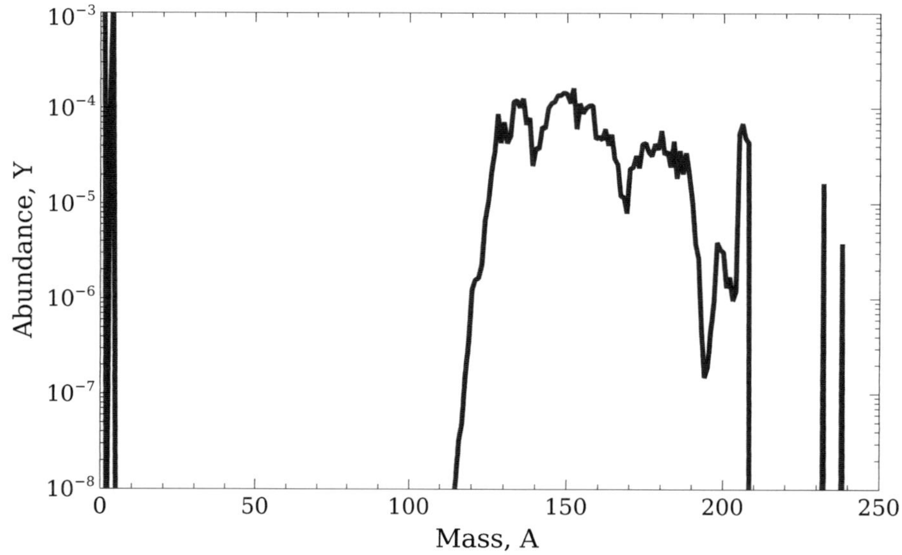

Figure 5. Results of an r-process simulation using SiRop from a QN which has ejected the NS crust (i.e. the QN ejecta). The parameters used are a QN ejecta of 10^{-3} $M_\odot$, an initial electron fraction $Y_e = 0.1$ and a 1% of total energy of the QN transformed into kinetic energy of the ejecta.

help bring consensus to the scientific community about the nature of hadronic matter and the conditions that allow it to undergo a QCD phase transition.

Burn-UD, our specialized in house software[1], offers the tools for scientists to study the time-dependent phase transition of hadronic-to-quark matter. The two new modules, neutrino reactions and QS evolution, have pointed at an electron-gradient induced instability and mass ejection due to $\nu\bar{\nu}$ annihilation. This mechanism seems to be possible in addition to others like core-collapse [45] and photon fireball [46]. The joint result of these mechanisms make us believe that the transition of hadronic-to-quark mater can result in an explosion that releases about $\sim 10^{52}$ ergs.

Such an energetic event will undoubtedly have consequences in astronomy. We believe that the explosive transition of a NS to a QS, a QN, is the mechanism powering the double humped light curve of SLSNe. Our model assumes the presence of a QN inside a massive binary system or in a WO type SN remnant and has successfully reproduced the light curves of over half a dozen SLSNe. Furthermore we have simulated the conditions of a QN site and the hot temperatures along with the rich neutron density expose the QN as a preferred site for the r-process.

References

[1] Bodmer A R 1971 *Phys. Rev. D* **4**(6) 1601–1606
[2] Witten E 1984 *Phys. Rev. D* **30**(2) 272–285
[3] Olinto A V 1987 *Physics Letters B* **192** 71–75
[4] Heiselberg H, Baym G and Pethick C 1991 *Nuclear Physics B-Proceedings Supplements* **24** 144–147
[5] Olesen M L and Madsen J 1991 *Nuclear Physics B-Proceedings Supplements* **24** 170–174
[6] Drago A, Lavagno A and Parenti I 2007 *The Astrophysical Journal* **659** 1519
[7] Bhattacharyya A, Ghosh S K, Joarder P S, Mallick R and Raha S 2006 *Physical Review C* **74** 065804

[1] Burn-UD and SiRop are publicly available at http://quarknova.ca.

[8] Benvenuto O, Horvath J and Vucetich H 1989 *International Journal of Modern Physics A* **4** 257–265
[9] Benvenuto O and Horvath J 1989 *Physical review letters* **63** 716
[10] Lugones G, Benvenuto O and Vucetich H 1994 *Physical Review D* **50** 6100
[11] Niebergal B, Ouyed R and Jaikumar P 2010 *Physical Review C* **82** 062801
[12] Herzog M and Röpke F K 2011 *Physical Review D* **84** 083002
[13] Manrique M A A and Lugones G 2015 *Brazilian Journal of Physics* **45** 457–466
[14] Ouyed A, Welbanks L, Koning N and Ouyed R 2015 *Conference Proceedings of the Compact Stars in the QCD Phase Diagram IV (CSQCD IV)*
[15] Salmonson J D and Wilson J R 1999 *The Astrophysical Journal* **517** 859
[16] Ouyed A, Ouyed R, Koning N and Jaikumar P *Neutrino-transport in hadron-quark combustion using Burn-UD: implications to explosive astrophysics. [In Preparation]*
[17] Ouyed A 2017 *Neutrino-transport in hadron-quark combustion using Burn-UD: Implications to explosive astrophysics.* Ph.D. thesis University of Calgary
[18] Ouyed R, Dey J and Dey M 2002 *Astronomy & Astrophysics* **390** L39–L42
[19] Ouyed R and Leahy D 2009 *The Astrophysical Journal* **696** 562
[20] Ouyed R, Leahy D, Koning N and Shand Z 2016 *arXiv preprint arXiv:1601.04236*
[21] Ouyed R, Leahy D, Koning N and Staff J E 2016 *arXiv preprint arXiv:1601.04235*
[22] Gal-Yam A, Mazzali P, Ofek E, Nugent P *et al.* 2009 *Nature* **462** 624–627
[23] Gal-Yam A 2012 *Science* **337** 927–932
[24] Hatano K, Branch D, Nomoto K, Deng J, Maeda K, Nugent P and Aldering G 2001 *Bulletin of the American Astronomical Society* **33** 838
[25] Silverman J M, Nugent P E, Gal-Yam A, Sullivan M *et al.* 2013 *The Astrophysical Journal Supplement Series* **207** 3
[26] Firth R, Sullivan M, Gal-Yam A, Howell D *et al.* 2015 *Monthly Notices of the Royal Astronomical Society* **446** 3895–3910
[27] Nugent P E, Sullivan M, Cenko S B, Thomas R C *et al.* 2011 *Nature* **480** 344–347
[28] Moriya T J, Liu Z W, Mackey J, Chen T W and Langer N 2015 *Astronomy & Astrophysics* **584** L5
[29] Dexter J and Kasen D 2013 *The Astrophysical Journal* **772** 30
[30] Woosley S 2010 *The Astrophysical Journal Letters* **719** L204
[31] Kasen D and Bildsten L 2010 *The Astrophysical Journal* **717** 245
[32] Ouyed R, Leahy D and Koning N 2015 *Monthly Notices of the Royal Astronomical Society* **454** 2353–2359
[33] Ouyed R, Leahy D and Koning N 2015 *The Astrophysical Journal* **809** 142
[34] Ouyed R, Leahy D and Koning N 2016 *The Astrophysical Journal* **818** 77
[35] Nicholl M, Smartt S, Jerkstrand A, Sim S *et al.* 2015 *The Astrophysical Journal Letters* **807** L18
[36] Leloudas G, Chatzopoulos E, Dilday B, Gorosabel J *et al.* 2012 *Astronomy & Astrophysics* **541** A129
[37] Dong S, Shappee B J, Prieto J L, Jha S W *et al.* 2016 *Science* **351** 257–260
[38] Godoy-Rivera D, Stanek K, Kochanek C, Chen P *et al.* 2017 *Monthly Notices of the Royal Astronomical Society* **466** 1428
[39] Ouyed R, Leahy D, Welbanks L and Koning N 2016 *Submitted [arXiv:1611.03657]*
[40] Burbidge E M, Burbidge G R, Fowler W A and Hoyle F 1957 *Reviews of modern physics* **29** 547
[41] Cameron A G W 1957 *Publications of the Astronomical Society of the Pacific* **69** 201–222
[42] Shand Z 2016 *The r-process Code SiRop: Applications to Nuclear Sensitivity Studies and High-energy Astrophysics* Master's thesis University of Calgary
[43] Kostka M, Koning N, Shand Z, Ouyed R and Jaikumar P 2014 *Astronomy & Astrophysics* **568** A97
[44] Charignon C, Kostka M, Koning N, Jaikumar P and Ouyed R 2011 *Astronomy & Astrophysics* **531** A79
[45] Keränen P, Ouyed R and Jaikumar P 2005 *The Astrophysical Journal* **618** 485
[46] Ouyed R, Rapp R and Vogt C 2005 *The Astrophysical Journal* **632** 1001

IOP Conf. Series: Journal of Physics: Conf. Series **861** (2017) 012009 doi:10.1088/1742-6596/861/1/012009

Unified description of nuclear matter properties within the CBF effective interaction approach

Omar Benhar[1] **and Alessandro Lovato**[2]
[1]INFN and Dipartimento di Fisica, Sapienza University, I-00185 Rome, Italy
[2]Physics Division, Argonne National Laboratory, Argonne, IL 60439
E-mail: [1]omar.benhar@roma1.infn.it
E-mail: [2]lovato@anl.gov

Abstract. We discuss the derivation of an effective interaction, obtained from a realistic nuclear Hamiltonian using the formalism of Correlated Basis Functions (CBF) and the cluster expansion technique. Unlike the bare nucleon-nucleon potential, the CBF effective interaction is well behaved, and suitable for carrying out perturbative calculations in the basis of eigenstates of the non interacting system. The results of studies of a variety of properties of nuclear matter, at both zero and nonzero temperatures, are reported and analyzed.

1. Introduction

This paper will focus on the tiny—although arguably most important—region of the QCD phase diagram corresponding to vanishing or very low temperatures and densities comparable to the central density of atomic nuclei, $\rho_0 = 0.16$ fm^{-3}. Under these conditions, the ground state of strongly-interacting matter is known to consist of protons and neutrons, the interactions of which can be described by a non relativistic Hamiltonian.

In the case of isospin-symmetric nuclear matter (SNM)—that is, a tanslation invariant system with equal numbers of protons and neutrons interacting through nuclear forces only—the binding energy per particle can be determined from the extrapolation of the nuclear mass formula, yielding the result $E_0 \approx -16$ MeV. Additional information can be extracted from measurements of nuclear properties and nuclear reactions, giving access to quantities such as the compression modulus, $K \approx 260 \pm 30$ MeV, and the symmetry energy, $E_{\rm sym} \approx 31.6 \pm 2.66$.

Effective interactions specifically designed to reproduce the available empirical information on SNM (see, e.g., Refs. [1, 2]), while being remarkably successful in a number of instances, are limited by the lack of a connection with nuclear dynamics at microscopic level. As a consequence, they are inherently unable to provide a quantitative account of nucleon-nucleon scattering processes—both in free space and in the nuclear medium—whose understanding is needed for the description of non-equilibrium properties [3, 4].

Early attempts to derive an effective interaction from a phenomenological nucleon-nucleon potential are described in Refs.[5, 6]. More recently, the authors of Refs. [7, 8] have improved on the approach of Ref. [5], and developed a procedure to determine the effective interaction using the Correlated Basis Function (CBF) formalism and the cluster expansion technique.

The results of extensive studies of the Fermi hard-sphere system strongly suggest that the CBF effective interaction approach provides accurate estimates of a variety of fundamental

IOP Conf. Series: Journal of Physics: Conf. Series **861** (2017) 012009 doi:10.1088/1742-6596/861/1/012009

quantities of interacting many-body systems other than the ground-state energy, such as the self-energy determining the two-point Green's function [9, 10].

The applications to nuclear matter, discussed in this paper, include calculations of a number of properties of great astrophysical interest, such as the equation of state, the quasiparticle spectrum and the chemical potentials of uniform matter at fixed baryon density and large neutron excess, at both zero and nonzero temperature.

The main features of the nuclear Hamiltonian and the derivation of the CBF effective interaction are described in Section 2, while Section 3 is devoted to the discussion of numerical results, including the ground-state energy at arbitrary proton fraction, the symmetry energy, the pressure and the single-nucleon spectrum. Finally, in Section 4 we summarize our findings, and outline the prospects for future applications of the approach.

2. Theoretical framework

In this section, we will outline the phenomenological model of nuclear dynamics employed to obtain the numerical results, and describe the procedure leading to the determination of the effective interaction.

2.1. The nuclear Hamiltonian

Within non relativistic many-body theory (NMBT) atomic nuclei, as well as infinite nuclear matter, are described in terms of point-like nucleons of mass m, whose dynamics are dictated by the Hamiltonian

$$H = \sum_i -\frac{\nabla_i^2}{2m} + \sum_{i<j} v_{ij} + \sum_{i<j<k} V_{ijk} \,. \tag{1}$$

The nucleon-nucleon (NN) potential v_{ij} is obtained from an accurate fit of the measured properties of the two-nucleon system—in both bound and scattering states—and reduces to the Yukawa one-pion-exchange potential at large distances. Coordinate-space NN potentials are usually written in the form

$$v_{ij} = \sum_p v^p(r_{ij}) O_{ij}^p \,, \tag{2}$$

where $r_{ij} = |\mathbf{r}_i - \mathbf{r}_j|$ is the distance between the interacting particles, and the sum includes up to eighteen terms. The most prominent contributions are those associated with the operators

$$O_{ij}^{p\leq 6} = [1, (\boldsymbol{\sigma}_i \cdot \boldsymbol{\sigma}_j), S_{ij}] \otimes [1, (\boldsymbol{\tau}_i \cdot \boldsymbol{\tau}_j)] \,, \tag{3}$$

where $\boldsymbol{\sigma}_i$ and $\boldsymbol{\tau}_i$ are Pauli matrices acting in spin and isospin space, respectively, while the operator

$$S_{ij} = \frac{3}{r_{ij}^2} (\boldsymbol{\sigma}_i \cdot \mathbf{r}_{ij})(\boldsymbol{\sigma}_j \cdot \mathbf{r}_{ij}) - (\boldsymbol{\sigma}_i \cdot \boldsymbol{\sigma}_j) \,, \tag{4}$$

accounts for the occurrence of non-spherically-symmetric interactions. The potentials obtained including the six operators of Eqs. (3)-(4) explain deuteron properties and the S-wave scattering phase shifts. In order to describe the P-wave, one has to include two additional components involving the momentum dependent operators

$$O_{ij}^{p=7,8} = (\boldsymbol{\ell} \cdot \mathbf{S}), \ (\boldsymbol{\ell} \cdot \mathbf{S})(\boldsymbol{\tau}_1 \cdot \boldsymbol{\tau}_2) \,, \tag{5}$$

where $\boldsymbol{\ell}$ denotes the angular momentum of the relative motion of the interacting particles.

IOP Conf. Series: Journal of Physics: Conf. Series **861** (2017) 012009 doi:10.1088/1742-6596/861/1/012009

The operators corresponding to $p = 7, \ldots, 14$ are associated with the non-static components of the NN nteraction, while those corresponding to $p = 15, \ldots, 18$ account for small violations of charge symmetry. All these terms are included in the state-of-the-art Argonne v_{18} (AV18) potential [11], providing a fit of the scattering data collected in the Nijmegen database, the low-energy nucleon-nucleon scattering parameters and deuteron properties with a reduced chi-square $\chi^2 \simeq 1$.

The results reported in this paper have been obtained using the so-called Argonne v_6' (AV6P) interaction, which is not simply a truncated version of the full AV18 potential—obtained neglecting the contributions with $p > 6$ in Eq. (2)—but rather its reprojection on the basis of the six spin-isospin operators of Eqs. (3)-(4) [12].

The inclusion of the additional three-nucleon (NNN) term, V_{ijk}, is needed to explain the binding energies of the three-nucleon systems and the saturation properties of SNM. The derivation of V_{ijk} was first discussed in the pioneering work of Fujita and Miyazawa [13]. They argued that its main component originates from two-pion-exchange processes in which a NN interaction leads to the excitation of one of the participating nucleons to a Δ resonance, which then decays in the aftermath of the interaction with a third nucleon. Commonly used phenomenological models of the NNN force, like the Urbana IX (UIX) potential [14], are written in the form

$$V_{ijk} = V_{ijk}^{2\pi} + V_{ijk}^{N} , \tag{6}$$

where $V_{ijk}^{2\pi}$ is the attractive Fujita-Miyazawa term, while V_{ijk}^{N} is a purely phenomenological repulsive term. The parameters entering the definition of the above potential are adjusted in such a way as to reproduce the ground state energy of the three-nucleon systems and the equilibrium density of SNM, when used in conjunction with the AV18 NN interaction.

Note that within the framework of NMBT the determination of the nuclear Hamiltonian implies very little theoretical bias, becaus the two- and three-nucleon systems are solved exactly, and the equilibrium properties of SNM can be computed with great accuracy.

It has to be emphasized that local NN potentials derived within the alternate framework of chiral perturbation theory are also written in the form of Eq. (2) [15, 16]. Since local versions of the chiral NNN potentials [17] have the same spin-isospin structure of the UIX force, the scheme described in this paper can be readily applied using chiral nuclear Hamiltonians.

2.2. *Derivation of the CBF effective interaction*

The formalism of Correlated Basis Functions (CBF) is based on the variational approach to the many-body problem with strong forces, first proposed by R. Jastrow back in the 1950s [18]. Within this scheme, the trial ground state of the nuclear hamiltonian is written in the form

$$|\Psi_0\rangle \equiv \frac{\mathcal{F}|\Phi_0\rangle}{\langle\Phi_0|\mathcal{F}^\dagger\mathcal{F}|\Phi_0\rangle^{1/2}} , \tag{7}$$

where $|\Phi_0\rangle$ is a Slater determinant built from single particle states $|\phi_\alpha\rangle$, with $\{\alpha\}$ being the set of quantum numbers of the states belonging to the Fermi sea. In the case of uniform matter at density $\rho = \nu k_F^3/(6\pi^2)$—where k_F and ν denote the Fermi momentum and the degeneracy of the momentum eigenstates, respectively—$|\phi_\alpha\rangle$ consists of a plane wave, with momentum $\mathbf{k}$ such that $|\mathbf{k}| \leq k_F$, and the Pauli spinors associated with the spin and isospin degrees of freedom.

The many-body operator $\mathcal{F}$, describing the effects of correlations among the nucleons, is written as a product of two-body operators, whose structure mirrors that of the AV6P potential, as

$$\mathcal{F} \equiv \mathcal{S} \prod_{i<j} F_{ij} , \tag{8}$$

with

$$F_{ij} = \sum_{p=1}^{6} f^p(r_{ij}) O^p_{ij} \ . \tag{9}$$

Note that the symmetrization operator $\mathcal{S}$ is needed to fulfill the requirement of antisymmetrization of $|\Psi_0\rangle$ under particle exchange, since, in general, $[O^p_{ij}, O^q_{jk}] \neq 0$.

The radial dependence of the correlation functions $f^p(r_{ij})$, determined from functional minimization of the expectation value of the Hamiltonian in the correlated ground state,

$$E_V = \langle \Psi_0 | H | \Psi_0 \rangle \ . \tag{10}$$

is largely shaped by the strongly repulsive core of the NN interaction, resulting in a drastic suppression of the probability to find two nucleons at relative distance $r_{ij} \lesssim 1$ fm. Longer range correlations, mainly due to the non-central components of the NN potential, are also important.

The calculation of the variational energy of Eq. (10) involves severe difficulties. It can be efficiently carried out expanding the right-hand side in a series, whose terms describe the contributions of clusters involving an increasing number of correlated particles [19]. The terms of the cluster expansion can be represented by diagrams, and classified according to their topological structures. Selected classes of diagrams can then be summed to all orders solving a set of coupled non-linear integral equations, referred to as Fermi Hyper-Netted Chain/Single-Operator Chain (FHNC/SOC) equations [20, 21], to obtain an accurate estimate of the ground state energy.

Under the assumption that the correlation structure of the ground and excited states of the system be the same, the operator $\mathcal{F}$ obtained from the variational calculation of E_V can be used to generate correlated excited states from Eq. (7) by replacing $|\Phi_0\rangle \to |\Phi_n\rangle$, with $|\Phi_n\rangle$ being any eigenstate of the non interacting Fermi gas. The resulting correlated states span a complete, although non orthogonal, set, that can be used to carry out perturbative calculations within the scheme developed in Ref. [22]. This approach, known as CBF perturbation theory, has been successfully applied to study a variety of nuclear matter properties, including the response functions [23, 24] and the two-point Green's function [25, 26].

In CBF perturbation theory, one has to evaluate matrix elements of the *bare* nuclear Hamiltonian, the effects of correlations being taken into account by the transformation of the basis states describing the non interacting system. However, the same result can in principle be obtained transforming the Hamiltonian, and using the Fermi gas basis. This procedure leads to the appearance of an *effective* Hamiltonian suitable for use in standard perturbation theory, thus avoiding the non trivial difficulties arising from the use of a non-orthogonal basis [27].

The CBF effective interaction is defined through the matrix element of the bare Hamiltonian in the correlated ground state, according to

$$\langle \Psi_0 | H | \Psi_0 \rangle = T_F + \langle \Phi_0 | \sum_{i<j} v^{\rm eff}_{ij} | \Phi_0 \rangle \ , \tag{11}$$

where T_F denotes the energy of the non interacting Fermi gas, and the effective potential is written in terms of the same spin-isospin operators appearing in Eq. (2)

$$v^{\rm eff}_{ij} = \sum_p v^{{\rm eff},p}(r_{ij}) O^p_{ij} \ . \tag{12}$$

From the above equation, it is apparent that $v^{\rm eff}_{ij}$ embodies the effect of correlations. As a consequence, it is well behaved at short distances, and can in principle be used to carry out perturbative calculations of any properties of nuclear matter.

The authors of Ref. [7] first proposed to obtain the effective interaction performing a cluster expansion of the left-hand side of Eq. (11) and keeping the two-body cluster contribution only. While leading to a very simple and transparent expression for $v^{\rm eff}_{ij}$, however, this scheme was seriously limited by its inability to take into account the NNN potential V_{ijk}. In Ref. [8] the effects of interactions involving more than two nucleons have been included through a density-dependent modification of the NN potential at intermediate range [28].

A significant improvement has been achieved by the authors of Refs. [29, 30], who explicitly took into account three-nucleon cluster contributions to the ground-state energy. This procedure allows to describe the effects of three-nucleon interactions at fully microscopic level using the UIX potential.

Note that the correlation functions $f^p(r_{ij})$ entering the definition of $v^{\rm eff}_{ij}$ are *not* the same as those obtained from the minimization of the variational energy of Eq. (10). They are adjusted so that the ground state energy computed at first order in $v^{\rm eff}_{ij}$—that is, in the Hartree-Fock approximation—reproduces the value of E_V resulting from the full FHNC/SOC calculation. In Refs. [29] and [30], this procedure was applied, *separately*, to SNM and pure neutron matter (PNM). The effective interaction employed to obtain the results discussed in this paper, on the other hand, simultaneously describes the density dependence of the energy per nucleon of *both* SNM and PNM. This feature is essential for astrophysical applications, because it allows to evaluate the properties of nuclear matter at fixed baryon density and large neutron excess, which is believed to make up a large region of the neutron star interior.

3. Nuclear Matter Properties

3.1. Ground state energy of cold nuclear matter

The procedure developed by the authors of Ref. [31, 32], taking into account the contributions of three-nucleon clusters, allows to use a realistic nuclear Hamiltonian, comprising two- and three-nucleon potentials. The results reported in this paper have been obtained combining the AV6P [33] and UIX [34] potentials.

The resulting effective interaction—derived using as baseline the FHNC variational estimates of the ground state energy of PNM and SNM—can be used to evaluate the energy per nucleon of neutron rich and spin polarized matter at fixed baryon density

$$\rho = \sum_\lambda \rho_\lambda = \rho \sum_\lambda x_\lambda \ , \tag{13}$$

where the index $\lambda = 1, 2, 3, 4$ labels spin-up protons, spin-down protons, spin-up neutrons and spin-down neutrons, respectively, the corresponding densities being $\rho_\lambda = x_\lambda \rho$. In SNM $x_1 = x_2 = x_3 = x_4 = 1/4$, while in PNM $x_1 = x_2 = 0$ and $x_3 = x_4 = 1/2$.

At first order in the CBF effective interaction, the energy per baryon can be written in the form

$$\frac{E}{A} = \frac{3}{5}\sum_\lambda x_\lambda \frac{{k_{F_\lambda}}^2}{2m} + \frac{\rho}{2}\sum_{\lambda\mu} x_\lambda x_\mu \int d^3r \left[v^{\rm eff,d}_{\lambda\mu}(r) - v^{\rm eff,e}_{\lambda\mu}(r)\ell(k_{F_\lambda} r)\ell(k_{F_\mu} r)\right] \tag{14}$$

where $v^{\rm eff,d}_{\lambda\mu}(r)$ and $v^{\rm eff,e}_{\lambda\mu}(r)$ denote the direct and exchange matrix elements of the effective interaction in spin-isospin space, $k_{F_\lambda} = (6\pi^2\rho_\lambda)^{1/3}$ is the Fermi momentum of the nucleons of type λ and

$$\rho_\lambda \ell(k_{F_\lambda} r) \equiv \frac{1}{V}\sum_{\mathbf{k}} e^{i\mathbf{k}\cdot\mathbf{r}} n_\lambda(\mathbf{k}) \ , \tag{15}$$

with $n_\lambda(\mathbf{k}) = \theta(k_{F_\lambda} - |\mathbf{k}|)$ and V being is the zero-temperature Fermi distribution and the normalization volume, respectively.

IOP Conf. Series: Journal of Physics: Conf. Series **861** (2017) 012009 doi:10.1088/1742-6596/861/1/012009

The solid lines of Figures 1 and 2 illustrate the density dependence of the energy per nucleon of PNM and SNM, respectively, obtained from Eq. (14) using the CBF effective interaction. The shaded regions show the FHNC results obtained using the bare Hamiltonian, with the associated theoretical uncertainty arising from the treatment of the kinetic energy [35]. For comparison, the results of a calculation carried out using the Auxiliary Field Diffusion Monte Carlo (AFDMC) [36] technique are also displayed. It clearly appears that the FHNC variational estimates provide a very accurate upper bound to the ground state energy. Note, however, that while the empirical equilibrium density of SNM is well reproduced, the corresponding energy is underestimated by ~ 5 MeV.

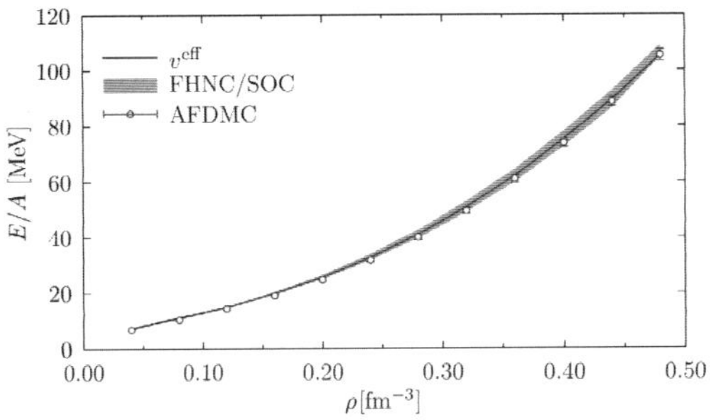

Figure 1. Density dependence of the energy per nucleon of PNM. The solid lines display results obtained using Eqs. (14)-(15) and the CBF effective interaction.The variational FHNC results are represented by the shaded regions, accounting for the uncertainty arising from the treatment of the kinetic energy, while the open circles correspond to the results obtained using the AFDMC technique.

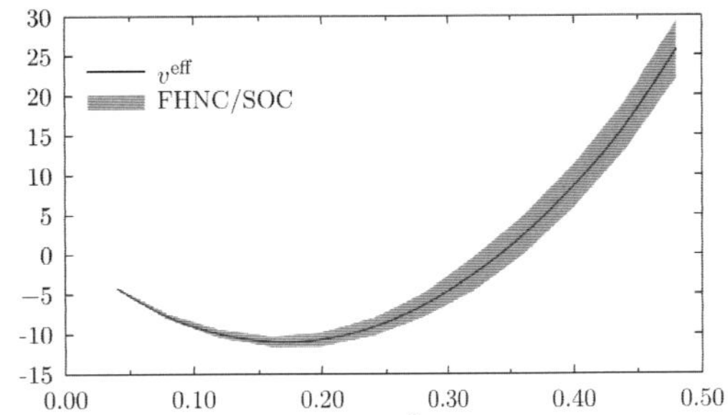

Figure 2. Density dependence of the energy per nucleon of SNM. The solid lines display results obtained using Eqs. (14)-(15) and the CBF effective interaction.The variational FHNC results are represented by the shaded regions, accounting for the uncertainty arising from the treatment of the kinetic energy.

The main advantage of the effective interaction approach is the possibility of carrying out calculations of a variety of nuclear matter properties of astrophysical interest using perturbation theory and the orthogonal basis of eigenstates of the non interacting system. Of great importance, in this context, is the ground state energy of matter at fixed baryon density and arbitrary proton fraction, $x_p = x_1 + x_2$, shown in Fig. 3 for $0 \leq x_p \leq 0.5$.

3.2. Symmetry energy

Consider again unpolarized matter with proton and neutron densities $\rho_p = x_p\rho$ and $\rho_n = (1-x_p)\rho$, respsctively. The ground-state energy per nucleon can be expanded in series of powers of the quantity $\delta = (\rho_n - \rho_p)/\rho$, providing a measure of neutron excess. The resulting expression reads (see, e.g., Ref. [37])

$$\frac{1}{A}E_0(\rho,\delta) = \frac{1}{A}E_0(\rho,0) + E_{\rm sym}(\rho)\delta^2 + O(\delta^4)\ , \tag{16}$$

where the symmetry energy

$$\begin{aligned} E_{\rm sym}(\rho) &= \left\{ \frac{\partial^2[E_0(\rho,\delta)/A]}{\partial\delta^2} \right\}_{\delta=0} \\ &\approx \frac{1}{A}E_0(\rho,1) - \frac{1}{A}E_0(\rho,0) \end{aligned} \tag{17}$$

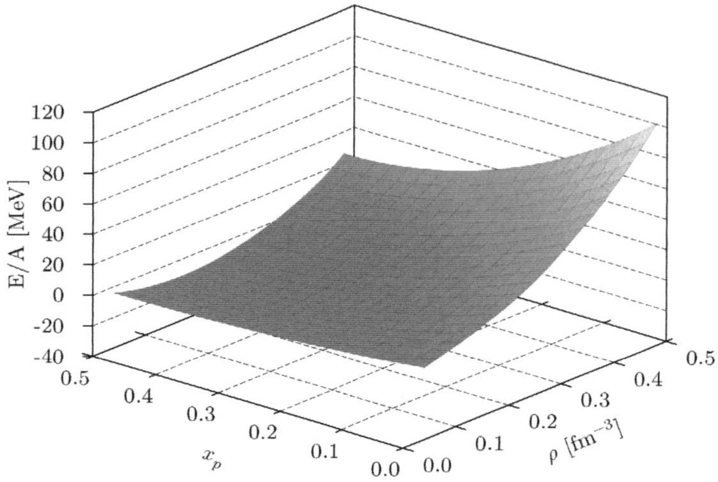

Figure 3. Energy per nucleon of uniform nuclear matter, computed as a function of baryon density, ρ, and proton fraction, $x_p = x_1 + x_2$, using Eqs. (14)-(15) and the CBF effective interaction.

can be interpreted as the energy required to convert SNM into PNM. The density dependence of $E_{\rm sym}(\rho)$, that can be obtained expanding around the equilibrium density, is conveniently characterized by the quantity

$$L = 3\rho_0 \left(\frac{dE_{\rm sym}}{d\rho} \right)_{\rho=\rho_0} . \tag{18}$$

Empirical information on $E_{\rm sym}(\rho_0)$ and L have been extracted from data collected by laboratory experiments and astrophysical observations [38]. The values resulting from our calculations, $E_{\rm sym}(\rho_0) = 30.9$ MeV and $L = 67.9$ MeV, turn out to be compatible with those obtained from a survey of 28 analyses, carried out by the authors of Ref. [38], yielding $E_{\rm sym}(\rho_0) = 31.6 \pm 2.66$ and $L = 58.9 \pm 16$ MeV.

The density dependence of the symmetry energy has been recently discussed in Ref. [39], whose authors combined the results of isospin-dependent flow measurements carried out by the ASY-EOS Collaboration at GSI with those obtained from analyses of low-energy heavy-ion collisions [40] and nuclear structure studies [41, 42, 43].

Figure 4 shows a comparison between $E_{\rm sym}(\rho)$ resulting from our calculations and the empirical information reported in Refs.[38, 39, 40, 41, 42, 43]. It appears that theoretical results are compatible with experiments at most densities.

3.3. Pressure

The pressure of nuclear matter, which plays a critical role in determining the properties of the equilibrium configurations of neutron stars, is simply related to the the ground-state energy through

$$P = -\left(\frac{\partial E_0}{\partial V} \right)_A = \rho^2 \frac{\partial (E_0/A)}{\partial \rho} , \tag{19}$$

where the derivative is taken keeping the number of nucleons constant.

The dashed line of Fig. 5 illustrates the density dependence of the pressure of SNM obtained from our approach. For comparison, the shaded area also shows the region consistent with the experimental flow data discussed in Ref. [44], providing a constraint on $P(\rho)$ at $\rho \geq 2\rho_0$. It appears while being within the allowed boundary at $2\rho_0 \leq \rho \leq 3\rho_0$, the calculated pressure exhibits a slope suggesting that a discrepancy may occur at higher density. However, in this

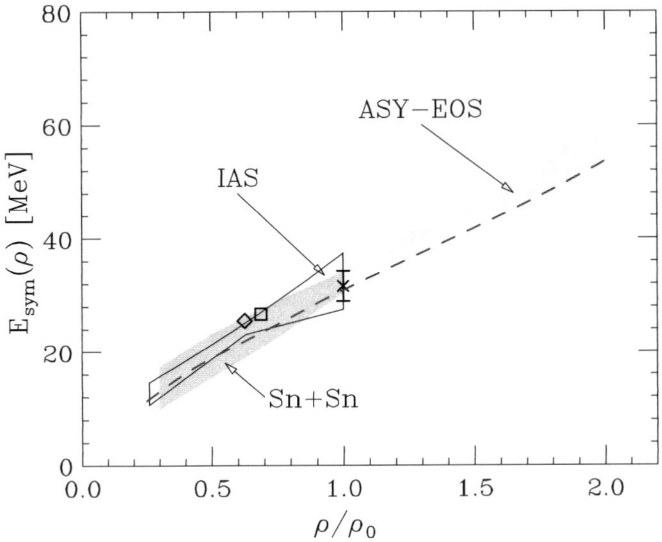

Figure 4. Density dependence of the symmetry energy of nuclear matter. The regions labelled ASY-EOS, Sn+Sn and IAS represent the results reported in Refs.[39], [40], and [41], respectively, while the symbols correspond to the analyses of Refs. [38] (cross with error bar), [42] (diamond), and [43] (square). The results of the CBF effective interaction approach are displayed by the dashed line.

context it has to be kept in mind that, being based on a non relativistic formalism, the CBF effective interaction approach is bound to predict a violation of causality—signaled by a value of the speed of sound in matter exceeding the speed of light—in the high-density limit.

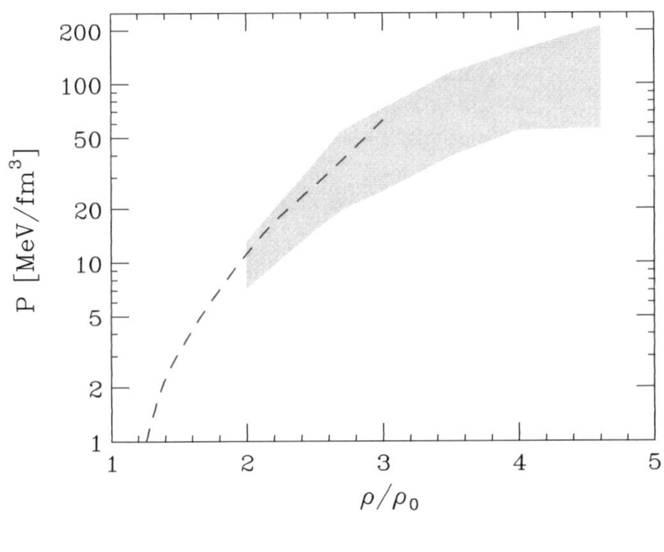

Figure 5. The dashed line illustrates the density dependence of the pressure of SNM obtained from the CBF effective interaction approach. The shaded area corresponds to the region consistent with the experimental flow data reported in Ref. [44].

3.4. Single particle spectrum and chemical potentials

At first order in the effective interaction, corresponding to the Hartree-Fock approximation, the energy spectrum of nucleons of type λ can be obtained from

$$e_\lambda(k) = \frac{k^2}{2m} + \rho \sum_\mu x_\mu \int d^3r \left[v^{\text{eff,d}}_{\lambda\mu}(\mathbf{r}) - j_0(kr)\ell(k_{F,\mu}r) v^{\text{eff,e}}_{\lambda\mu}(\mathbf{r}) \right] , \tag{20}$$

with $j_0(x) = \sin x/x$. The above expression is often parametrized in terms of the effective mass $m^\star$, defined as

$$\frac{1}{m^\star} = \frac{1}{k}\left(\frac{de_\lambda}{dk}\right) \tag{21}$$

For vanishing temperature, equation (20) also gives the chemical potentials, which are simply related to the energy spectrum through $\mu_\lambda = e_\lambda(k_{F,\lambda})$.

Figures 6 and 7 show the density dependence of the ratios $m^\star(k_F)/m$ and of the nucleon chemical potential in SNM (solid lines) and PNM (dashed lines).

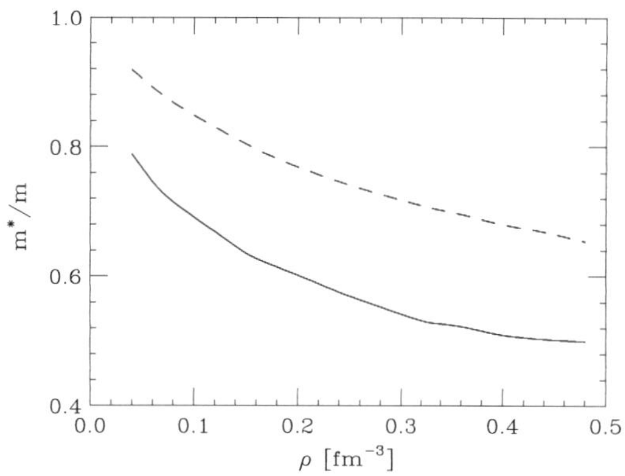

Figure 6. Density dependence of the ratio between the nucleon effective mass at $k = k_F$ and the bare nucleon mass. The solid and dashed lines correspond to SNM and PNM, respectively.

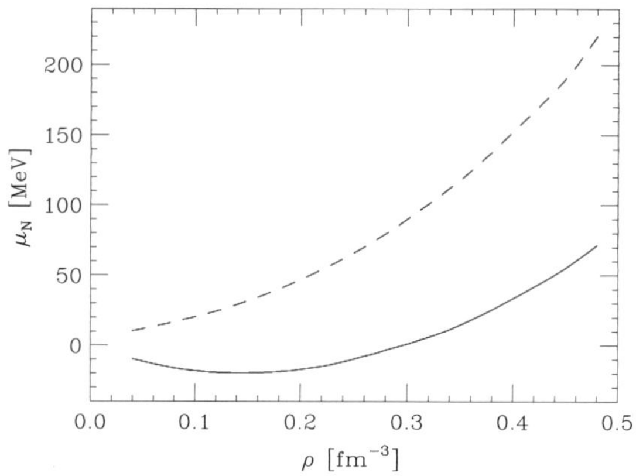

Figure 7. Density dependence of the nucleon chemical potential. The solid and dashed lines correspond to SNM and PNM, respectively.

3.5. Extension to nonzero temperature

To the extent to which thermal effects do not lead to modifications of the underlying nuclear dynamics, the approach described in this paper can be readily generalized to treat nuclear matter at nonzero temperature, by replacing the $T = 0$ Fermi distribution appearing in Eq. (15) with the corresponding distribution at temperature $T > 0$

$$n_\lambda(k, T) = \{1 + e^{\beta[e_\lambda(k) - \mu_\lambda]}\}^{-1} , \tag{22}$$

where $\beta = 1/T$, $e_\lambda(k)$ is the energy of a nucleon of type λ carrying momentum k, and the chemical potential μ_λ is determined by the constraint

$$\frac{1}{V}\sum_{\mathbf{k}\lambda} n_\lambda(k, T) = \rho_\lambda . \tag{23}$$

As an example, Figs. 8 and 9 show the temperature- and density-dependence of the Gibbs free energy per baryon of PNM and SNM, respectively, defined as

$$\frac{F}{A} = \frac{E_0 - TS}{A} , \tag{24}$$

where S denotes the entropy.

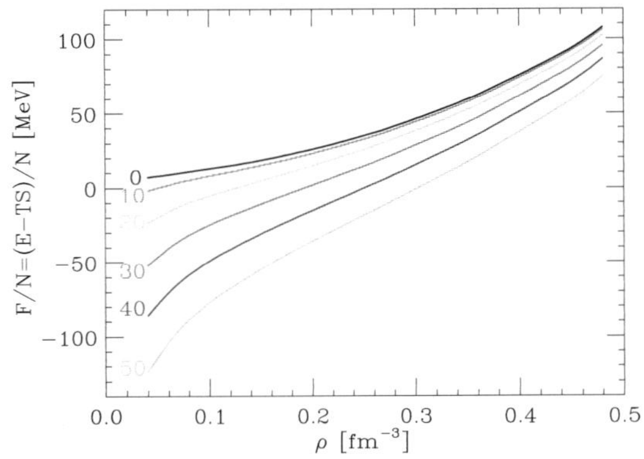

Figure 8. Density dependence of the Gibbs free energy per nucleon of PNM. The lines are labelled according to the temperature, T, expressed in units of MeV.

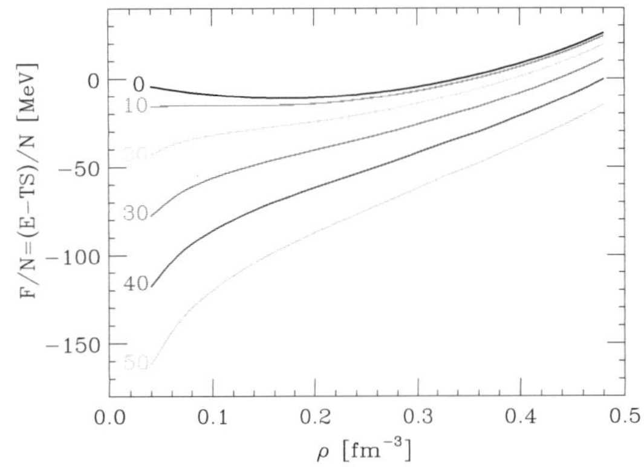

Figure 9. Same as in Fig. 8, but for SNM.

4. Summary and outlook

The results discussed in this paper provide convincing evidence that the effective interaction obtained from a realistic nuclear Hamiltonian using the CBF formalism and the cluster expansion technique is a powerful tool to carry out perturtative calculations of a variety of nuclear matter properties, ranging from the ground-state energy to the quasiparticle spectrum, the in-medium collision probability and the transport coefficients.

Unlike other advanced many-body approaches based on realistic models of nuclear dynamics at microscopic level—the applications of which are largely limited to calculations of the equation of state of SNM and PNM—the CBF effective interaction approach allows to study the properties of nuclear matter at large neutron excess, which are known to play a critical role in many astrophysical processes. In addition, under the assumption that thermal effects do not significantly affect strong interaction dynamics—which may be expected to be applicable at temperatures $T \ll m_\pi$, $m_\pi \sim 140$ MeV being the pion mass—it can be readily generalized to treat nuclear matter at nonzero temperature.

Ongoing and future applications to neutron star matter include the calculations of the shear and bulk viscosity coefficients, the superfluid gaps and the neutrino emission and absorption rates.

Acknowledgments

OB wishes to express his deepest gratitude to the organizers of "Compact Stars in the QCD phase diagram V", as well as to the Gran Sasso Scientific Institute (GSSI). The work of AL is supported by the U.S. Department of Energy, Office of Science, Office of Nuclear Physics, under contracts DE-AC02-06CH11357.

References

[1] Rikovska Stone J, Miller J C, Koncewicz R, Stevenson P D and Strayer M R 2003 *Phys. Rev. C* **68** 034324
[2] Chabanat E, Bonche P, Haensel P, Meyer J and Schaeffer R 1997 *Nucl. Phys. A* **627** 710
[3] Benhar O, Polls A, Valli M and Vidaña I 2010 *Phys. Rev. C* **81** 024305
[4] Zhang H F, Lombardo U and Zuo W 2010 *Phys. Rev. C* **82** 015805
[5] L Amundsen and E Østgaard 1985 *Nucl. Phys. A* **437** 487
[6] J Wambach, TL Aisworth and D Pines 1993 *Nucl. Phys. A* **555** 128
[7] Cowell S and Pandharipande V 2006 *Phys. Rev. C* **73** 025801
[8] Benhar O and Valli M 2007 *Phys. Rev. Lett.* **99** 232501

IOP Conf. Series: Journal of Physics: Conf. Series **861** (2017) 012009 doi:10.1088/1742-6596/861/1/012009

[9] Mecca A, Lovato A, Benhar O and Polls A 2015 *Phys. Rev. C* **91** 034325
[10] Mecca A, Lovato A, Benhar O and Polls A 2016 *Phys. Rev. C* **93** 035802
[11] Wiringa R B, Stoks V G J and Schiavilla R 1995 *Phys. Rev. C* **51** 38–51
[12] Wiringa R B and Pieper S C 2002 *Phys. Rev. Lett.* **89** 182501
[13] Fujita J and Miyazawa H 1957 *Prog. Theor. Phys.* **17** 360–365
[14] Pudliner B S, Pandharipande V R, Carlson J and Wiringa R B 1995 *Phys. Rev. Lett.* **74** 4396–4399
[15] A Gezerlis *et al* 2013 *Phys. Rev. Lett.* **111** 032501
[16] M Piarulli *et al* 2015 *Phys. Rev.* **C91** 024003
[17] J E Lynn *et al* 2016 *Phys. Rev. Lett.* **116** 062501
[18] Jastrow R 1955 *Phys. Rev.* **98**(5) 1479–1484 URL `http://link.aps.org/doi/10.1103/PhysRev.98.1479`
[19] Clark J W 1979 *Prog. Part. Nucl. Phys.* **2** 89 – 199
[20] Fantoni S and Rosati S 1974 *Nuovo Cim.* **A20** 179–193
[21] Pandharipande V R and Wiringa R B 1979 *Rev. Mod. Phys.* **51** 821–859
[22] Fantoni S, Friman B L and Pandharipande V R 1982 *Nucl. Phys. A* **399** 51
[23] Fantoni S and Pandharipande V R 1987 *Nucl. Phys. A* **473** 234
[24] Fabrocini A and Fantoni S 1989 *Nucl. Phys. A* **503** 375
[25] Benhar O, Fabrocini A and Fantoni S 1989 *Nucl. Phys. A* **505** 267
[26] Benhar O, Fabrocini A and Fantoni S 1992 *Nucl. Phys. A* **550** 201
[27] Fantoni S and Pandharipande V R 1988 *Phys. Rev. C* **37** 1697
[28] Lagaris I and Pandharipande V R 1981 *Nucl. Phys. A* **359** 349
[29] Lovato A, Losa C and Benhar O 2013 *Nucl. Phys.* **A901** 22–50
[30] Lovato A, Benhar O, Gandolfi S and Losa C 2014 *Phys. Rev.* **C89** 025804
[31] A Lovato, C Losa, and O Benhar 2013 *Nucl. Phys. A* **901** 22
[32] A Lovato, O Benhar, S Gandolfi and C Losa 2002 *Phys. Rev. C* **89** 182501
[33] RB Wiringa and SC Pieper 2014 *Phys. Rev. Lett.* **89** 025804
[34] BS Pudliner, VR Pandharipande, J Carlson and RB Wiringa 1995 *Phys. Rev. Lett.* **74** 4396
[35] Clark J W 1979 *Prog. Part. Nucl. Phys.* **2** 89
[36] KE Schmidt and S Fantoni 1999 *Phys. Lett. B* **446** 99
[37] M Baldo and and F Burgio 216 *Prog. Part. Nucl. Phys.* **91** 203
[38] B-A Li and X Han 2013 *Phys. Lett. B* **727** 276
[39] P Russotto *et al* 2016 *Phys. Rev. C* **94** 034608
[40] M B Tsang *et al* 2009 *Phys. Rev. Lett.* **102** 122701
[41] P Danielewicz and J Lee 2002 *Nucl. Phys. A* **298** 1592
[42] Brown B A 2014 *Phys. Rev. Lett.* **922** 1
[43] Zhang Z and Chen L W 2013 *Phys. Lett. B* **726** 234 – 238
[44] P Danielewicz, R Lacey, and W G Lynch 2002 *Science* **298** 1592

The stifness of the supranuclear equation of state (once again)

J E Horvath[1] and R A de Souza

Departamento de Astronomia, IAG-USP, São Paulo University, 05579-090 São Paulo SP, Brazil

E-mail: [1]`foton@iag.usp.br`

Abstract. We revisit the present status of the stiffness of the supranuclear equations of state, particularly the solutions that increase the stiffness in the presence of hyperons, the putative transition to a quark matter phase and the robustness of massive compact star observations.

1. The advent of high masses

Astrophysics is not a precisely a fairy tale, but if it was, one could read a future imaginary book on compact stars starting as:

"Once upon a time, everybody believed that a neutron star was composed of neutrons and its typical mass was not very different from $1.4M_{\odot}$..."

This belief actually lasted until the first years of the 21st century, it can be found in many books and review articles and was part of the "conventional wisdom" in the high-energy field. However, as early as 1970's the possibility of an exotic composition (quarks) was considered. In its mildest version, quarks should be the relevant degrees of freedom at $\sim$ several times the nuclear saturation density and constitute an inner central core. In its extreme version, a form of cold quark matter (strange quark matter) was hypothesized to be the true ground state of strong interactions and grow to compose most of the "neutron" stars [1-3] Many other phases were postulated (i.e. pion and kaon condensates, etc.) and the task of both theoreticians and observational astrophysics was to disentangle the composition through a combination of calculations and observations of actual compact stars, a task which is still in the beginning (see [4,5] for reviews).

The last few years yield a definite advance in the determination of neutron stars masses (although radii are still controversial and elusive [5]). As we shall discuss, there is now conclusive evidence for heavy ($\leq 2M_{\odot}$) neutron stars, and therefore the softest equations of state became unrealistic. However, it is not yet clear from a theoretical point of view where do we stand, and what does it really mean for hyperon and quark models, two of the best studied alternatives for the internal composition. We shall review here some of these possibilities without the pretension to be extensive, but rather to exemplify the type of behavior of the equation of state that can comply with the observations, and what type of physical elements are required. We conclude with a brief critical appraisal of this problem as it stands in 2016.

IOP Conf. Series: Journal of Physics: Conf. Series **861** (2017) 012010 doi:10.1088/1742-6596/861/1/012010

2. Observations and masses

It is well-known that high masses (substantially above $1.4M_{\odot}$) can be produced by stiff equations of state only, therefore the state-of-the-art pre-2000 or so did not discriminate among the several possibilities on this basis only. This is why in the last years firm evidence for masses well above the former "canonical" value was received by the community as an actual novelty, possibly allowing a better comprehension of the equation of state itself in the long run.

The first important evidence for high masses was presented by Demorest et al [6] with the univocal detection of the Shapiro delay in the system containing the source PSR J1614-2230. The precision of the fit allowed a robust (and clean) determination of a value $1.97 \pm 0.04\, M_{\odot}$ well accepted today. A more conventional approach (optical observations of the companion, allowing a determination of radial velocities + modeling of the companion WD) allowed Antoniadis et al. [7] to calculate a mass of $2.01 \pm 0.04\, M_{\odot}$ for PSR J0348+0432, reinforcing this high mass findings. Very recent works have also contributed to this problem, for instance, Falanga et al. [8] reported a mass of $2.12 \pm 0.16\, M_{\odot}$ for the HMXB Vela X-1 and $1.96 \pm 0.19\, M_{\odot}$ for 4U 1700-37. The quest for the highest mass continues because of its enormous implications, presently the record is held by PSR J1748-2021B in NGC 6440 [9] with a whopping $2.74 \pm 0.21\, M_{\odot}$, although this extreme value needs confirmation.

The whole issue of the mass distribution has been a subject of interest recently. The inclusion of high mass objects and application of Bayesian tools prompted Valentim, Rangel and Horvath [10] to claim at least a bimodal distribution with peaks at $1.37M_{\odot}$ and $1.73M_{\odot}$. Somewhat different values, but nevertheless firm evidence for more than one peak has been presented by Zhang et al [11], Özel et al [12] and Kiziltan, Kottas and Thorsett [13]. Using the latest data Ozel et al. updated their values to $1.39M_{\odot}$ and $1.8M_{\odot}$, and also inferred a possible maximum mass of $2.15M_{\odot}$. The important point here is that a bin with massive NS has been identified, even though its exact shape is still uncertain. Recent reviews on this issue can be found in [5] and [14]. While still far from the Rhoades-Ruffini limit [15] there is no question that masses above $2M_{\odot}$ should be accommodated by theoretical proposals, and this new limit brings considerable discussion for the stiffness of the equation of state, as we discuss below.

3. Hyperons? Which hyperons?

The inclusion of hyperons in dense matter is an anathema lasting at least 40 years. Hyperons are known to exist (!) and a natural part of the baryonic sector. Within a non-relativistic approach and even allowing a considerable uncertainty about their actual coupling to neutrons and protons, hyperons are believed to appear at around $2 - 3\rho_0$, and soften considerably the equation of state (see M. Baldo's talk in these Proceedings). Thus, hyperons are considered "nasty beasts" in nuclear physics at least regarding this issue. Many equations of state present in the literature did not include hyperons at all or neglect their interaction with neutrons and protons and/or treat the matter in the mean-field approach, which is known to produce stiffer equations of state almost by construction.

With the confirmation of high-mass neutron stars the issue of the presence of the hyperons gained a new twist, and became known as the "hyperon puzzle". In a few words this name refers to the quandary posed by the (inevitable?) presence of hyperons which would preclude precisely the high masses.

The so-called hyperon puzzle has some sources which can be readily identified as a source of problems. The first is to check what is really included and what is not when a definite scheme of calculations is given, including all the interaction terms and three-body forces (see Pederiva's talk in these Proceedings). In particular, it is important to check whether three-body forces can be "tuned" to give more repulsion ($C_T < 1$ in nuclear physics standard notation). The second issue is the old problem of computed vs. measured hypernuclei binding energies. While calculations are not particularly difficult, experiments are scarce and probe a limited range only.

This lack of reliable input is a source of considerable uncertainty regarding the influence of interactions on the hyperon appearance in dense matter, noting also that for the interesting regime the constraints are even weaker than for the nuclear matter case (see I. Bombaci review for a full appraisal of this issue [16]).

3.1. A model calculation as an example

As a definite example treating the hyperon fields we shall discuss our own results employing the $\sigma - \omega - \rho - \delta - \phi$ model presented in Gomes et al. [17]. The model includes the full baryonic octet and meson interactions that bear its name. The presence of derivative couplings somehow "sums up" the many body forces discussed above. The most relevant part of the Lagrangian reads

$$\sum_b \overline{\psi}_b \left(1 + \frac{g_{\sigma b}\sigma + g_{\sigma^* b}\sigma^* + \frac{1}{2} g_{\delta b}\tau.\delta}{\zeta m_b}\right)^{-\zeta} m_b \psi_b \qquad (1)$$

The effective masses become

$$m^*_{\lambda b} = \left(1 + \frac{g_{\sigma b}\sigma + g_{\sigma^* b}\sigma^* + \frac{1}{2} g_{\delta b}\tau.\delta}{\lambda m_b}\right)^{-\lambda} \qquad (2)$$

with $\lambda = \xi, \kappa, \eta, \zeta$ and b denotes the components of the baryonic octet.. This model attempts to describe many-body forces by parametrizing them in terms of non-linear interactions. It also uses the concept of "naturalness" for the coupling constants. Thus, its results go beyond the usual treatment of H-n interactions and may display a solution to the hyperon puzzle.

The equation of state derived by Gomes et al. [17] and employed here is shown in Fig. 1. Here ζ has been identified as the most important parameter of all for the behavior (slope) of the equation of state, once the properties of nuclear matter at the saturation point have been employed to fix the remaining set. Decreasing ζ produces a stiffening of the equation of state for a range of acceptable values of the former.

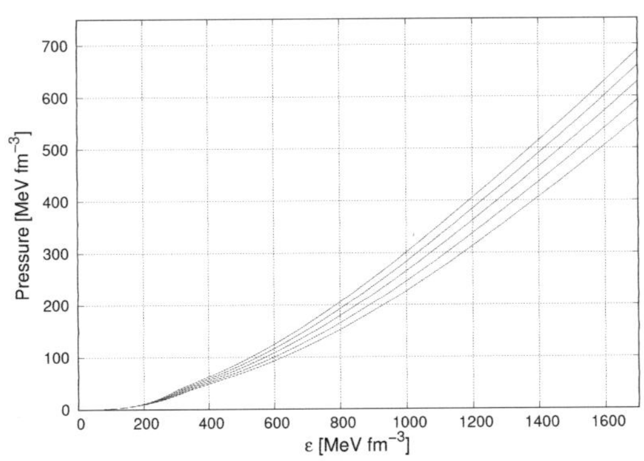

Figure 1. The equation of state of the $\sigma - \omega - \rho - \delta - \phi$ model. The curves have been generated by fixing all parameters but ζ and varying the latter. From bottom to top the adopted values were 8.5, 7.1, 5.9, 4.9 and 4 $\times 10^{-2}$.

With the aim of investigating the appearance of quark matter in the central core, we have constructed equilibrium phase transitions employing the stiff equation of state for cold quark matter (three flavors in chemical equilibrium) denominated MFTQCD due to Franzon et al. [18]. The latter begins by performing a mean-field approximation and separating soft and hard gluons according to Fogaça and Navarra [19]. A dynamical mass of the gluon m_G is generated by the mean field expectation value of the condensates of order 2, while those of the oder 4 produce a constant term identified as a contribution to the vacuum energy. The value of the latter and the quotient $\xi = g/m_G$ are the parameters of the model. The important point here

IOP Conf. Series: Journal of Physics: Conf. Series **861** (2017) 012010 doi:10.1088/1742-6596/861/1/012010

is that the equation of state is "bag-like" and stiff enough to have some chance to be present at very high pressures.

The equation of state is shown in Fig. 2. For the more stiff versions we have attempted to construct a phase transition *à la Maxwell* ignoring the conservation of many charges, as suggested by the work of Yasutake et al. [20] who showed that it is almost impossible to produce a large coexistence region within realistic models of the interface.

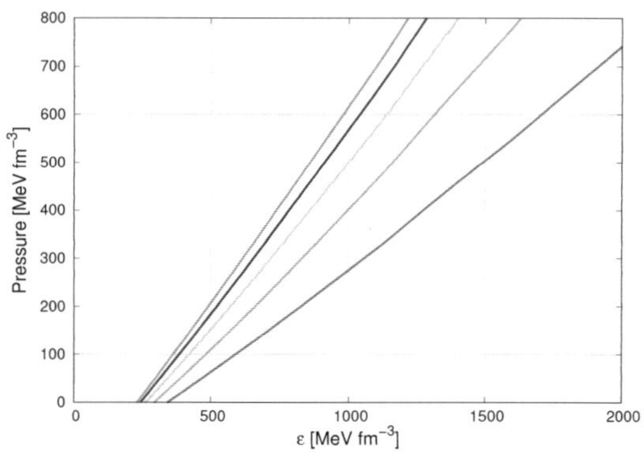

Figure 2. The equation of state of the MFTQCD cold quark matter phase. The vacuum term has been fixed as $B_{MFTQCD} = 90MeV/fm^3$, which is consistent with known hadronic physics and the quotient ξ varied from bottom to top to be, 1.5, 3, 4.5, 6 and 7.5 $\times 10^{-4}$. Note that the stiffness of the equation of state *decreases* with ξ.

This attempt is shown in Figs. 3 and 4. As seen, the existence of a quark phase is not easy, precisely because the $\sigma - \omega - \rho - \delta - \phi$ produces enough repulsion to postpone the appearance of quark matter, at least for most values of the parameter ζ.

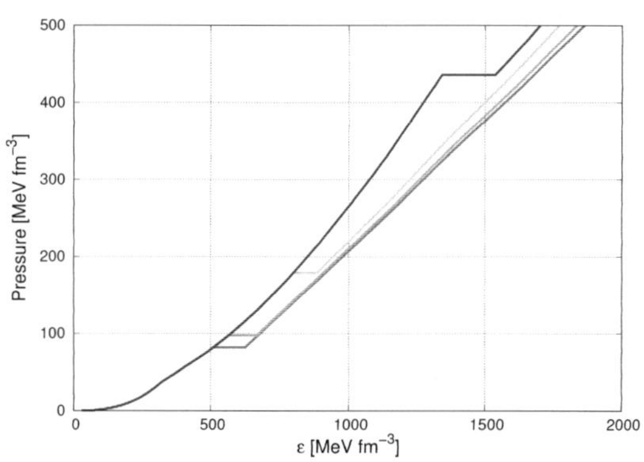

Figure 3. Maxwell construction for the transition between the $\sigma - \omega - \rho - \delta - \phi$ and the MFTQCD quark phase. For definiteness the $B_{MFTQCD} = 90MeV/fm^3$ and $\zeta = 5.9 \times 10^{-2}$ and ξ has been varied from 4 $\times 10^{-4}$ (lowest coexistence pressure) to 9 $\times 10^{-4}$ (highest coexistence pressure). The stiffness of the quark phase controlled by the latter parameter and determining the appearance or absence of the phase is apparent.

The equations of state just described were employed to integrate the TOV equations and generate theoretical non-rotating stellar sequences. As a result (Fig. 5), we verified that the maximum mass M_{max} is barely above the last determined value $\sim 2M_{\odot}$ for a limited set of the parameters. In other words, this class of hyperonic equations of state is in some trouble to reproduce the highest masses, and even more, if they happen to be correct, it is almost impossible that a quark core can develop inside them. While one may think of constructing an even stiffer EoS for the quark phase, it is difficult to imagine that vacuum+interaction terms in cold quark matter can change this feature.

4. Conclusions

We revisited the issue of the stiffness of the equation of state within a limited set of possibilities prompted by the study of the hyperonic sector. We do not yet know how large an actual compact

IOP Conf. Series: Journal of Physics: Conf. Series **861** (2017) 012010 doi:10.1088/1742-6596/861/1/012010

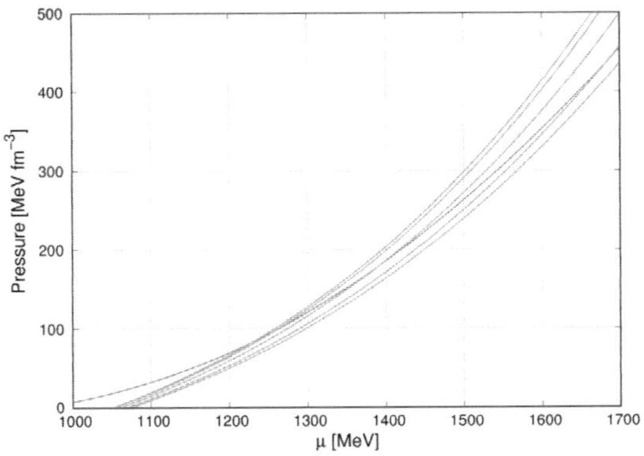

Figure 4. An example of equilibrium between $\sigma - \omega - \rho - \delta - \phi$ and the quark phase described by the MFTQCD for a favorable set of parameters in which the curves cross.

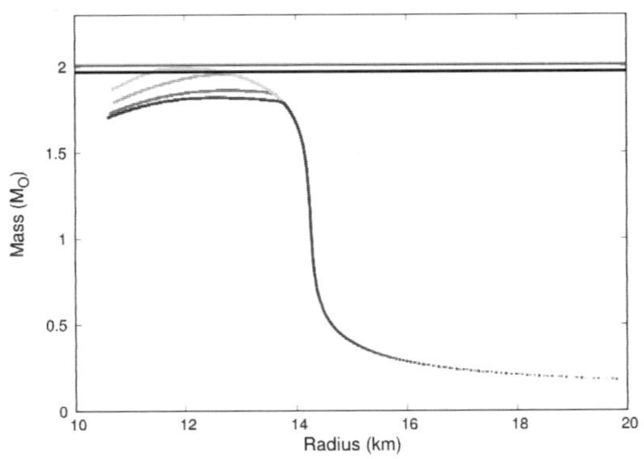

Figure 5. The stellar sequences emerging from the integration of the equations of state. The constraint $M_{max} > 2M_\odot$ is barely satisfied

star mass could be, just that $\sim 2M_\odot$ objects have been confirmed. Higher masses would be even more difficult to accommodate (of course), and the physics of the ultradense matter could be approaching the Rhoades-Ruffini limit dangerously, without a clear physical reason for that.

Since we know that without the inclusion of hyperons the maximum masses of the sequence can go above $\sim 2M_\odot$ quite easily, one strange solution to the hyperon puzzle could be the "hyperonless" hyperonic matter, in which hyperons are suppressed by some convenient mechanism. A conventional alternative to this would be to have enough repulsion from them. However, and stated quite boldly but pointing to the trend which is clearly present in our calculations, the latter possibility would relegate the study of cold quark matter to the status of an academic problem (unless that SQM composes all compact objects and we have to admit from scratch that we never experienced the true ground state of hadronic interactions in the lab...).

Acknowledgments

We would like to acknowledge the Local Organizing Committee for the organization of a nice CSQCD Meeting. JEH has been supported by the Fapesp Agency (Process 2013/26258-4) and CNPq, Brazil (Process 305328/2013-1).

References

[1] Bodmer A 1971 *Phys. Rev.* D **4** 1601
[2] Terazawa H 1979 *Tokyo U. Report* INS336
[3] Witten E 1984 *Phys. Rev.* D **30** 272
[4] Lattimer J 2012 *Annu. Rev. Nucl. Part. Sci.* **62** 485
[5] Özel F and Freire P 2016 *Annu. Rev. Astron. Astrophys.* **54** 401

[6] Demorest P, Pennucci T, Ransom S M, Roberts MSE and Hessels JWT 2010 *Nature* **467** 1081
[7] Antoniadis J et al. 2013 *Science* **340** 448
[8] Falanga M, Bozzo E, Lutovinov A, Bonnet-Bidaud JM, Fetisova Y and Puls J 2015 *Astron. & Astrophys.* **577** A130
[9] Freire PCC, Ransom SM, Bégin S, Stairs IH, Hessels JWT, Frey LH and Camilo F 2008 *Astrophys. J.* **675** 670
[10] Valentim R, Rangel E and Horvath JE 2011 *Mon. Not. R. Astron. Soc.* **414** 1427
[11] Zhang CM, Wang J, Zhao YH, Yin HX, Song LM, Menezes DP, Wickramasinghe DT, Ferrario L and Chardonnet P 2011 *Astron. & Astrophys.* **527** A83
[12] Özel F, Psaltis D, Narayan R and Santos Villareal A 2012 *Astrophys. J.* **757** A55
[13] Kiziltan B, Kottas A, De Yoreo M and Thorsett SE 2013 *Astrophys. J.* **778** A66
[14] Horvath JE and Valentim R 2016 *Preprint* arXiv:1607.06981
[15] Rhoades, CE, Ruffini, R 1974 *Phys. Rev. Lett.* **32** 324
[16] Bombaci I 2016 *Preprint* arXiv:1601.05339
[17] Gomes RO, Dexheimer V, Schramm S and Vasconcellos CAZ 2015 *Astrophys. J.* **808** 8
[18] Franzon B, Fogaça DA, Navarra FS and Horvath, JE 2012 *Phys. Rev.* D **86** 065031
[19] Fogaça DA and Navarra FS 2011 *Phys. Lett.* B **700** 236
[20] Yasutake N, Chen H, Maruyama T and Tatsumi T 2016 *Journal of Physics: Conference Series* **665** 012068

IOP Conf. Series: Journal of Physics: Conf. Series **861** (2017) 012011 doi:10.1088/1742-6596/861/1/012011

The CSS parametrization for Hybrid Stars with the Field Correlator Method

G F Burgio

INFN Sezione di Catania. Via S. Sofia 64, I-95123 Catania, Italy

E-mail: `fiorella.burgio@ct.infn.it`

Abstract. We explore the structure of hybrid stars based on a nuclear matter equation of state (EoS) built with the microscopic Brueckner-Hartree-Fock many-body theory, and a quark matter EoS derived with the Field Correlator Method (FCM), which can be accurately represented by the CSS (constant speed of sound) parametrization. We find that the main features of the hadron-quark phase transition are directly related to the FCM parameters, i.e. the quark-antiquark potential V_1, the gluon condensate G_2 and the color-flavour superconducting gap Δ, whose values range can be determined by the observational data on neutron star (NS) masses.

1. Introduction

The appearance of quark matter (QM) in neutron stars (NS) interiors and the corresponding equation of state are among the mostly debated issues in the physics of compact objects. Unfortunately, while the microscopic theory of the nucleonic EoS has reached a high degree of sophistication [1, 2, 3, 4, 5], the quark matter EoS is still poorly known at zero temperature and at the high baryonic density typical for NS. In fact the appropriate theoretical tool, i.e. the lattice formulation of the quantum chromodynamics (QCD) is inapplicable at large baryon densities and small temperature due to the so-called Sign Problem [6], due to its complicated nonlinear and nonperturbative nature. On the other hand, in the large temperature and small density region lattice QCD simulations have provided controlled results for the EoS as well as for the nature of the transition [7, 8].

The value of the maximum mass of NS is probably one of the physical quantities that is most sensitive to the presence of quark matter in NS. The recent observation of a large NS mass in PSR J0348+0432 [9] with mass $\mathrm{M} = 2.01 \pm 0.04\,\mathrm{M}_\odot$ ($M_\odot = 2 \times 10^{33}$g) implies that the EoS of NS matter has to be stiff enough to keep the maximum mass at these large values. Purely nucleonic EoS are able to accommodate such large masses [10], but the likely appearance of non-nucleonic degrees of freedom, like hyperons and quarks, will eventually soften the EoS with respect to the purely nucleonic case. The resulting low maximum mass value would in this case be incompatible with observations. The large value of the mass could then be explained only if both hyperonic and quark matter EoS are stiffer than expected.

Many models of quark matter do exist, and they all contain a high degree of uncertainty. The best one can do is to compare the predictions of different models and to estimate the uncertainty of the results for the NS matter as well as for the NS structure and mass. In this paper we use the Field Correlator Method model for the quark EoS [11], which contains *ab initio* the property of confinement, which is expected to play a role as far as the stability of a neutron star is concerned

IOP Conf. Series: Journal of Physics: Conf. Series **861** (2017) 012011 doi:10.1088/1742-6596/861/1/012011

[12], and in principle is able to cover the full temperature-chemical potential plane [13]. In ref. [14] we tested this model against NS observations, which could seriously constrain the model parameters, i.e. the quark-antiquark potential V_1, the gluon condensate G_2 and, more recently, also the color-flavour superconducting gap Δ [15]. In Sect.2 we illustrate the FCM model at finite temperature and density, with the inclusion of the color-flavour locking effect.

As far as the hadronic phase is concerning, we use two definite EoS, based on the Brueckner-Hartree-Fock many-body theory for nuclear matter, both in its non-relativistic and relativistic version. Those are briefly reviewed in the Sect.3. Sect.4 contains some numerical results on the hadron-quark phase transition as described by the CSS parametrization, and a discussion of the FCM mapping onto the CSS parametrization. Moreover, results for the mass-radius-central density relation for hybrid stars will be illustrated, while effects due to the inclusion of hyperons will be briefly mentioned. Conclusions are reported in Sect.5.

2. Quark Matter EoS : the Field Correlator Method

The approach based on the FCM provides a natural treatment of the dynamics of confinement in terms of the Color Electric (D^E and D_1^E) and Color Magnetic (D^H and D_1^H) Gaussian correlators, being the former one directly related to confinement, so that its vanishing above the critical temperature implies deconfinement [11]. The extension of the FCM to finite temperature T and chemical potential $\mu_q = 0$ gives analytical results in reasonable agreement with lattice data thus allowing to describe correctly the deconfinement phase transition [16, 17]. In this work, we are interested in the physics of neutron stars, and therefore we extended the FCM to finite values of the chemical potential [16] and zero temperature.

Within the FCM, the quark pressure for a single flavour is simply given by [16]

$$P_q/T^4 = \frac{1}{\pi^2}\left[\phi_\nu\left(\frac{\mu_q - V_1/2}{T}\right) + \phi_\nu\left(-\frac{\mu_q + V_1/2}{T}\right)\right] \tag{1}$$

where

$$\phi_\nu(a) = \int_0^\infty du \frac{u^4}{\sqrt{u^2+\nu^2}} \frac{1}{(\exp[\sqrt{u^2+\nu^2} - a] + 1)}. \tag{2}$$

being $\nu = m_q/T$, and V_1, which indicates the large distance limit of the static $q\bar{q}$ potential, i.e.

$$V_1 = \int_0^{1/T} d\tau(1-\tau T)\int_0^\infty d\chi\chi D_1^E(\sqrt{\chi^2+\tau^2}), \tag{3}$$

Therefore its value is an effective measure of the strength of the interaction when the particles are infinitely separated. The potential V_1 in Eq.(3) is assumed to be independent on the chemical potential, and this is partially supported by lattice simulations at very small chemical potential. The EoS is completely specified once the gluon contribution is added to the quark pressure, i.e.

$$P_g/T^4 = \frac{8}{3\pi^2}\int_0^\infty d\chi\chi^3 \frac{1}{\exp\left(\chi + \frac{9V_1}{8T}\right) - 1} \tag{4}$$

and therefore

$$P_{qg} = P_g + \sum_{j=u,d,s} P_q^j + \Delta\epsilon_{vac} \tag{5}$$

where P_g and P_q^j are respectively given in Eq. (1) and (4), and

$$\Delta\epsilon_{vac} \approx -\frac{(11 - \frac{2}{3}N_f)}{32}\frac{G_2}{2} \tag{6}$$

corresponds to the difference of the vacuum energy density between deconfined and confined phase, being N_f the flavour number ($N_f = 3$ in this paper). G_2 is the gluon condensate whose numerical value, determined by the QCD sum rules, is known with large uncertainty [18]

$$G_2 = 0.012 \pm 0.006 \text{ GeV}^4 \tag{7}$$

It is interesting to notice that G_2 appears only in the vacuum contribution to the pressure (6), and plays the same role of the bag constant in the MIT bag model. In addition, if one turns V_1 off, P_q becomes the pressure of free quarks, and in this case the FCM reduces to the simplest version of the bag model. Therefore V_1 can be regarded as the main correction to the free quarks dynamics inside the bag. Therefore the EoS in Eq.(5) essentially depends on two parameters, namely the quark-antiquark potential V_1 and the gluon condensate G_2. A comparison with the lattice calculations of the Wuppertal-Budapest [19] and hotQCD [20] collaborations provides clear indications about their specific values at finite temperature and $\mu_B = 0$, but no direct relation is found with the corresponding values at finite μ_B. Therefore, in our analysis we choose to keep V_1 and G_2 as a free parameters, and check what kind of indications can be extracted from the determination of the maximum mass of neutron stars. Additional effects that could induce changes in the FCM EoS are due to the colour superconductivity pairing mechanism, which is expected to take place for extremely large values of the chemical potential [21, 22, 23]. We included these effects in our analysis [15] by adding to the full FCM pressure P_{qg} the color-flavour locked (CFL) pressure contribution, i.e. $P_{cfl} = \Delta^2\mu_B^2/3\pi^2$ only when the chemical potential is greater than $\mu_B = 3m_s^2/(4\Delta)$, as discussed for instance in [23]. The gap Δ is expected to be in the range $10 - 100$ MeV in the region of interest of μ_B for the NS, and we treat it as a third free parameter in the FCM model.

3. Hadronic Phase: EoS in the Brueckner-Bethe-Goldstone theory

The BHF method for the nuclear matter EoS is based on the Brueckner-Bethe-Goldstone (BBG) many-body theory, which is a linked cluster expansion of the energy per nucleon of nuclear matter [24]. In this approach one systematically replaces the bare nucleon-nucleon (NN) interaction V by the Brueckner reaction matrix G, which is the solution of the Bethe-Goldstone equation

$$G(\rho;\omega) = V + V \sum_{k_a k_b} \frac{|k_a k_b\rangle Q \langle k_a k_b|}{\omega - e(k_a) - e(k_b)} G(\rho;\omega), \tag{8}$$

where ρ is the nucleon number density, ω is the starting energy, and $|k_a k_b\rangle Q\langle k_a k_b|$ is the Pauli operator. $e(k) = e(k;\rho) = \frac{\hbar^2 k^2}{2m} + U(k;\rho)$ is the single particle energy, and U is the single-particle potential,

$$U(k;\rho) \quad = \quad \sum_{k' \leq k_F} \langle kk'|G(\rho; e(k) + e(k'))|kk'\rangle_a \tag{9}$$

The subscript "a" indicates antisymmetrization of the matrix element. In the BHF approximation the energy per nucleon is

$$\frac{E}{A}(\rho) = \frac{3}{5}\frac{\hbar^2\, k_F^2}{2m} + \frac{1}{2A} \sum_{k,k' \leq k_F} \langle kk'|G(\rho; e(k) + e(k'))|kk'\rangle_a \tag{10}$$

The inclusion of nucleonic three- body forces (TBFs) in this approach allows to reproduce correctly the nuclear matter saturation point $\rho_0 \approx 0.16$ fm^{-3}, $E/A \approx -16$ MeV, and gives values of incompressibility and symmetry energy at saturation compatible with those extracted

IOP Conf. Series: Journal of Physics: Conf. Series **861** (2017) 012011 doi:10.1088/1742-6596/861/1/012011

from phenomenology [1]. In this work we choose the Argonne v_{18} nucleon-nucleon potential [25], supplemented by the so-called Urbana model [26] as three-body force. In the following we will show results obtained with the relativistic counterpart, i.e. the Dirac-Brueckner-Hartree-Fock scheme [27] where the Bonn A potential is used as NN interaction.

In this paper we adopt a conventional description of stellar matter as composed by neutrons, protons, and leptons in beta equilibrium. The chemical potentials of each species are the fundamental input for solving the equations of chemical equilibrium, charge neutrality and baryon number conservation, yielding the equilibrium fractions of all species. Once the composition of the β-stable, charge neutral stellar matter is known, one can calculate the equation of state, i.e., the relation between pressure P and energy density ϵ as a function of the baryon density ρ. It can be easily obtained from the thermodynamical relation

$$P = -\frac{dE}{dV} = P_B + P_l \tag{11}$$

$$P_B = \rho^2 \frac{d(\epsilon_B/\rho)}{d\rho}, \quad P_l = \rho^2 \frac{d(\epsilon_l/\rho)}{d\rho} \tag{12}$$

with E the total energy and V the total volume. The total nucleonic energy density ϵ_B is obtained by adding the energy densities of each species ϵ_i. As far as leptons are concerned, at those high densities electrons are a free ultrarelativistic gas, whereas muons are relativistic. Hence their energy densities ϵ_l are well known from textbooks [28]. The numerical procedure has been often illustrated in papers and textbooks, and therefore it will not be repeated here.

4. The hadron-quark phase transition

Recently, we found that the FCM model can be expressed in the language of the "Constant Speed of Sound" (CSS) parametrization [29, 30], and we showed how its parameters can be mapped onto the CSS parameter space. We remind that the CSS scheme is a general parametrization suitable for classifying different models of quark matter and establishing connections among them, and for expressing experimental constraints in a model-independent way. We remind that it is applicable to high-density equations of state for which: (a) there is a sharp interface between nuclear matter and quark matter, (b) the speed of sound in the high-density phase is pressure-independent in the range between the first-order transition pressure up to the maximum central pressure of neutron stars. Given the nuclear matter EoS $\epsilon_{\rm NM}(p)$, the high-density EoS can be expressed as

$$\epsilon(p) = \begin{cases} \epsilon_{\rm NM}(p) & p < p_{trans} \\ \epsilon_{\rm NM}(p_{trans}) + \Delta\epsilon + c_{\rm QM}^{-2}(p - p_{trans}) & p > p_{trans} \end{cases} \tag{13}$$

where the three parameters: the pressure p_{trans} at the transition, the discontinuity in energy density $\Delta\epsilon$ at the transition, and the speed of sound c_{QM} characterize completely the high-density phase. It has also been shown [29] that the CSS parametrization gives four different topologies of the mass-radius curve for compact stars, i.e. a hybrid branch connected to the nuclear branch (C), or disconnected (D), or both present (B) or neither (A). We use the term "hybrid branch" to refer to the part of the mass-radius relation of hybrid stars whose central pressure is above $p_{\rm trans}$, and so they contain a core of the high-density phase. The occurrence of these as a function of the CSS parameters $p_{\rm trans}/\varepsilon_{\rm trans}$ and $\Delta\epsilon/\varepsilon_{\rm trans}$ at fixed $c^2_{\rm QM}$ is shown schematically in Fig. 1 (taken from Ref. [29]). The mass-radius curve in each region is depicted in inset plots, in which the thick green line is the hadronic branch, the thin solid red lines are stable hybrid stars, and the thin dashed red lines are unstable hybrid stars. In the phase diagram the solid red line shows the threshold value ϵ_{crit} below which there is always a stable hybrid star branch connected to the neutron star branch. This critical value is given by [31, 32, 33]

$$\frac{\Delta\epsilon_{crit}}{\varepsilon_{\rm trans}} = \frac{1}{2} + \frac{3}{2}\frac{p_{\rm trans}}{\varepsilon_{\rm trans}} \tag{14}$$

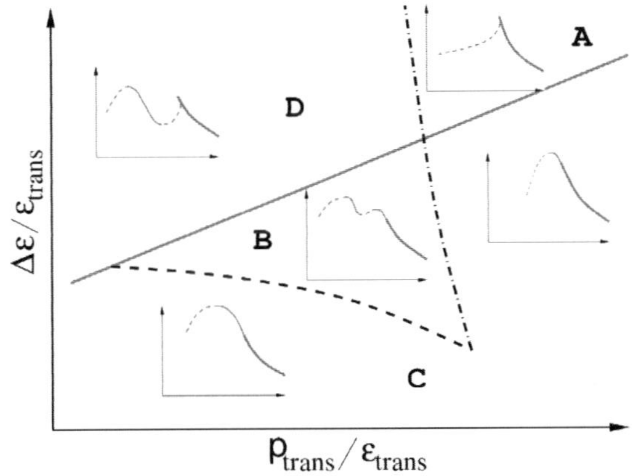

Figure 1. Schematic phase diagram [29] for hybrid star branches in the mass-radius relation of compact stars. The four regions are (A) no hybrid branch ("absent"); (B) both connected and disconnected hybrid branches; (C) connected hybrid branch only; (D) disconnected hybrid branch only.

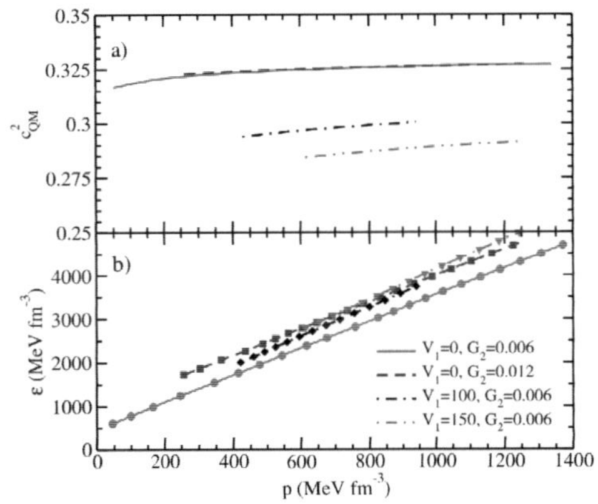

Figure 2. (Color online) The squared speed of sound $c^2_{\rm QM}$ (panel (a)) is displayed vs. quark matter pressure for several values of V_1 (in MeV) and G_2 (in GeV^4). In panel (b), the FCM energy density is represented by full symbols, whereas the full lines denote the CSS parametrization given by Eq.(13).

and was obtained by performing an expansion in powers of the size of the core of high-density phase. Eq.(14) is an analytic result, independent of $c^2_{\rm QM}$ and the nuclear matter EoS. The dashed and dot-dashed black lines mark the appearance-disappearance of the connected or disconnected hybrid star branch. The position of these lines depends on the value of $c^2_{\rm QM}$ and (weakly) on the accompanying nuclear matter EoS [29].

In Fig. 2 we show that the CSS parametrization is applicable to the FCM EoS, since the speed of sound depends only weakly on the density or pressure. The upper panel shows the speed of sound vs. pressure for different values of the FCM parameters, displayed in the lower panel. We see that the speed of sound varies by less than 5% over the considered range of pressures along each curve, and lies in the interval $0.28 < c^2_{\rm QM} < 1/3$. The value of $c^2_{\rm QM}$ shows a weak dependence on V_1 and extremely weak on G_2. The transition pressure is more sensitive to the FCM parameters, increasing rapidly with V_1 and with G_2, whereas the energy density at a given pressure increases slightly with V_1 or G_2.

Let us now discuss the main features of the hadron-quark phase transition. Fig. 3 shows numerical results for the pressure as a function of the baryon chemical potential μ_B in hadronic matter and quark matter in beta equilibrium. In particular, the green (red) solid curves represent the BHF (DBHF) EoS, whereas the remaining curves are the results for the FCM model with different choices of the quark-antiquark potential V_1 (expressed in MeV). In the left, middle and right panels the value assumed for the gap Δ is respectively equal to 0, 50 and 100 MeV. All calculations shown in Fig. 3 are performed taking $G_2 = 0.006$ GeV^4. We notice that with increasing the value of V_1 the transition point is shifted to larger values of the chemical potential, hence of the baryon density. However, the exact value depends also on the stiffness of the hadronic EoS at those densities. In this case, being the DBHF EoS stiffer than the BHF, the transition takes place at smaller values of the density. We notice that the transition point is affected also by the value of the gap Δ, and is shifted toward smaller μ_B for larger value of the gap.

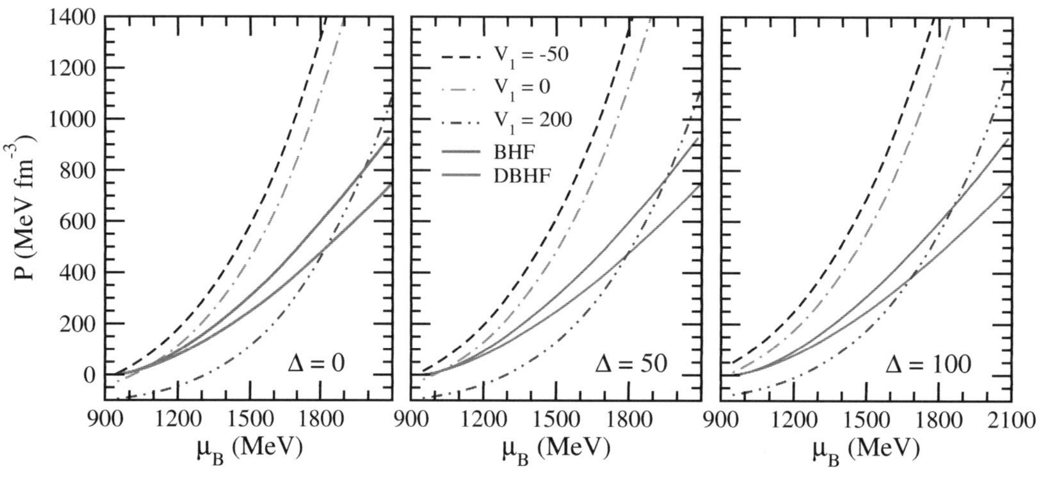

Figure 3. The pressure is displayed vs. the baryon chemical potential μ_B for the FCM quark matter and the purely hadronic matter. All calculations for FCM have been performed for $G_2 = 0.006$ GeV^4, and several values of V_1. The solid curves represent the BHF (green) and DBHF (red) EoS. In each panel results for different values of the gap Δ, i.e. 0, 50, and 100 MeV are shown.

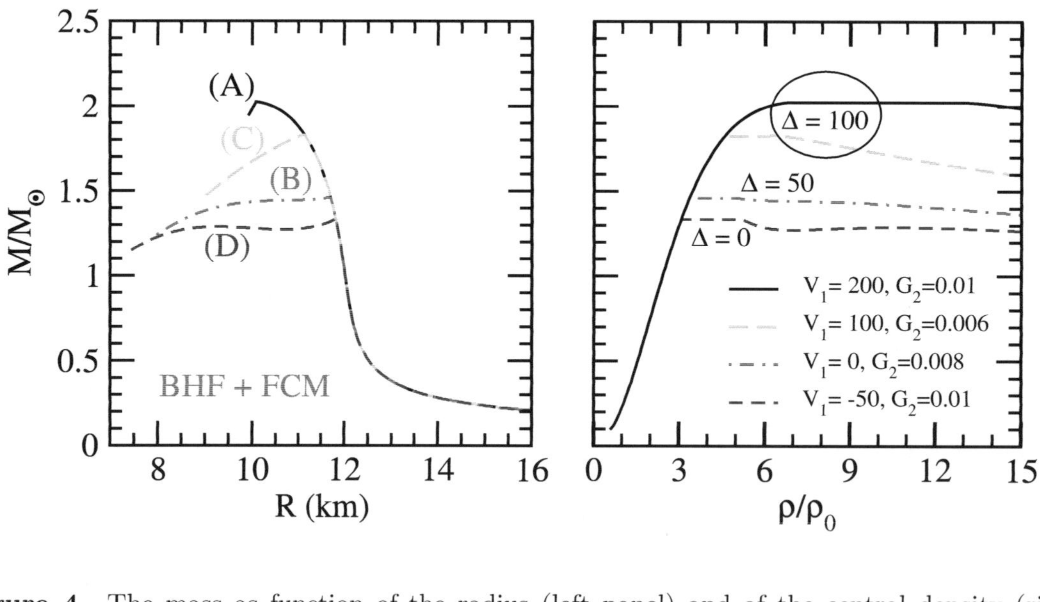

Figure 4. The mass as function of the radius (left panel) and of the central density (right panel) is displayed for several values of V_1, G_2 and Δ. The labels (A), (B), (C) and (D) indicate the specific topologies of the hybrid star branch. The BHF EoS is used for the hadronic phase.

IOP Conf. Series: Journal of Physics: Conf. Series **861** (2017) 012011 doi:10.1088/1742-6596/861/1/012011

The resulting EoS, for the several cases discussed, is the main input for solving the well-known Tolman-Oppenheimer-Volkoff equations [28] for spherically symmetric NS, thus obtaining the mass-radius-central density relation. This is illustrated in Fig. 4, where the relation between mass and radius (left panel) and central density (right panel) in units of the saturation density ρ_0 is displayed. Results are plotted for different values of V_1, G_2 and Δ and the BHF EoS is used for hadronic matter. The largest value of the maximum mass in this example is obtained for $V_1 = 200$ MeV, $\Delta = 100$ and $G_2 = 0.01$, and it is compatible with the largest mass observed up to now, i.e. $(2.01 \pm 0.04)M_\odot$ in PSR J0348+0432 [9]. In Fig. 4 we show also the topology of the mass-radius curve. For example, for unpaired quark matter and $V_1 = -50$ MeV, the mass-radius relation obtained with $G_2 = 0.01$ GeV^4 (blue dashed line) exhibits a branch of stable hybrid stars disconnected (D) by the hadronic branch. With increasing V_1 the transition point moves to larger values of the pressure and the energy density, and as a consequence we explore regions of the phase diagram where the topology changes. For instance, for $V_1 = 0$ we can get both (B) connected and disconnected hybrid star branches, whereas for $V_1 = 100$ MeV connected (C) hybrid star branches are present and, for the largest value of $V_1 = 200$ MeV the hybrid branch is absent (A). This is clearly shown by a cusp in the mass-radius relation, and all configurations with radii smaller than the one characterizing the cusp are unstable. Therefore only purely nucleonic stars do exist in this case. We observe a similar topology when the DBHF EoS is used for the hadronic phase. In both cases we see that the largest possible values of the maximum mass are obtained only for values of $V_1 > 100$ MeV, and that only in the DBHF case maximum masses can be well above the observational limit. In fact, the heaviest BHF+FCM hybrid star has a mass of 2.03 $M_\odot$, and the heaviest DBHF+FCM hybrid star has a mass of 2.31 $M_\odot$ [15]. Those values are indicated by an orange cross in Fig. 5, where we display the mapping between the FCM and CSS parameters. In the upper (lower) panels we show results for the BHF (DBHF) hadronic EoS, whereas in the left, middle and right panels calculations are displayed for different values of the gap $\Delta = 0,\ 50,\ 100$ MeV respectively. Along each line we keep V_1 constant and vary G_2. For the BHF case the maximum hybrid star mass has value $V_1 = 240$ MeV, $G_2 = 0.0024$ GeV^4, and for the DBHF case $V_1 = 255$ MeV, $G_2 = 0.0019$ GeV^4. In Fig. 5 the dashed black contour delimits the region accessible by the FCM calculation, and the symbols connected by solid lines show the CSS parametrization of the FCM quark matter EoS. The vertical black dashed lines indicate the parameter regions accessible by the FCM and consistent with the measurement of a $\mathrm{M} = 2\,\mathrm{M}_\odot$. Hybrid stars with mass heavier than 2 $\mathrm{M}_\odot$ lie on a very small connected branch on the right side of the vertical black dashed lines, and cover a small range of central pressures, having a very tiny quark core, with mass and radius similar to those of the heaviest purely hadronic star, as was already discussed in Ref. [30]. We have found that for a hybrid star with $\mathrm{M} = 2\,\mathrm{M}_\odot$ the hadronic layer occupies the largest portion of the star, and is characterized by a radius of about 10 km, whereas the crust radius is always smaller than 1 km.

Finally we briefly discuss the effects of adding the hyperon degrees of freedom in the BHF approach, which should produce a softening of the EoS with a strong decrease of the maximum mass [34, 35], with the maximum mass remaining well below the $2\mathrm{M}_\odot$, at variance with the observational constraint. In ref.[15] we found that when V_1 is small, the masses remain below 1.5 $\mathrm{M}_\odot$, whereas with larger V_1, the mass of the NS grows up to 1.95 $\mathrm{M}_\odot$, which is reasonably close to the observational constraint of 2 $\mathrm{M}_\odot$. Therefore only with a special choice of the FCM parameters, we are able to reach sufficiently high NS masses in the presence of hyperon degrees of freedom.

5. Conclusions

We have studied the structure of hybrid NS employing for the hadronic phase the non-relativistic BHF EoS and its relativistic counterpart, the DBHF EoS. For the quark matter phase we have

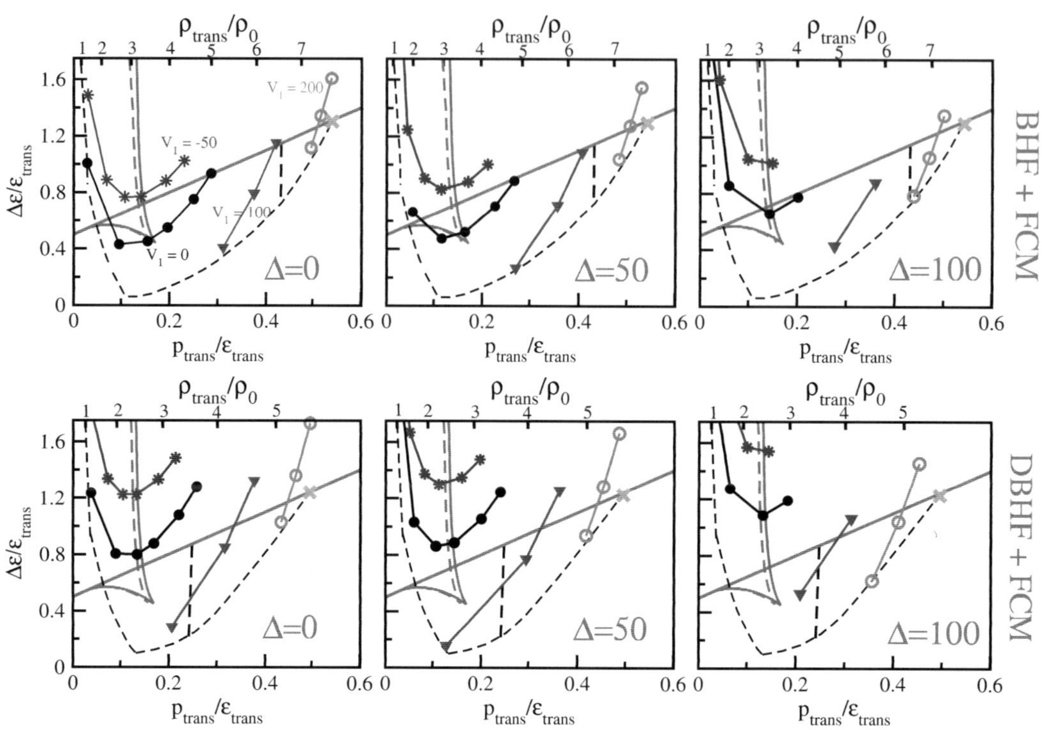

Figure 5. The mapping of the FCM quark matter model onto the CSS parametrization. The green curves are the phase boundaries for the occurrence of connected and disconnected hybrid branches. The dashed black line delimits the region yielded by the FCM model, and the symbols give CSS parameter values as G_2 is varied at constant V_1 (given in MeV). The (orange) cross denotes the parameters for the heaviest FCM hybrid star. Results are obtained using the BHF (upper panels) and DBHF (lower panels) nuclear matter EoS.

used the FCM extension at finite T and μ_B, which provides a very simple description of the quark dynamics in terms of two parameters, namely the potential V_1, which summarizes the interaction corrections to the free quark and gluon pressure, and the gluon condensate G_2, that parametrizes the vacuum pressure and energy density. For completeness, we have included the effects of color superconductivity through the CFL mechanism, which amounts to the addition of a new free energy contribution in terms of the gap Δ.

With this new set of more refined calculations, we confirm the trend already observed in [14, 36, 30], i.e. the maximum mass of hybrid stars grows with the two parameters V_1 and G_2 while it decreases when Δ is increased. We have also extended the mapping developed in [30] among the parameters of the FCM and those defining the CSS parametrization, by displaying the effect of the gap Δ, thus concluding that a particular configuration with mass around or above two solar masses can be realized in the FCM by different pairs of G_2 and V_1, depending on the specific value assigned to Δ. The inclusion of the hyperons induces dramatic changes in this picture. In fact, a regular transition from nuclear to quark matter with a stable quark phase up to very high chemical potential requires a particular tuning of the FCM parameters that leads to values of the maximum mass up to 1.95 $M_{\odot}$.

Acknowledgments
Partial support comes from "NewCompStar", COST Action MP1304.

6. References

[1] Baldo M, Bombaci I and Burgio G F 1997 *Astron.Astrophys.* **328** 274–282
[2] Akmal A, Pandharipande V R and Ravenhall D G 1998 *Phys.Rev.* **C58** 1804–1828
[3] Zhou X R, Burgio G F, Lombardo U, Schulze H J and Zuo W 2004 *Phys.Rev.* **C69** 018801
[4] Baldo M, Polls A, Rios A, Schulze H J and Vidaña I 2012 *Phys.Rev.* **C86** 064001
[5] Li Z, Lombardo U, Schulze H J and Zuo W 2008 *Phys.Rev.* **C77** 034316
[6] de-Forcrand P 2009 Simulating QCD at finite density *Symposium on Lattice Field Theory* p 10
[7] Aoki Y, Endrődi G, Fodor Z, Katz S D and Szabó K K 2006 *Nature* **443** 675–678
[8] Bazavov A *et al.* 2009 *Phys. Rev.* **D80** 014504
[9] Antoniadis J, Freire P, Wex N, Tauris T, Lynch R and et al 2013 *Science* **340** 1233232
[10] Taranto G, Baldo M and Burgio G F 2013 *Phys.Rev.* **C87** 045803
[11] Di Giacomo A, Dosch H G, Shevchenko V and Simonov Y 2002 *Phys.Rept.* **372** 319–368
[12] Baldo M, Burgio G F, Castorina P, Plumari S and Zappala D 2007 *Phys. Rev.* **C75** 035804
[13] Logoteta D and Bombaci I 2013 *Phys. Rev.* **D88** 063001
[14] Baldo M, Burgio G F, Castorina P, Plumari S and Zappala D 2008 *Phys.Rev.* **D78** 063009
[15] Burgio G F and Zappal D 2016 *Eur. Phys. J.* **A52** 60
[16] Simonov Y and Trusov M 2007 *JETP Lett.* **85** 598–601
[17] Simonov Y and Trusov M 2007 *Phys.Lett.* **B650** 36–40
[18] Shifman M A, Vainshtein A and Zakharov V I 1979 *Nucl.Phys.* **B147** 385–447
[19] Borsanyi S, Fodor Z, Hoelbling C, Katz S D, Krieg S, Ratti C and Szabo K K 2010 *JHEP* **09** 073
[20] Bazavov A *et al.* 2012 *Phys. Rev.* **D85** 054503
[21] Alford M G, Schmitt A, Rajagopal K and Schfer T 2008 *Rev.Mod.Phys.* **80** 1455–1515
[22] Rajagopal K and Wilczek F 2001 *Phys.Rev.Lett.* **86** 3492–3495
[23] Alford M, Braby M, Paris M and Reddy S 2005 *Astrophys.J.* **629** 969–978
[24] Baldo M (ed) 1999 *Nuclear Methods and The Nuclear Equation of State* (Singapore: World Scientific)
[25] Wiringa R B, Stoks V and Schiavilla R 1995 *Phys.Rev.* **C51** 38–51
[26] Carlson J, Pandharipande V and Wiringa R B 1983 *Nucl.Phys.* **A401** 59–85
[27] Gross-Boelting T, Fuchs C and Faessler A 1999 *Nucl.Phys.* **A648** 105–137
[28] Shapiro S L and Teukolsky S A 1986 *Black Holes, White Dwarfs and Neutron Stars: The Physics of Compact Objects*
[29] Alford M G, Han S and Prakash M 2013 *Phys.Rev.* **D88** 083013
[30] Alford M G, Burgio G F, Han S, Taranto G and Zappalà D 2015 *Phys. Rev.* **D92** 083002
[31] Seidov Z F 1971 *Sov. Astron.* **15** 347
[32] Schaeffer R, Zdunik L and Haensel P 1983 *Astron.Astrophys.* **126** 121–145
[33] Lindblom L 1998 *Phys.Rev.* **D58** 024008
[34] Baldo M, Burgio G F and Schulze H 1998 *Phys.Rev.* **C58** 3688–3695
[35] Baldo M, Burgio G F and Schulze H 2000 *Phys.Rev.* **C61** 055801
[36] Plumari S, Burgio G F, Greco V and Zappala D 2013 *Phys.Rev.* **D88** 083005

IOP Conf. Series: Journal of Physics: Conf. Series **861** (2017) 012012 doi:10.1088/1742-6596/861/1/012012

A self-consistent study of magnetic field effects on hybrid stars

V Dexheimer[1], B Franzon[2] and S Schramm[2]

[1]Department of Physics, Kent State University, Kent OH 44242 USA
[2]FIAS, Ruth-Moufang 1, 60438 Frankfurt am Main, Germany

E-mail: [1]`vdexheim@kent.edu`

Abstract. It is understood that strong magnetic fields affect the structure of neutron stars. Nevertheless, many calculations for magnetized neutron stars are still being performed using symmetric solutions of Einstein's equations. In this conference proceeding, we review why this is not the correct procedure and we also discuss the effects of magnetic fields on the stellar population and temperature profiles.

1. Introduction

In addition to being dense and, to some extent, hot, neutron stars rotate extremely fast and possess strong magnetic fields. According to observations of star periods and period derivatives, magnetars can have a surface magnetic field as large as $10^{14} - 10^{15}$ G [1] and one expects to find even stronger magnetic fields inside these stars. One calculation estimates the internal magnetic field of a star to be of the order of 10^{16} G [2] but, according to virial theorem arguments, neutron stars can have central magnetic fields of the order of 10^{18} G depending on their composition, among other factors [3, 4, 5, 6]. Unfortunately, the origin of strong magnetic fields in compact stars is still unclear.

2. Modelling

A self-consistent study of strong magnetic field effects on neutron stars must contain two parts:

(i) Effects of magnetic fields on the equation of state (EoS)

(ii) Effects of magnetic fields on the solutions of Einstein's equations

To illustrate the first item, we make use of the non-linear realization of the SU(3) sigma model (also referred to as Chiral Mean Field or CMF model) [7, 8, 9]. This is an effective quantum relativistic model within the mean field approximation that describes hadrons interacting via meson exchange. The model is constructed from symmetry relations, which allows it to be chirally invariant at high densities and/or temperatures.

In addition, the model also contains quarks and an effective potential for the field related to deconfinement, which has non-zero values for any density/chemical potential and/or temperature as long as the quarks are present [10, 11]. In this way, different kinds of phase transitions (from smooth crossovers to strong first order ones) can be reproduced and the interactions with the medium determine what are the degrees of freedom present at a certain temperature and baryon

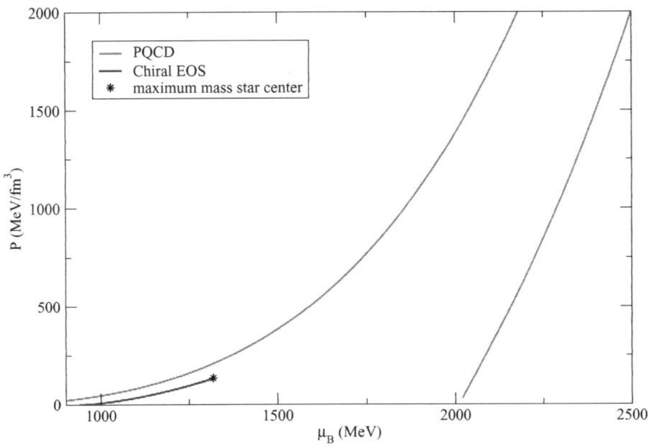

Figure 1. CMF Model equation of state (pressure versus baryon chemical potential) for neutron stars in comparison with PQCD results. The error band reflects uncertainties due to renormalization. PQCD data from Ref [12].

chemical potential. In the low temperature case, a first order phase transition is found and, due to the assumption of global charge neutrality, a mixed phase is reproduced.

At low densities, the CMF Model is constrained by nuclear physics data, while at high temperatures and densities it is fitted to lattice QCD results and astrophysical observations, respectively. As an additional test of the formalism, we show in Fig. 1 the EoS for neutron star matter and compare it to perturbative QCD (PQCD) results for 3-flavor quark-gluon plasma at zero temperature including beta-equilibrium and charge neutrality [12].

The effects of magnetic fields in the EoS are included in the CMF model though Landau quantization, the anomalous magnetic moment (AMM) for hadrons and the magnetization. The first one increases the amount of charged particles in the system [13]. The third one modifies the pressure in the direction perpendicular to the magnetic field and creates an asymmetry in the EoS (independent of the pure magnetic field asymmetry), which is larger at low temperatures (see Ref. [14] and references therein for details).

The anomalous magnetic moment induces a difference in the amount of spin up and spin down fermions in the star [15]. Particles with spin projection opposite to their charge sign cannot have a zeroth Landau level, even when their orbital angular momentum is equal to zero, causing them to have a smaller density when summing over all levels. Particles with spin projection equal to their charge sign are enhanced. This causes, for example, the electrons to be fully polarized in the quark phase. For non-charged baryons, the ones with negative spin projection and negative AMM coupling are suppressed due to the increase in their effective masses. The baryons with negative spin projection and positive coupling are enhanced by the AMM. Particles with spin up have the exactly opposite behavior. For a figure showing the particle population for different spin projections see Ref. [16].

For the solutions of Einstein's equations in the presence of strong magnetic fields, an axisymmetric configuration must be used. In addition, Maxwell's equations must be fulfilled in all parts of the star. To guarantee these, a macroscopic poloidal magnetic field is generated through currents in a numerical code which is part of the LORENE C++ class library for numerical relativity [17, 18]. In this formalism, the strength of the magnetic field throughout the star is controlled by either the dipole magnetic moment μ or the dimensionless current function f_0.

3. Stellar Structure

To include the CMF EoS in the LORENE code, we calculate the EoS as a function of baryon chemical potential and magnetic field, as explained in Refs. [19, 20, 21]. Then, we solve the

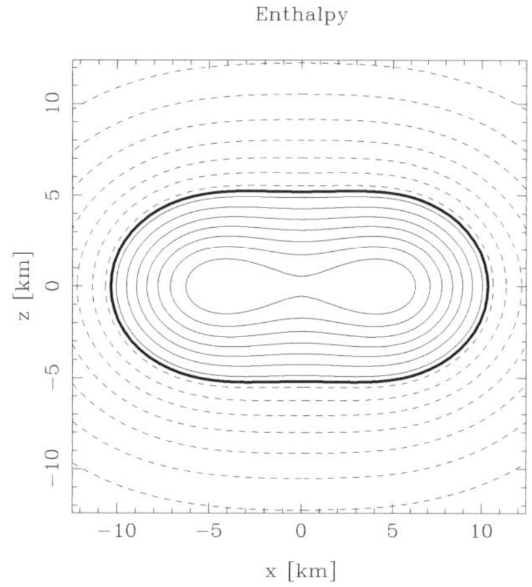

Figure 2. Baryon chemical potential isocontour lines for the most massive star reproduced by $\mu = 3.5 \times 10^{32}$ A m^2.

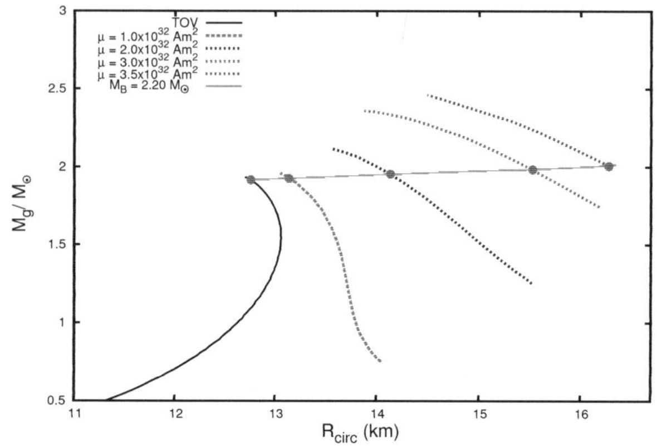

Figure 3. Mass-radius diagram for non-magnetized and magnetized models (with fixed magnetic moments). The gray line shows an equilibrium sequence for a fixed baryon mass of 2.2 M$_\odot$.

coupled Einstein-Maxwell's equations and determine the magnetic field in each part of a star. As a result, larger dipole magnetic moments μ (reproducing larger magnetic fields throughout the star) significantly increase the mass of the star. Using the maximum μ value allowed by the code, $\mu = 3.5 \times 10^{32}$ A m^2, a maximum magnetic field of $\sim 2 \times 10^{18}$ G is found in the stellar center. This configuration has a maximum stellar mass of about 0.6 M$_\odot$ larger than the field-free case. Note that the effects solely due to magnetic field effects from the equation of state on the star mass are mainly due to the magnetization and are relatively small, being visible only for $\mu \gtrsim 3 \times 10^{32}$ A m^2. Fig. 2 shows the deformation of the most massive star reproduced by $\mu = 3.5 \times 10^{32}$ A m^2. In this case, the ratio between the magnetic pressure and the matter pressure in the center for this star is 0.79.

When we fix the baryon mass in order to reproduce the time evolution of an isolated star, the magnitude of magnetic field effects decreases considerably. In this case, a maximum magnetic field of $\sim 1 \times 10^{18}$ G in the stellar center is reproduced by using $\mu \gtrsim 3.5 \times 10^{32}$ A m^2 and a fixed baryon mass of 2.2 M$_\odot$. This configuration has a maximum stellar mass of about 0.1 M$_\odot$ larger than the field-free case. Nevertheless, the radius of stars with fixed baryon mass is still very different for different magnetic field strengths, as can be seen in Fig. 3. The circumferencial radius is a coordinate-independent characterization of the stellar equator $C/(2\pi)$, with C being the proper length of the circumference of the star in the equatorial plane.

4. Stellar Population

Strong magnetic fields have a very large impact on the population of neutron stars. In addition to the effects already discussed in Section 2 related to the EoS, there are magnetic field effects related to the star structure. First, the central density (and respective central baryon chemical potential) in magnetized stars is substantially reduced. This can be seen in Fig. 4, where the different panels show that the population is less exotic (less hyperons or quarks) for larger dipole magnetic moments. This induces a much lower amount of strangeness in the center of the star. Looking at this in another way, exotic degrees of freedom appear over time (and in a larger portion of the star) as the stellar magnetic field decays.

The second effect due to strong magnetic fields that is related to the star structure has to

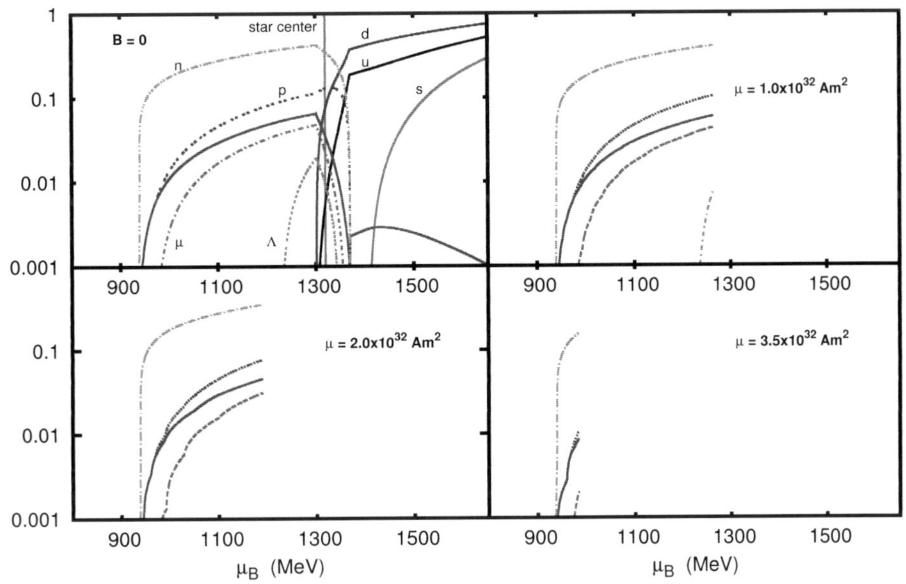

Figure 4. Stellar particle population as a function of the baryon chemical potential. Each panel shows a star with a different dipole magnetic moment. All figures represent an equilibrium sequence at fixed baryon mass $M_B = 2.2\ \mathrm{M}_\odot$. For the non-magnetized case (B=0), the vertical red curve represents the chemical potential reached at the center of the star.

do with deformation. Quantities like the amount of trapped neutrinos and temperature, both associated with the cooling of stars, are quite different in deformed stars [22]. To study these, we look at hot stars with fixed entropy per baryon $S/A = s/n_B = 2$ and trapped neutrinos $Y_L = (n_e + n_{\nu_e})/n_B = 0.4$, corresponding to the first evolutionary stage of neutron stars. Fig. 5 shows how the neutrinos are distributed along the equatorial ($\theta = \pi/2$) and polar ($\theta = 0$) directions in the star for different current functions. At the center of the star, it can easily be noted that the amount of neutrinos is reduced by higher magnetic field distributions. Towards the outside part of the star, the situation is more complicated. A higher magnetic field distribution implies a larger amount of neutrinos at a certain equatorial radius. This is related to the fact that the star becomes more oblate for larger f_0 and that the Lorenz force reverses its direction in the stellar equatorial plane, acting, therefore, differently at different radii.

Fig. 6 shows the temperature distribution in the star. Again, higher magnetic field distribution means a lower temperature at the center of the star, but a larger temperature at a larger equatorial radius of the stellar core. For a discussion on the relation of strong magnetic fields and star surface thermal distribution see, for example, Ref. [23]. The knowledge of the correct temperature distribution in proto-neutron and neutron stars is crucial for modeling the cooling of these stars. Thus, models that include the presence of strong magnetic fields should be reconsidered, not only to investigate the effects of the anisotropy of the energy-momentum tensor due to the magnetic field, but also to include the asymmetric temperature distribution in these objects.

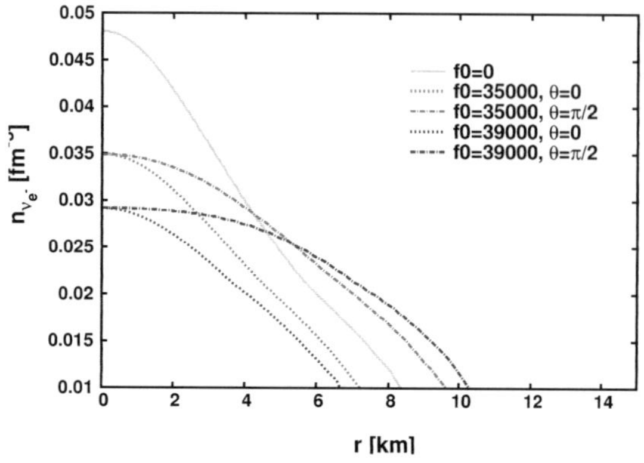

Figure 5. Electron neutrino density profile as a function of the equatorial and polar coordinate radii for one proto-neutron star with M_B = 2.35 $M_\odot$ shown for different current functions.

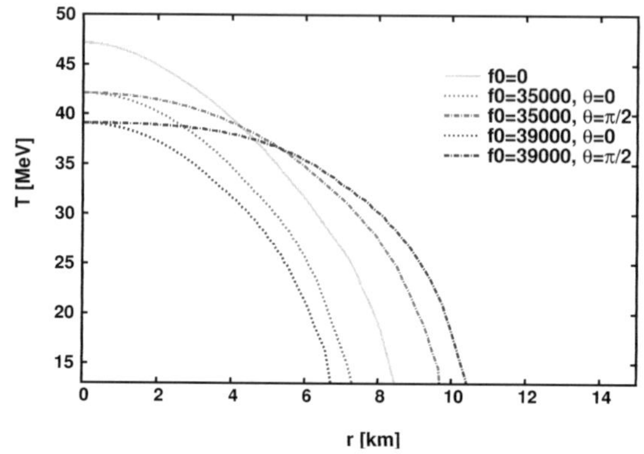

Figure 6. Temperature profile as a function of the equatorial and polar coordinate radii for one proto-neutron star with M_B = 2.35 $M_\odot$ shown for different current functions.

5. Conclusions

Stars that possess strong magnetic fields need to be modeled very carefully. This includes direct effects from the magnetic field on the equation of state and also on the star structure. For example, we showed that stars that have a central magnetic field $\sim 10^{18}$ G are already completely deformed and cannot be described through simple spherical solutions of Einstein's equations [24, 25]. In this case, their population must be calculated using the magnetic field values and central density obtained from the axisymmetric solution of Einstein-Maxwell's equations. In our case, we model the microscopic EoS by using the CMF Model, which is an extended version of the non-linear realization of the SU(3) sigma model that also contains quarks.

Although the increase of stellar mass due to the magnetic field is small for a fixed baryon mass, the polar and equatorial radii are significantly modified, together with the amount of neutrinos and temperature profile, which now have to be calculated separately in different directions in the star. When we further assume that the stellar magnetic field decays over time throughout the star, we find that the appearance of exotic matter, such as hyperons, and phase transitions to quark matter take place at later evolution stages, when the central stellar density increases.

References

[1] Melatos A 1999 *Astrophys. J.* **519** L77
[2] Makishima K, Enoto T, Hiraga J S, Nakano T, Nakazawa K, Sakurai S, Sasano M and Murakami H 2014 *Phys. Rev. Lett.* **112** 171102
[3] Fushiki I, Gudmundsson E and Pethick C 1989 *The Astrophysical Journal* **342** 958–975
[4] Lai D and Shapiro S L 1991 *The Astrophysical Journal* **383** 745–751
[5] Cardall C Y, Prakash M and Lattimer J M 2001 *Astrophys. J.* **554** 322–339
[6] Ferrer E J, de la Incera V, Keith J P, Portillo I and Springsteen P L 2010 *Phys. Rev.* **C82** 065802 (*Preprint* 1009.3521)
[7] Papazoglou P, Zschiesche D, Schramm S, Schaffner-Bielich J, Stoecker H and Greiner W 1999 *Phys. Rev.* **C59** 411–427
[8] Dexheimer V and Schramm S 2008 *Astrophys. J.* **683** 943–948
[9] Dexheimer V, Negreiros R and Schramm S 2015 *Phys. Rev.* **C92** 012801
[10] Dexheimer V A and Schramm S 2010 *Phys. Rev.* **C81** 045201
[11] Hempel M, Dexheimer V, Schramm S and Iosilevskiy I 2013 *Phys. Rev.* **C88** 014906
[12] Fraga E S, Kurkela A and Vuorinen A 2014 *Astrophys. J.* **781** L25
[13] Broderick A, Prakash M and Lattimer J M 2000 *Astrophys. J.* **537** 351

[14] Strickland M, Dexheimer V and Menezes D P 2012 *Phys. Rev.* **D86** 125032
[15] Broderick A E, Prakash M and Lattimer J M 2002 *Phys. Lett.* **B531** 167–174
[16] Dexheimer V, Negreiros R, Schramm S and Hempel M 2013 *AIP Conf. Proc.* **1520** 264–269
[17] Bonazzola S, Gourgoulhon E, Salgado M and Marck J A 1993 *Astron. Astrophys.* **278** 421–443
[18] Bocquet M, Bonazzola S, Gourgoulhon E and Novak J 1995 *Astron. Astrophys.* **301** 757
[19] Dexheimer V, Negreiros R and Schramm S 2012 *Eur. Phys. J.* **A48** 189
[20] Chatterjee D, Elghozi T, Novak J and Oertel M 2015 *Mon. Not. Roy. Astron. Soc.* **447** 3785
[21] Franzon B, Dexheimer V and Schramm S 2016 *Mon. Not. Roy. Astron. Soc.* **456** 2937–2945
[22] Franzon B, Dexheimer V and Schramm S 2016 *Phys. Rev.* **D94** 044018
[23] Aguilera D N, Pons J A and Miralles J A 2008 (*Preprint* 0803.0486)
[24] Oppenheimer J R and Volkoff G M 1939 *Phys. Rev.* **55** 374–381
[25] Tolman R C 1939 *Phys. Rev.* **55** 364–373

IOP Conf. Series: Journal of Physics: Conf. Series **861** (2017) 012013 doi:10.1088/1742-6596/861/1/012013

Neutron star structure with chiral interactions

Domenico Logoteta[1] **and Ignazio Bombaci**[2]

[1]INFN, Sezione di Pisa, Largo Bruno Pontecorvo 3, I-56127 Pisa, Italy
[2]Dipartimento di Fisica, Universitá di Pisa, Largo Bruno Pontecorvo 3, I-56127 Pisa, Italy and INFN, Sezione di Pisa, Largo Bruno Pontecorvo 3, I-56127 Pisa, Italy

E-mail: [1]domenico.logoteta@pi.infn.it

E-mail: [2]ignazio.bombaci@unipi.it

Abstract. We use two-body and three-body nuclear interactions derived in the framework of chiral perturbation theory (ChPT) with and without the explicit Δ isobar contributions to calculate the energy per particle of symmetric nuclear matter and pure neutron matter employing the microscopic Brueckner–Hartree–Fock approach. In particular, we present nuclear matter calculations using the new fully local in coordinate-space two-nucleon interaction at the next-to-next-to-next-to-leading-order (N3LO) of ChPT with Δ isobar intermediate states (N3LOΔ) recently developed by Piarulli *et al.* [1]. We compute the β-equilibrium equation of state and determine the neutron star mass-radius and mass-central density sequences. We find that the adopted interactions are able to provide satisfactory properties of nuclear matter at saturation density as well as to fulfill the limit of two-solar mass for the maximum mass configuration as required by recent observations.

1. Introduction

Chiral effective field theory (EFT) opened a new avenue for a description of nuclear interactions [2, 3, 4]) and nuclear systems consistent with quantum chromodynamics (QCD), the fundamental theory of the strong interaction. The considerable advantage of using such method lies in the fact that two-body, three-body as well as many-body nuclear interactions can be calculated perturbatively, i.e. order by order, according to a well defined scheme based on a low-energy effective QCD Lagrangian which retains the symmetries of QCD, and in particular the approximate chiral symmetry. Within this chiral perturbation theory (ChPT) the details of the QCD dynamics are contained in parameters, the so called low-energy constants (LECs), which are fixed by low-energy experimental data.

Recently Piarulli et al. [1] have developed a fully local in coordinate-space two-nucleon chiral potential which includes the Δ isobar intermediate state. This new potential represents the fully local version of the minimally non-local chiral interaction reported in Ref. [5]. It has been shown by various authors [6, 7] that a Δ-full ChPT has an improved convergence with respect to the Δ-less ChPT. In addition, the Δ-full ChPT naturally leads to three-nucleon forces (TNFs) induced by two-pion exchange with excitation of an intermediate Δ (the celebrated Fujita–Miyazawa three-nucleon force [8]).

In this work, we present microscopic calculations of the equation of state (EoS) of symmetric nuclear matter and pure neutron matter using the local chiral potential of Ref. [1] and employing the Brueckner–Bethe–Goldstone (BBG) [9, 10] many body theory within the Brueckner–Hartree–Fock (BHF) approximation. The present work contains some of the results published in

Ref. [11] and represents a development with respect to our previous works [12, 13] where ChPT nuclear interactions have been used in BHF calculations of nuclear matter properties.

2. Chiral nuclear interactions

Let us now focus on the specific interactions we have employed in the present work. Among the large variety of nucleon-nucleon (NN) interactions derived in the framework of ChPT, as the two-body nuclear interaction, we have used the fully local chiral potential at N3LO including Δ isobar excitations in intermediate state (hereafter N3LOΔ) recently proposed in Ref. [1]. Originally this potential was presented in Ref. [5] in a minimal non-local form. Notice that Ref. [1] reports different parametrizations of the local potential obtained by fitting the low energy NN experimental data using different long- and short-range cutoffs. In the calculations presented in this work, we use the *model b* described in Ref. [1] (see their Tab. II) which fits the Granada database [14] of proton-proton (pp) and neutron-proton (np) scattering data up to an energy of 125 MeV in the laboratory reference frame and has a χ^2/datum$\sim$ 1.07.

To compare with other NN interactions, we have also employed the N3LO chiral NN potential by Entem and Machleidt (EM) [15], considering two different values of the cutoff, $\Lambda = 500$ MeV and $\Lambda = 450$ MeV, employed to regularize the high momentum components of the interaction. Notice that for consistency reasons, the same value of the cutoff has been employed in each calculation, both in the two- and three-nucleon interactions. However the assumed shape of the cutoff in the two-body and in the three-body interaction is in general different (see Ref. [11] for more details).

Concerning the TNF, we have used the N2LO potential by Epelbaum et al. [16] in its local version given by Navratil [17]. We note that the non locality of the N2LO three-nucleon interaction depends only on the cutoff used to regularize the potential. The N2LO TNF depends on factors c_1, c_3, c_4, c_D and c_E which are the so called low energy constants. The N2LO interaction keeps the same operatorial structure both including or not the Δ degrees of freedom ([7]). We note that the constants c_1, c_3 and c_4 are already fixed at the two-body level by the N3LO interaction. However when including the Δ isobar in the three-body potential, the parameters c_3 and c_4 take additional contribution from the Fujita–Miyazawa diagram. Such a diagram appears at the NLO and is clearly not present in the theory without the Δ (see discussion in Ref. [11] about this issue). The values of the constants c_i for the TNFs that we have considered in the present work are reported in Tab. 1.

The remaining parameters c_D and c_E are not determined by the two-body interaction and have to be fixed constraining some specific observables in few-nucleon systems or to reproduce the empirical saturation point of symmetric nuclear matter. In particular, for the interaction model N3LO+N2LO(450), following reference [18], we have set $c_D = -0.24$ and $c_E = -0.11$ in order to reproduce the binding energies of ^{3}H and ^{3}He and the Gammow-Teller matrix element for the ^{3}H β-decay considering contributions to the axial nuclear current up to order N3LO [18]. For the interaction model N3LO+N2LO(500), we have adopted a recent constraint on c_D and c_E employing the same strategy of Ref. [18] but considering contributions to the axial nuclear current up to order N4LO [19]. We note that this parametrization has also the valuable property to reproduce the neutron–deuteron doublet scattering length.

Finally, for the very recent model N3LOΔ+N2LOΔ [1] no calculation for few-body nuclear systems has been done so far. Thus we have fitted the LECs c_D and c_E to get a good saturation point for symmetric nuclear matter.

3. The Brueckner–Hartree–Fock approach with averaged three-body forces

The BHF approach is the lowest order of the BBG many-body theory [9, 10]. In this theory, the ground state energy of nuclear matter is evaluated in terms of the so-called hole-line expansion, where the perturbative diagrams are grouped according to the number of independent hole-lines.

IOP Conf. Series: Journal of Physics: Conf. Series **861** (2017) 012013 doi:10.1088/1742-6596/861/1/012013

Table 1. Values of the low energy constants (LECs) of the TNFs models used in the present calculations. In the first row, we report the parametrizations of the N2LO three-body force with the Δ isobar excitations [1]. Notice that the values c_1, c_3 and c_4 have been kept fixed. In the third and in the forth rows we report the N2LO TNF parametrizations obtained in conjunction with the EM [15] N3LO two-nucleon potential with $\Lambda = 500$ MeV (third row) and with $\Lambda = 450$ MeV (forth row). The LECs c_1, c_3 and c_4 are expressed in GeV^{-1}, whereas c_D and c_E are dimensionless.

TNF	c_D	c_E	c_1	c_3	c_4
N2LOΔ	-0.10	1.30	-0.057	-3.63	3.14
N2LO500	-1.88	-0.48	-0.810	-3.20	5.40
N2LO450	-0.11	-0.24	-0.810	-3.40	3.40

The expansion is derived by means of the in-medium two-body scattering Brueckner G-matrix which describes the effective interaction between two nucleons in presence of the surrounding nuclear medium. In the case of asymmetric nuclear matter [1] with neutron density ρ_n, proton density ρ_p, total nucleon density $\rho = \rho_n + \rho_p$ and isospin asymmetry $\beta = (\rho_n - \rho_p)/\rho$ (asymmetry parameter), one has different G-matrices describing the nn, pp and np in medium effective interactions. They are obtained by solving the well known Bethe–Goldstone equation [9].

We make use of the so-called continuous choice [22] for the single-particle potential $U_\tau(k)$ when solving the Bethe–Goldstone equation. As shown in Refs. [23, 24], the contribution of the three-hole-line diagrams to the energy per particle E/A is minimized in this prescription and a faster convergence of the hole-line expansion for E/A is achieved [23, 24, 25] with respect to the so-called gap choice for $U_\tau(k)$.

As it is well known, within the most advanced non-relativistic quantum many-body approaches [26], it is not possible to reproduce the empirical saturation point of symmetric nuclear matter, $\rho_0 = 0.16 \pm 0.01$ fm^{-3}, $E/A|_{\rho_0} = -16.0 \pm 1.0$ MeV, when using two-body nuclear interactions only. In addition, TNFs are crucial in the case of dense β-stable nuclear matter to obtain a stiff equation of state (EoS) [27, 28] compatible with the measured masses, $M = 1.97 \pm 0.04\, M_\odot$ [29] and $M = 2.01 \pm 0.04\, M_\odot$ [30] of the neutron stars in PSR J1614-2230 and PSR J0348+0432 respectively.

Within the BHF approach TNFs cannot be used directly in their original form. This is because it would be necessary to solve the three-body Faddeev equations in the nuclear medium (Bethe–Faddeev equations) [31, 32] and currently this is a task still far to be achieved. To circumvent this problem an effective density dependent two-body force is built starting from the original three-body one by averaging over one of the three nucleons [33, 34].

In the present work, we consider the in medium effective NN force derived in Ref. [35] (see Ref. [11] for more details on the average).

4. Results and discussion

In this section we present and discuss the results of our calculations for the equation of state (EoS), i.e. the energy per particle E/A as a function of the density ρ, for symmetric nuclear matter (SNM) and pure neutron matter (PNM) using the chiral nuclear interaction models and the BHF approach described in the previous two sections. In all the calculations performed

[1] In the present work we consider spin unpolarized nuclear matter. Spin polarized nuclear matter within the BHF approach has been considered, for example, in Ref. [20, 21].

Table 2. Nuclear matter properties at saturation point for the models described in the text. In the first column we report the model name; in the other columns we give the saturation density (ρ_0) of symmetric nuclear matter, the corresponding value of the energy per particle E/A, the symmetry energy, its slope parameter L and the incompressibility K_∞. All these values refer to the calculated saturation density.

Model	$\rho_0(\mathrm{fm}^{-3})$	E/A (MeV)	E_{sym} (MeV)	L (MeV)	K_∞ (MeV)
N3LOΔ+N2LOΔ	0.171	-15.23	35.39	76.0	190
N3LO+N2LO(500)	0.135	-12.12	25.89	38.3	153
N3LO+N2LO(450)	0.156	-14.32	29.20	39.8	205

in this work, we have considered partial wave contributions up to a total two-body angular momentum $J_{max} = 8$.

In Fig. 1 we show the energy per particle of PNM [panel (a)] and SNM [panel (b)] for the considered interaction models. The dashed lines, in both panels, refer to the calculations performed employing the two-body potential without any TNF, whereas the continuous lines refer to the calculations where the contribution of the TNFs to the energy per nucleon has been included.

Focusing first on the case of PNM (Fig. 1(a)), we note sizable differences between the three energy per nucleon curves produced by the different NN interactions. The model N3LOΔ (upper (black) dashed line) gives indeed a much stiffer EoS than the N3LO ones for both cutoff values, $\Lambda = 500$ MeV (middle (red) dashed line) and $\Lambda = 450$ MeV (lower (blue) dashed line). This behaviour is both due to the local form of the potential and to the inclusion of Δ isobar.

In addition, looking at Tab. 1, we see that the values of c_1 and c_3 are very similar for the considered models. Thus we expect a comparable effect of TNFs on the EoS for PNM as confirmed by our results.

The EoS for symmetric nuclear matter is shown in Fig. 1(b). When only two-body interactions are included, the models based on the EM N3LO potential [15] give unsatisfactory nuclear matter saturation properties. More specifically the model N3LO(500) (middle (red) dashed line) gives a saturation point ($\rho_0 = 0.41$ fm^{-3}, $E/A|_{\rho_0} = -24.25$ MeV), whereas the EoS curve for the model N3LO(450) (lower (blue) dashed line) shows no saturation point up to density of ~ 0.5 fm^{-3}. The EoS for the N3LOΔ NN interaction [1] (upper (black) dashed line) has instead a very different trend. In this case the saturation point turns out to be (0.24 fm^{-3}, -18.27 MeV).

The overall repulsive effect introduced by the inclusion of TNFs produces a significant improvement of the calculated SNM saturation point (see the continuous lines in Fig. 1(b)) with respect to the results described above for the case with no TNFs.

These results clearly show that in the case of Δ-full chiral nuclear interactions the contribution to the energy per particle generated by the TNFs is strongly reduced in comparison to the case where the EoS is obtained from Δ-less chiral interactions.

In Tab. 2 we report the calculated values of the saturation points of SNM for the interaction models considered in the present work. All the models, with the exception of the N3LO+N2LO(500) one, provide reasonable saturation points.

The energy per nucleon of asymmetric nuclear matter can be accurately reproduced [36] using the so called parabolic (in the asymmetry parameter β) approximation

$$\frac{E}{A}(\rho,\beta) = \frac{E}{A}(\rho,0) + E_{sym}(\rho)\beta^2 \,. \tag{1}$$

IOP Conf. Series: Journal of Physics: Conf. Series **861** (2017) 012013 doi:10.1088/1742-6596/861/1/012013

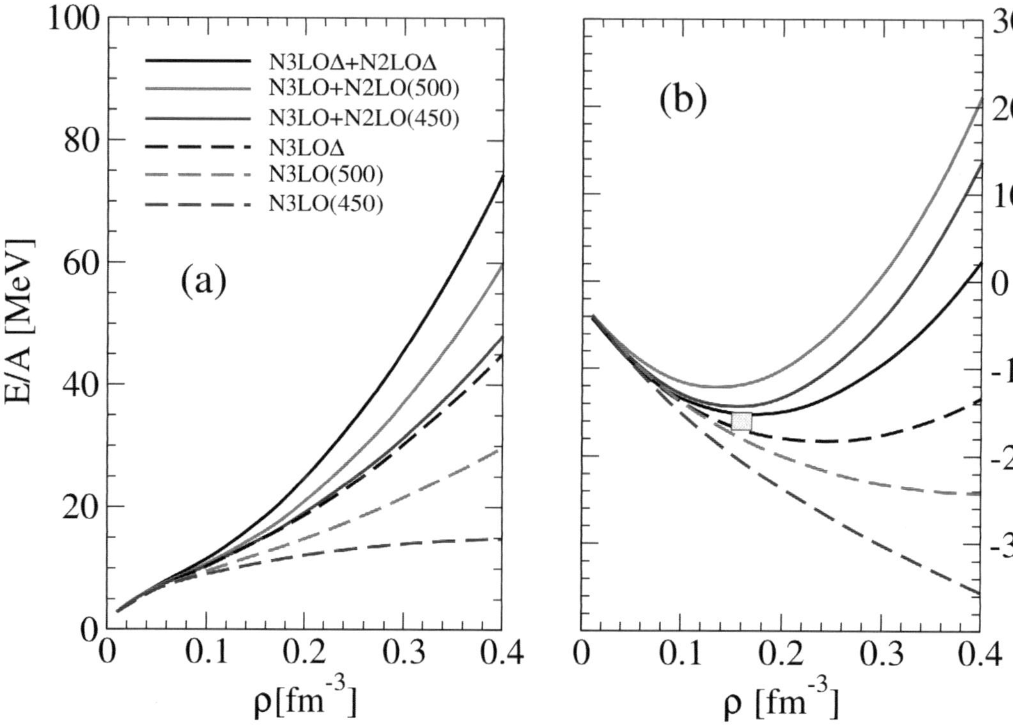

Figure 1. (Color online) Energy per particle of pure neutron matter [panel (a)] and symmetric nuclear matter [panel (b)] as a function of the nucleonic density for the models described in the text. Continuous lines have been obtained using two- plus three-body interactions, while the dashed lines have been obtained considering only the two-body interaction. The empirical saturation point of nuclear matter $\rho_0 = 0.16 \pm 0.01$ fm^{-3}, $E/A|_{\rho_0} = -16.0 \pm 1.0$ MeV is denoted by the grey box in the panel (b).

where $E_{sym}(\rho)$ is the nuclear symmetry energy [37]. Using Eq. (1), the symmetry energy can be calculated as the difference between the energy per particle of pure neutron matter ($\beta = 1$) and symmetric nuclear matter ($\beta = 0$).

In Tab. 2 we report the symmetry energy and the so called slope parameter

$$L = 3\rho_0 \frac{\partial E_{sym}(\rho)}{\partial \rho}\Big|_{\rho_0} \tag{2}$$

at the calculated saturation density ρ_0 (second column in Tab. 2) for the interaction models considered in the present work. As we can see our calculated $E_{sym}(\rho_0)$ and L are in a satisfactory agreement with the values obtained by other BHF calculations with two- and three-body interactions (see e.g. [38, 39]) and with the values extracted from various experimental data, $E_{sym}(\rho_0) = 29.0 - 32.7$ MeV, and $L = 40.5 - 61.9$ MeV, as summarized in Ref. [40].

The incompressibility K_∞ of symmetric nuclear matter at saturation density is given by:

$$K_\infty = 9\rho_0^2 \frac{\partial^2 E/A}{\partial \rho^2}\Big|_{\rho_0} . \tag{3}$$

The incompressibility K_∞ is usually extracted from experimental data of giant monopole resonance (GMR) energies in medium-mass and heavy nuclei. This analysis gives $K_\infty = 210 \pm 30$

IOP Conf. Series: Journal of Physics: Conf. Series **861** (2017) 012013 doi:10.1088/1742-6596/861/1/012013

MeV [41] or more recently $K_\infty = 240 \pm 20$ MeV [42]. Recently the authors of Ref. [43] performed a re-analysis of GMR data finding 250 MeV$< K_\infty <$ 315 MeV. The incompressibility K_∞, at the calculated saturation point for the various interaction models used in the present work, is reported in the last column of Tab. 2. These calculated values for K_∞ are rather low when compared with the empirical values extracted from GMR in nuclei. This is a common feature with many other BHF nuclear matter calculations with two- and three-body interactions (see e.g. [39].

5. Neutron star structure

The composition of the inner core of a neutron star cannot be completely constrained by observations and thus different possibilities are currently under investigation. The appearence of hyperons ([44]) and the transition to a deconfined quark phase ([45, 46, 47, 48]) are among the most popular possibilities. In this work we consider the simplest case of nucleonic matter. Our aim is to establish if nuclear matter EoSs derived from modern chiral interactions can fulfill the constraints put by astrophysical observations. This is an essential step in order to consider more sophisticated possibilities. We thus apply our new EoS model based on the N3LOΔ+N2LOΔ chiral interaction to determine the structure of neutron stars [49]. To this end we calculate the EoS for β-stable nuclear matter including the lepton contribution. The composition of β-stable matter is fixed by the relations between the chemical potentials of the various species. As we are considering pure nucleonic neutrino-free matter one has:

$$\mu_n - \mu_p = \mu_e \ , \qquad \mu_e = \mu_\mu . \tag{4}$$

In Eq. (4) μ_n, μ_p, μ_e and μ_μ are the chemical potentials of neutrons, protons, electrons and muons (with muons appearing above a threshold density given by $\mu_e = m_\mu$) Finally charge neutrality requires:

$$\rho_p = \rho_e + \rho_\mu \tag{5}$$

The various chemical potentials are determined through:

$$\mu_N = \frac{\partial \epsilon}{\partial \rho_N} \ , \qquad \mu_l = \frac{\partial \epsilon}{\partial \rho_l} \tag{6}$$

where $\epsilon = \epsilon_{NN} + \epsilon_L$ is the total energy density which sums up the lepton contribution ϵ_L and the nucleonic one ϵ_{NN}.

The last one is calculated in BHF approximation from $\epsilon_{NN} = \rho E/A(\rho, \beta)$ where we have used the parabolic approximation to determine $E/A(\rho, \beta)$ for asymmetric matter. We have self consistently solved the equations (4), (5), (6) and obtained the EoS for $\beta-$stable matter.

Next to calculate the structure of neutron stars, we have integrated the hydrostatic equilibrium equations in general relativity [49]. For nucleonic density ≤ 0.08 fm^{-3} we have matched our EoS (which describe the neutron star core) with the Baym–Pethick–Sutherland [50] and Negele–Vautherin [51] EoS for the stellar crust. The results are shown in Fig. 2 where we plot the mass-radius (left panel) and mass-central density (right panel) relations for the considered EoS model. The hatched region in the left panel of Fig. 2 represents the mass-radius constraints based on the analysis of recent observations of both transient and bursting X-ray sources obtained in [52, 53].

The maximum mass predicted by our model, $M_{max} = 2.07\ M_\odot$, is compatible with the accurate measurement of the masses, $M = 1.97 \pm 0.04\, M_\odot$ [29] and $M = 2.01 \pm 0.04\, M_\odot$ [30], of the neutron stars in PSR J1614-2230 and PSR J0348+0432 respectively. In addition our EoS model is also able to fulfill the empirical constraints on mass-radius relationship obtained in Refs. [52, 53].

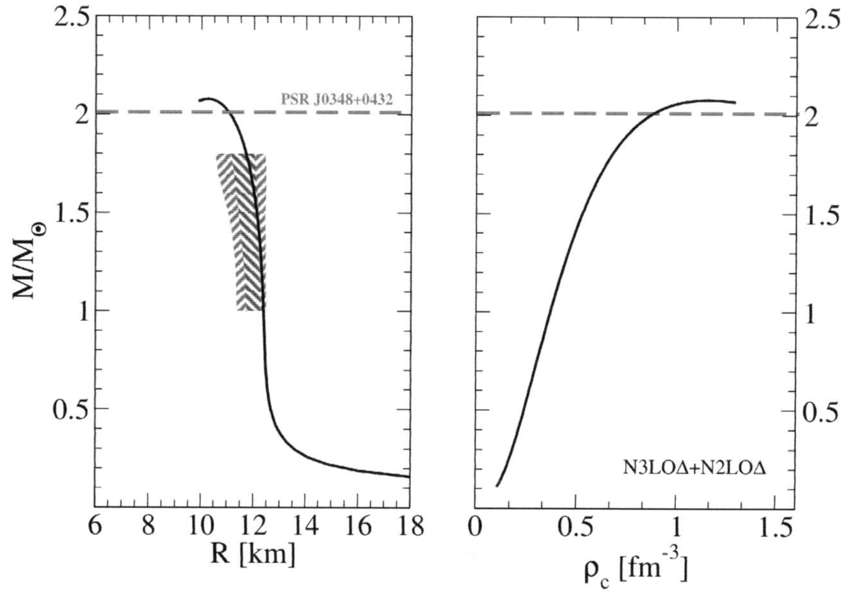

Figure 2. Mass-radius relationship (left panel) and mass-central density relationship (right panel) for the N3LOΔ+N2LOΔ model. The hatched region in the left panel represents the mass-radius constraints obtained in [52, 53]. The red dashed line stands for the measured mass of PSR J0348+0432 [30]. $M_\odot = 1.9885 \times 10^{33}$ g is the mass of the sun.

The values for the gravitational maximum mass M_{max} and the correspondig radius R and central density ρ_c are reported in Tab. 3 and are in good agreement with the results of other calculations [27, 28] based on microscopic approaches.

Table 3. Properties of the maximum mass configuration for the N3LOΔ+N2LOΔ EoS model.

Model	M_{max} ($M_\odot$)	R (km)	ρ_c (fm^{-3})
N3LOΔ+N2LOΔ	2.07	10.26	1.15

6. Summary

We have investigated the behavior and the properties of nuclear matter using three microscopic models fully based on interactions derived in chiral effective field theory, in the framework of the Brueckner–Hartree–Fock many-body approach. In particular we have tested, the new fully local chiral potential at order N3LO which includes the Δ isobar contributions in the intermediate states of the NN interaction [1]. We have also considered two versions of the N3LO chiral NN potential by Entem and Machleidt [15], which differ in the value of the cutoff employed in the calculations. All the two-nucleon interactions have been supplemented with TNFs required to satisfactorily reproduce the empirical saturation point of symmetric nuclear matter. Our results for various nuclear matter properties at saturation density are in good agreement with the available experimental data except for the incompressibility K_∞ which is underestimated with respect to the highly uncertain empirical value [41, 42, 43].

IOP Conf. Series: Journal of Physics: Conf. Series **861** (2017) 012013 doi:10.1088/1742-6596/861/1/012013

Finally for the N3LOΔ+N2LOΔ chiral interaction, we have calculated the EoS of β-stable nuclear matter and the corresponding neutron star properties. We have found that the maximum mass obtained with our EoS model is compatible with present measured neutron star masses. In addition, our calculated neutron star masses and radii are in good agreement with the empirical constraints on these quantities based on the analysis of recent observations of both transient and bursting X-ray sources obtained in [52, 53].

Acknowledgments
This work has been partially supported by "NewCompstar", COST Action MP1304.

References

[1] Piarulli M, Girlanda L, Schiavilla R, Kievsky A, Lovato A, Marcucci L E, Pieper S C, Viviani M and Wiringa R B 2016 Phys. Rev. C **94** 054007
[2] Weinberg S 1979 Physica A **96** 327; 1990 Phys. Lett. B **251** 288; 1991 Nucl. Phys. B **363** 3; 1992 Phys. Lett. B **259** 114
[3] Epelbaum E, Hammer H-W and Meißner U G 2009 Rev. Mod. Phys. **81** 1773
[4] Machleidt R and Entem D R 2011 Phys. Rep. **503** 1
[5] Piarulli M, Girlanda L, Schiavilla R, Navarro Perez R, Amaro J E and Ruiz Arriola E 2015 Phys. Rev. C **91** 024003
[6] Kaiser N, Gerstendörfer S and Weise W 1998 Nucl. Phys. A **637** 395
[7] Krebs H, Epelbaum E and Meißner U G 2007 Eur. Phys. J. A **32** 127
[8] Fujita J and Miyazawa H 1957 Prog. Theor, Phys. **17** 360
[9] Day B D 1967 Rev. Mod. Phys. **39** 719
[10] Baldo M and Burgio G F 2012 Rep. Progr. Phys. **75** 026301
[11] Logoteta D, Bombaci I and Kievsky A 2016 Phys. Rev. C **94** 064001
[12] Logoteta D, Vidaña I, Bombaci I and Kievsky A 2015 Phys. Rev. C **91** 064001
[13] Logoteta D, Bombaci I and Kievsky A 2016 Phys. Lett. B **758** 449
[14] Navarro Pérez R, Amaro J E and Ruiz Arriola E 2013 Phys. Rev. C **88** 064002; Erratum 2015 Phys. Rev. C 91 029901
[15] Entem D R and Machleidt R 2003 Phys. Rev. C **68** 041001(R)
[16] Epelbaum E, Nogga A, Glöckle W, Kamada H and Meißner U G and Witała H 2002 Phys. Rev. C **66** 064001
[17] Navratil P 2007 Few-Body Syst. **41** 117
[18] Marcucci L E, Kievsky A, Rosati S, Schiavilla R and Viviani M 2012 Phys. Rev. Lett. **108** 052502
[19] Baroni A, Girlanda L, Kievsky A, Marcucci L E and Viviani M 2016 Phys. Rev. C **94** 024003.
[20] Vidaña I and Bombaci I 2002 Phys. Rev. C **66** 045801
[21] Bombaci I, Polls A, Ramos A, Rios A and Vidaña I 2006 Phys. Lett. B **632** 638
[22] Jeukenne J P, Lejeunne A and Mahaux C 1976 Phys. Rep. **25** 83
[23] Song H Q, Baldo M, Giansiracusa G and Lombardo U 1998 Phys. Rev. Lett. **81** 1584
[24] Baldo M, Giansiracusa G, Lombardo U and Song H Q 2000 Phys. Lett. B **473** 1
[25] Baldo M, Bombaci I, Giansiracusa G and Lombardo U 1990 J. Phys. G: Nucl. Part. Phys. **16** L263
[26] Bombaci I, Fabrocini A, Polls A and Vidaña I 2005 Phys. Lett. B **609** 232
[27] Baldo M, Bombaci I and Burgio G F 1997 Astron. and Astrophys. **328** 274
[28] Akmal A, Pandharipande V R and Ravenhall D G 1998 Phys. Rev. C **58** 1804
[29] Demorest P, Pennucci T, Ransom S, Roberts M and Hessels J 2010 Nature **467** 1081
[30] Antoniadis J et al. 2013 Science **340** 1233232
[31] Bethe H A 1965 Phys. Rev. **138** 804B
[32] Rajaraman R and Bethe H A 1967 Rev. Mod. Phys. **39** 745
[33] Loiseau B A, Nogami Y and Ross C K 1971 Nucl. Phys. **A401** 601
[34] Grangé P, Lejeunne A, Martzolff B and Mathiot J-F 1989 Phys. Rev.C **40** 1040
[35] Holt J W, Kaiser N and Weise W 2010 Phys. Rev. C **81** 024002
[36] Bombaci I and Lombardo U 1991 Phys. Rev. C **44** 1892
[37] Eur. Phys. J. A **50** (2) 2014, Topical issue on Nuclear Symmetry Energy, edited by Li B A, Ramos A, Verde G and Vidaña I
[38] Li Z H, Lombardo U, Schulze H-J, Zuo W, Chen L W and Ma H R 2006 Phys. Rev. C **74** 047304
[39] Li Z H and Schulze H-J 2008 Phys. Rev. C **78** 028801
[40] Lattimer J M 2014 Gen. Rel. Grav. **46** 1713

[41] Blaizot J P, Gogny D and Grammaticos B 1976 Nucl. Phys. A **265** 315
[42] Sholmo S, Kolomietz V K and Colò G 2006 Eur. Phys. J. A **30** 23
[43] Stone J R, Stone N J and Moszkowski S A 2014 Phys. Rev. C **89** 044316
[44] Vidaña I, Logoteta D, Providência C, Polls A and Bombaci I 2011 EPL **94** 11002
[45] Bombaci I, Logoteta D, Vidaña I and Providência C 2016 EPJA **52** 58
[46] Logoteta D, Bombaci I, Providência C and Vidaña I 2012 Phys. Rev. C **85** 055807
[47] Logoteta D, Providência C and Vidaña I 2013 Phys. Rev. C **88** 055802
[48] Bombaci I and Logoteta D 2013 Mon Not R Astron Soc Lett **433** L79
[49] Shapiro S L and Teukolsky S A 1983 *Black Holes, White Dwarfs andNeutron Stars* (Wiley, New York, 1983).
[50] Baym G, Pethick C and Sutherland D 1971 Astrophys. J. **170** 299
[51] Negele J W and Vautherin D 1973 Nucl. Phys. A **207** 298
[52] Steiner A W, Lattimer J M and Brown E F 2010 Astrophys. J. **722** 33
[53] Steiner A W, Lattimer J M and Brown E F 2013 Astrophys. J Lett. **765** L5

IOP Conf. Series: Journal of Physics: Conf. Series **861** (2017) 012014 doi:10.1088/1742-6596/861/1/012014

Rotating NSs/QSs and recent astrophysical observations

Ang Li[1] **and Jianmin Dong**[2]

[1] Department of Astronomy, Xiamen University, Xiamen, Fujian 361005, China
[2] Institute of Modern Physics, Chinese Academy of Sciences, Lanzhou 730000, China

E-mail: [1]`liang@xmu.edu.cn`

E-mail: [2]`dongjm07@impcas.ac.cn`

Abstract. Both fast and slow configurations of rotating neutron stars (NSs) are studied with the recently-constructed unified NS EoSs. The calculations for pure quark stars (QSs) and hybrid stars (HSs) are also done, using several updated quark matter EoSs and Gibbs construction for obtaining hadron-quark mixed phase. All three types of EoSs fulfill the recent 2-solar-mass constraint. By confronting the glitch observations with the theoretical calculations for the crustal moment of inertia (MoI), we find that the glitch crisis is still present in Vela-like pulsars. An upcoming accurate MoI measurement (eg., a possible 10% accuracy for pulsar PSR J0737-3039A) allows one to distinguish QSs from NSs, since the MoIs of QSs are generally $\gtrsim 1.5$ times larger than NSs and HSs, no matter the compactness and the mass of the stars. Using tabulated EoSs, we compute stationary and equilibrium sequences of rapidly rotating, relativistic stars in general relativity from the well-tested *rns* code, assuming the matter comprising the star to be a perfect fluid. All three observed properties of the short gamma-ray bursts (SGRBs) internal plateaus sample are simulated using the rotating configurations of NSs/QSs as basic inputs. We finally argue that for some characteristic SGRBs, the post-merger products of NS-NS mergers are probably supramassive QSs rather than NSs, and NS-NS mergers are a plausible location for quark deconfinement and the formation of QSs.

1. Introduction

Neutron stars (NSs) are by far one of the most interesting observational objects, since many mysteries remain on them due to their complexity. Future observations with advanced telescopes such as SKA, Advanced LIGO/VIRGO, AXTAR, FAST, NICER, LOFT, and eXTP, will certainly improve our current knowledge of such stellar objects, by providing precise measurements of their masses, radii, spin frequencies, glitches, surface temperatures, luminosities, magnetic strengthes, etc. Theoretically, since no first-principle theory can be used due to the non-perturbative nature of quantum chromodynamics at finite chemical density, their inner structures are still unclear. There is also the hypothesis of pure quark stars (QSs) and their transitions to NSs are both intriguing and important open problems. Therefore, for the study of NSs/QSs, particular attention should be paid on combining realistic enough theoretical models with accumulated observations. The present contribution is along this line.

For this purpose, we select unified NS EoSs that satisfy up-to-date experimental constraints from both nuclear physics [1] and astrophysics [2, 3, 4], based on modern nuclear many-body theories, including microscopic methods (e.g. BCPM [5]) starting from a realistic nucleon-nucleon

IOP Conf. Series: Journal of Physics: Conf. Series **861** (2017) 012014 doi:10.1088/1742-6596/861/1/012014

two-body force (usually accompanied with a nucleonic three body force) and phenomenological ones (e.g. BSk20, BSk21 [6], Shen [7]), starting from a nuclear effective force with parameters fitted to finite nuclei experiments and/or nuclear saturation properties; we also find typical parameter sets for QS EoSs in developed confinement density-dependent mass (CDDM) model, labelled as CIDDM, CDDM1, CDDM2 (see [8] and references therein). Hadron-quark mixed phase is obtained by Gibbs construction [9, 10] and the calculations of hybrid stars (HSs) are done in this framework. All three types of EoSs fulfill the recent 2-solar-mass constraint [2, 3, 4]. For simplicity, several exotic phases possible in NSs' cores are not included here, for example, hyperons (see, e.g., [11]), kaon meson condensation (see, e.g., [12]), Delta excitation (see, e.g., [13]). In the following, we introduce briefly the theoretical framework and discussions of our results in Sects. $2-4$, before providing the summary and future perspectives in Sect. 5. More details can be found in Refs. [8, 14, 15] and references therein.

2. Structures of the Vela pulsar and the glitch crisis

The Vela pulsar (PSR B0833-45) is among the most studied pulsars, which are usually regarded as highly-magnetized, rapidly rotating NSs. By far there is an accurate determination of several pulsar properties, such as the distance, the spin period (89.33 milliseconds), and 17 quasi-periodically produced glitches (sudden increases in spin frequency). Those observations will be used here for a detailed study of its global properties and inner structures. In particular, the "glitch crisis" problem (see eg: [16, 17]), relating to the moment of inertia (MoI) of the stellar crust(see eg: [18, 19]), has generated a great debate recently and will be discussed as well.

Assuming the pulsar mass ranging from $1.0M_\odot$ to $2.0M_\odot$, we calculate the detailed structures of this pulsar, namely the central density, the core/crust radii, the core/crustal mass, the core/crustal thickness, the total/crustal MoI. They are collected in Tables $2-4$ of Ref. [14]. We show here the results from BSk20 EoS as an example.

Furthermore, we analyse the "glitch crisis" problem based on the calculated MoI results. In Fig. 1, the fractional MoIs I_c/I are plotted as a function of the stellar mass for three unified NS EoSs (BCPM, BSk20, BSk21), to be compared with one matched EoS "BHF + NV + BPS". We first notice that for the previous glitch constraint $I_c/I \geq 1.6\%$, the mass of the Vela pulsar should only be smaller than $1.8\mathrm{M}_\odot$ based on selected NS EoS models here. This could be easily fulfilled for an isolated pulsar. However, the discovered crustal entrainment shifts the constraint to around $I_c/I \geq 7\%$, which could hardly be fulfilled by a pulsar heavier than $1.0M_\odot$, except a narrow range of $1.0 \sim 1.06M_\odot$ in the matched "BHF + NV + BPS" EoS case. Actually the predicted fractional moments of inertia based on three unified NS EoSs (BCPM, BSk21, BSk20) are no more than 6.48%. Therefore, we argue that the crisis could be still present in the standard two-component model.

3. NSs or QSs? MoI as a probe

The coexistence of NSs and QSs has been a great puzzle. Recently, it has been suggested to measure MoI rather accurately for pulsar PSR J0737-3039A, with a 10% accuracy (or a constraint on $R \sim 5\%$), using the double pulsar system within the next 20 years. This would significantly constrain the EoS for dense stellar matter and may serve as a tool to justify/discriminate NSs/QSs, see references [20, 21, 22, 23, 24, 25] and references therein for more details.

In Ref. [9], we developed our previous study [10] of the transition to deconfined quark phase in NSs, by including the interaction in the quark EoS to the leading order in the perturbative expansion within the CDDM model. Using the Gibbs conditions the hadron-quark mixed phase is constructed matching the latter with the hadron EoS derived from the microscopic Brueckner theory. High-mass HSs consistent with 2-solar-mass constraint are possible with mixed-phase core with typical parameter sets. Together with EoS models based on the same theories for pure NSs and QSs, respectively, we are ready to evaluate the MoIs of three types of compact stars,

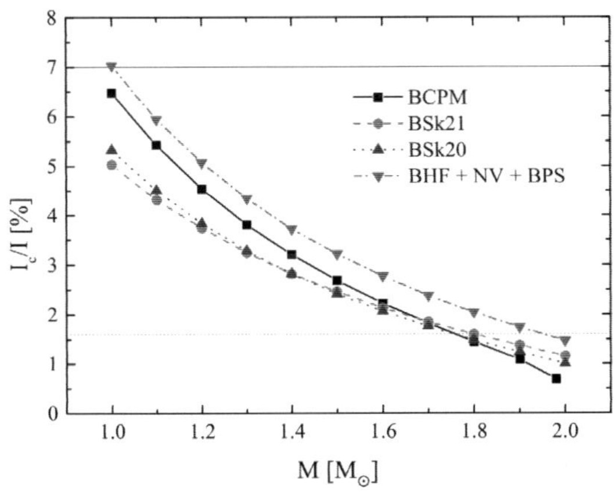

Figure 1. Fractional moments of inertia I_c/I as a function of the stellar mass, with three unified NS EoSs (BCPM, BSk21, BSk20) and one of the matched EoSs (BHF + NV +BPS). Taken from Ref. [14] and see there for details.

Table 1. Predictions for the properties of the Vela pulsar (PSR B0833-45) with a spin period of 89.33 milliseconds, taking the pulsar mass ranging from $1.0M_\odot$ to $2.0M_\odot$. The central densities are in units of fm^{-3}, the masses in solar masses $M_\odot$ (except the mass of the outer crust M_{ocrust} in $10^{-5}M_\odot$), the radii in kilometers, the total moments of inertia in $10^{45}\mathrm{g\ cm}^2$, and the fractional moments of inertia in percentage. Calculations are done with the BSk20 NS EoS, which is based on the widely-used Skyrme nuclear energy density functional, and the parameters were fitted to reproduce with high accuracy almost all known nuclear masses, and to various physical conditions including the nuclear matter EoS from microscopic calculations. Taken from Ref. [14] and see there for details and more results on other unified EoSs (eg: BCPM, BSk21).

Mass	Cent.	Mass			Radius				Moment of inertia	
		core	icrust	ocrust	total	core	icrust	ocrust	total	fraction
1.0	0.328	0.97	0.025	5.57	12.50	10.93	0.84	0.73	1.001	5.03
1.1	0.346	1.08	0.023	5.08	12.54	11.13	0.76	0.65	1.154	4.32
1.2	0.365	1.18	0.022	4.68	12.58	11.30	0.69	0.58	1.315	3.75
1.3	0.384	1.28	0.021	4.25	12.60	11.44	0.63	0.53	1.482	3.25
1.4	0.405	1.38	0.019	3.93	12.62	11.57	0.57	0.48	1.657	2.82
1.5	0.428	1.48	0.018	3.57	12.62	11.67	0.52	0.43	1.839	2.46
1.6	0.453	1.58	0.016	3.27	12.61	11.74	0.48	0.39	2.026	2.13
1.7	0.480	1.68	0.015	2.93	12.58	11.79	0.43	0.35	2.220	1.85
1.8	0.511	1.79	0.014	2.63	12.52	11.82	0.39	0.31	2.420	1.60
1.9	0.548	1.89	0.012	2.34	12.44	11.81	0.35	0.28	2.624	1.38
2.0	0.593	1.99	0.011	2.05	12.32	11.76	0.31	0.25	2.833	1.16

and study whether or not the upcoming MoI constraint might serve as probe for high-density phase of stellar matter. In addition, previous studies (eg: [20, 21]) found large spread in the results for self-bound QSs. The reason might be relatively free parameters in QS models which determine the surface density of strange quark matter at zero pressure. This situation could be largely improved by the new 2-solar-mass constraint used in the present study.

Fig. 2 presents the stars' gravitational masses as functions of central density for these EoSs models (NSs in solid black line, SSs in solid red and blue lines, and HSs in dash-dotted blue

IOP Conf. Series: Journal of Physics: Conf. Series **861** (2017) 012014 doi:10.1088/1742-6596/861/1/012014

line), together with two recent massive stars (PSR J1614-2230 [2, 3] and PSR J0348+0432 [4]) whose masses are accurately measured. The horizontal line represents the pulsar mass $1.337M_{\odot}$ of PSR J0737-3039A. The calculated MoI of QSs are generally $>\sim 1.5$ times larger than those of NSs and HSs, no matter the compactness and the mass of the stars. This is consistent with the calculations in Ref. [23]. Fig. 3 shows the scaled MoIs as a function of stellar mass, in comparison with an upcoming accurate MoI measurement (indicated by a shadowed area like Ref. [22]). One sees that such measurement could have the ability to distinguish QSs from NSs/HSs. More model calculations should be done to justify this argument as a model-independent one.

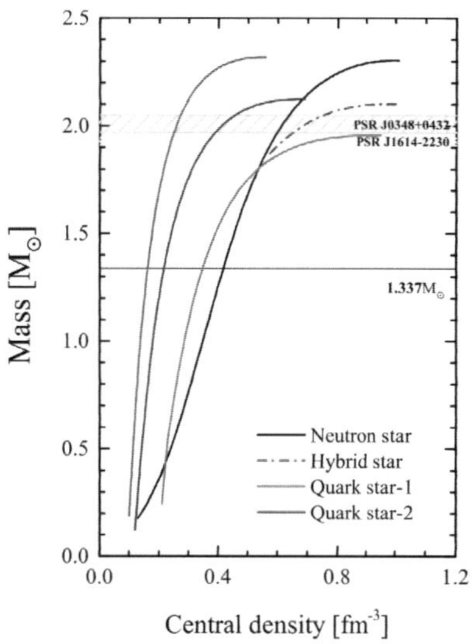

Figure 2. Gravitational masses as functions of the stars' central densities for NSs, HSs and QSs, together with two recently-measured massive stars. The horizontal line represents the pulsar mass $1.337M_{\odot}$ of PSR J0737-3039A. Taken from Ref. [15] and see there for details.

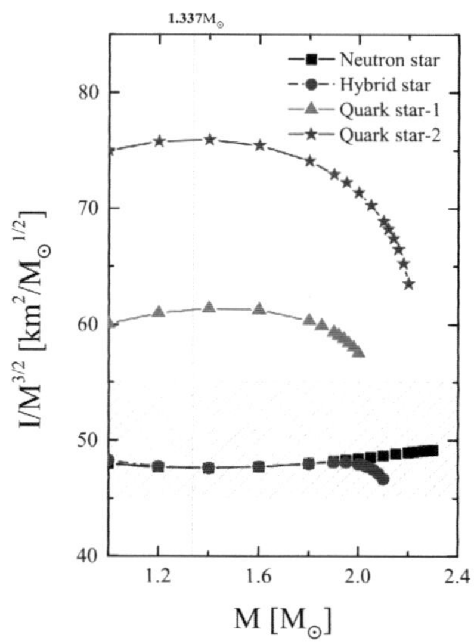

Figure 3. Scaled MoI as a function of stellar mass, based on EoS models in Fig. 2. Shadowed area indicates an upcoming accurate MoI measurement [22]. The vertical line represents the pulsar mass $1.337M_{\odot}$ of PSR J0737-3039A. Taken from Ref. [15] and see there for details.

4. Internal X-ray plateau in short GRBs: Signature of supramassive fast-rotating QSs?

Short gamma-ray bursts (SGRBs) are generally believed to originate from the mergers of two NSs or one NS and one black hole (NS-BH). The traditional view is that NS-NS mergers produce a BH promptly or less than 1 second after the merger, and accretion of remaining debris into the BH launches a relativistic jet that powers the SGRB. Such an engine is also naturally expected within NS-BH merger systems. *Swift* observations of SGRBs, on the other hand, showed extended central engine activities in the early X-ray afterglow phase, in particular the so-called internal plateau, characterized by a nearly flat light curve plateau extending to ~ 300 seconds followed by a rapid $t^{-(8\sim9)}$ decay, as shown in Fig. 4.

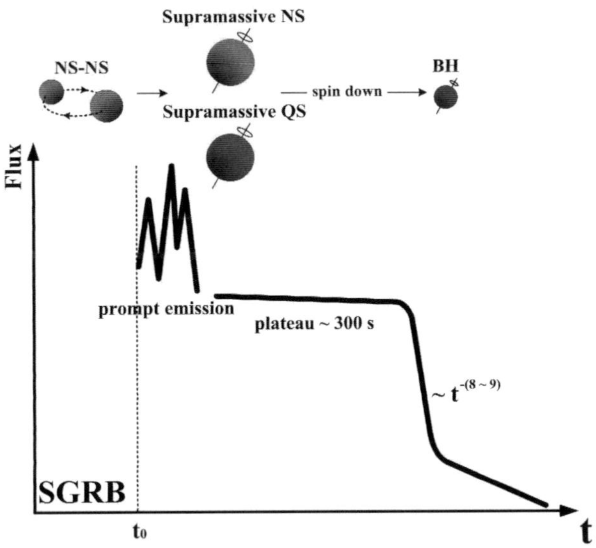

Figure 4. Illustration of the "internal plateau" observation and the NS/QS central engine model in SGRBs.

Table 2. NS/QS EoSs investigated in this study. Here $P_{\rm K}$, $I_{\rm K,max}$ are the Keplerian spin limit and the corresponding maximum moment of inertia, respectively; $M_{\rm TOV}$, $R_{\rm eq}$ are the static gravitational maximum mass and the corresponding equatorial radius, respectively; α, β are the fitting parameters for $M_{\rm max}$ in Eq. (1); A, B, C are the fitting parameters for $R_{\rm eq,max}$ in Eq. (2); a, q, k are the fitting parameters for $I_{\rm max}$ in Eq. (3). Taken from Ref. [8] and see there for details.

	EoS	$P_{\rm K}$ (ms)	$I_{\rm K,max}$ (10^{45}g cm^2)	$M_{\rm TOV}$ ($M_\odot$)	$R_{\rm eq}$ (km)	α ($P^{-\beta}$)	β	A (P^{-B})	B	C (km)	a	q (ms)	k (P^{-1})
	BCPM	0.5584	2.857	1.98	9.941	0.03859	-2.651	0.7172	-2.674	9.910	0.4509	0.3877	7.334
NS	BSk20	0.5391	3.503	2.17	10.17	0.03587	-2.675	0.6347	-2.638	10.18	0.4714	0.4062	6.929
	BSk21	0.6021	4.368	2.28	11.08	0.04868	-2.746	0.9429	-2.696	11.03	0.4838	0.3500	7.085
	Shen	0.7143	4.675	2.18	12.40	0.07657	-2.738	1.393	-3.431	12.47	0.4102	0.5725	8.644
	CIDDM	0.8326	8.645	2.09	12.43	0.16146	-4.932	2.583	-5.223	12.75	0.4433	0.8079	80.76
QS	CDDM1	0.9960	11.67	2.21	13.99	0.39154	-4.999	7.920	-5.322	14.32	0.4253	0.9608	57.94
	CDDM2	1.1249	16.34	2.45	15.76	0.74477	-5.175	17.27	-5.479	16.13	0.4205	1.087	55.14

Since it is very difficult for a BH engine to power such a plateau, one attractive interpretation is that NS-NS mergers produce a rapidly-spinning, supramassive NS rather than a BH, with the rapidly decaying phase signifying the epoch when the supramassive NS collapses to a BH after the NS spins down due to dipole radiation or gravitational wave radiation. Whether the current modelling of a NS could reproduce reasonably all three observed quantities (the break time $t_{\rm b}$, the break time luminosity $L_{\rm b}$ and the total energy in the electromagnetic channel $E_{\rm total}$) is crucially related to the underlying EoS of dense nuclear matter and the calculated rotating configurations.

While a previous analysis shows a qualitative consistency between this suggestion and the *Swift* SGRB data, the distribution of observed break time $t_{\rm b}$ is much narrower than the distribution of the collapse time of supramassive NSs for the several NS EoSs investigated. We here include developed QS EoSs (labelled as CIDDM, CDDM1, CDDM2), as well as recently-constructed unified NS EoSs (BCPM, BSk20, BSk21, Shen). The corresponding star properties are collected in the first four rows of Table 2. Rotating configurations ($M_{\rm max}$, $R_{\rm eq}$, $I_{\rm max}$) from

IOP Conf. Series: Journal of Physics: Conf. Series **861** (2017) 012014 doi:10.1088/1742-6596/861/1/012014

Table 3. Simulated parameter ranges for supramassive NS/QS properties from the *Swift* internal plateaux sample with EoS models (BSk20, BSk21, Shen, CIDDM, CDDM1, CDDM2). Here ϵ, P_i, B_p, η are the ellipticity, the initial spin period, the surface dipole magnetic field, and the radiation efficiency, respectively. $P_{\rm best}(t_b)$ in the last column is the best P value only for the t_b distribution. It is taken from Ref. [8] and see there for details.

EoS	ϵ	P_i (ms)	B_p (G)	η	$P_{\rm best}$ (t_b)
BSk20	0.002	$0.70-0.75$	$N(\mu_{\rm Bp}=10^{14.8-15.4}, \sigma_{\rm Bp}\leq 0.2)$	$0.5-1$	0.20
BSk21	0.002	$0.60-0.80$	$N(\mu_{\rm Bp}=10^{14.7-15.1}, \sigma_{\rm Bp}\leq 0.2)$	$0.7-1$	0.29
Shen	$0.002-0.003$	$0.70-0.90$	$N(\mu_{\rm Bp}=10^{14.6-15.0}, \sigma_{\rm Bp}\leq 0.2)$	$0.5-1$	0.41
CIDDM	0.001	$0.95-1.05$	$N(\mu_{\rm Bp}=10^{14.8-15.4}, \sigma_{\rm Bp}\leq 0.2)$	$0.5-1$	0.44
CDDM1	$0.002-0.003$	$1.00-1.40$	$N(\mu_{\rm Bp}=10^{14.7-15.1}, \sigma_{\rm Bp}\leq 0.3)$	$0.5-1$	0.65
CDDM2	$0.004-0.007$	$1.10-1.70$	$N(\mu_{\rm Bp}=10^{14.8-15.3}, \sigma_{\rm Bp}\leq 0.4)$	$0.5-1$	0.84

the *rns* code can be fitted well as a function of the spin period (P) (in millisecond) as follows:

$$\frac{M_{\rm max}}{M_\odot} = \frac{M_{\rm TOV}}{M_\odot}\left[1+\alpha\left(\frac{P}{\rm ms}\right)^{\beta}\right]; \tag{1}$$

$$\frac{R_{\rm eq,max}}{\rm km} = C + A\left(\frac{P}{\rm ms}\right)^{B}; \tag{2}$$

$$\frac{I_{\rm max}}{10^{45}{\rm g\ cm^2}} = \frac{M_{\rm max}}{M_\odot}\left(\frac{R_{\rm eq}}{\rm km}\right)^2 \frac{a}{1+e^{-k\left(\frac{P}{\rm ms}-q\right)}}, \tag{3}$$

where the parameters $(\alpha, \beta, A, B, C, a, q, k)$ are collected in the last eight columns of Table 2.

Before comparing our results with detailed SGRB observations, it is necessary to first check if the chosen NS/QS EoSs could reproduce the current fraction of supramassive stars, $\sim 22\%$, based on the mass distribution observation of Galactic NS-NS binary systems. The details are in Ref. [8] and the results are shown in Fig. 5. One can see that all except the BCPM NS EoS can reproduce the 22% fraction constraint (with slightly different required P_i). In the following we omit the BCPM EoS. The allowed static gravitational maximum mass $M_{\rm TOV}$ is then in the range of $2.17M_\odot - 2.45M_\odot$.

We finally apply our previous Monte Carlo simulations to the new EoSs for both NSs and QSs, trying to simultaneously reproduce all three distributions (t_b, L_b, $E_{\rm total}$) available from the observational data and revealing the post-merger supramassive stars' physics. The results are shown in Table 3 and Fig. 6.

In Table 3, we list the constrained ranges for the NSs'(QSs') parameters: an ellipticity ϵ as low as 0.002 (0.001), an initial spin period P_i commonly close to the Keplerian limit $P_{\rm K}$, a surface dipole magnetic field of $B_p \sim 10^{15}$ G [satisfying normal distributions $N(\mu_{\rm Bp}, \sigma_{\rm Bp})$], and an efficiency of $\eta = 0.5-1$ related to the conversion of the dipole spin-down luminosity to the observed X-ray luminosity. In the last column we show $P_{\rm best}(t_b)$, the best values only for the t_b distribution. It is clear that the KS test for the t_b distribution is indeed improved. In particular, as one can see from Fig. 6, the t_b distributions in the QS scenarios are more concentrated, which provide a better agreement with the observed ones. Some quantitative differences would exist for other QS model EoSs, but the same conclusions should hold because of the similar QS rotational properties obtained (more pronounced increases of $M_{\rm max}$ with frequency than those with the NS ones: $\sim 40\%$ vs. $18-20\%$ in the present calculation). We therefore argue that a supramassive

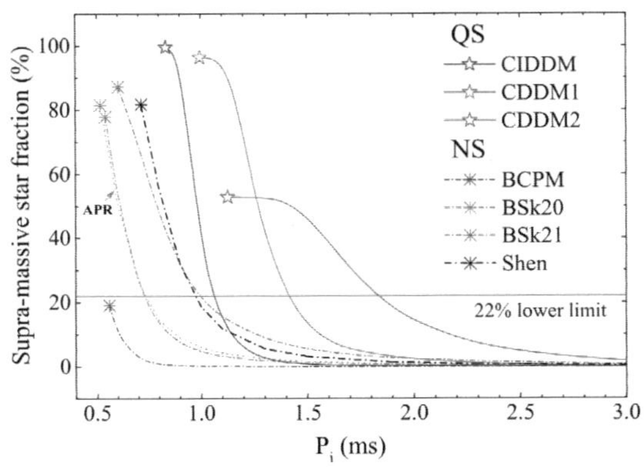

Figure 5. Theoretical estimations of the supramassive star fraction based on four cases of unified NS EoSs and three cases of QS EoSs, as compared with the observed 22% constraint. Previous calculations using the APR NS EoS model [26] are also shown for comparison. Taken from Ref. [8] and see there for details.

QS is favored than a supramassive NS to serve as the central engine of SGRBs with internal plateaux. NS-NS mergers are a plausible location for quark deconfinement and the formation of QSs.

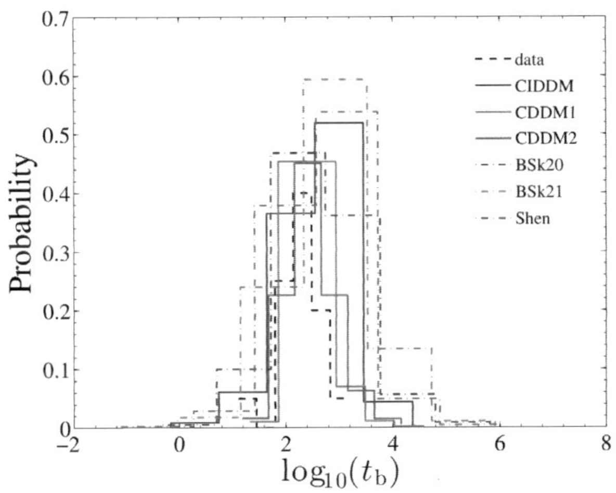

Figure 6. Simulated collapse time distributions with three unified NS EoSs and three QS EoSs, as compared with the observed one (dashed curve). Taken from Ref. [8] and see there for details.

5. Summary and future perspectives

In the present proceeding we show our three recent calculations on rotating compact stars, and compare them with available/upcoming accurate astrophysical observations, including pulsar glitch, MoI, and SQRB.

In terms of future methods, useful constraints could be possible by combining gravitational wave and electromagnetic observations from both coalescing NS-NS binaries and isolated NSs. It would be interesting and meaningful to study whether NS EoS could be determined from gravitational wave emission from elliptically deformed pulsars and NS-NS binaries.

Acknowledgments

The work was supported by the National Natural Science Foundation of China (Nos 11405223, U1431107).

References

[1] Danielewicz P, Lacey R and Lynch W G 2002 *Science* **298**, 1592
[2] Demorest P B, Pennucci R, Ransom S M, Roberts M S E and Hessels J W T 2010 *Nature* **467** 1081
[3] Antoniadis J, et al. 2013 *Science* **340** 6131
[4] Fonseca E *et al* arXiv:1603.00545
[5] Sharma B K, Centelles M, Vinas X, Baldo M and Burgio G F 2015 *Astron. Astrophys* **584** A103
[6] Potekhin A Y, Fantina A F, Chamel N, Pearson J M and Goriely S 2013 *Astron. Astrophys* **560** A48
[7] Shen H, Toki H, Oyamatsu K and Sumiyoshi K 1998 *Nucl. Phys. A* **637** 435
[8] Li A, Zhang B, Zhang N B, Gao H, Qi B and Liu T 2016 *Phys. Rev. D* **94** 083010
[9] Li A, Zuo W and Peng G X 2015 *Phys. Rev. C* **91** 035803
[10] Peng G X, Li A and Lombardo, U. 2008 *Phys. Rev. C* **77** 065807
[11] Hu J N, Li A, Toki H and Zuo W 2014 *Phys. Rev. C* **89** 025802; Burgio G F, Schulze H-J and Li A 2011 *Phys. Rev. C* **83** 025804
[12] Li A, Zhou X R, Burgio G F and Schulze H J 2010 *Phys. Rev. C* **81** 025806; Li A, Burgio G F, Lombardo U and Zuo W 2006 *Phys. Rev. C* **74** 055801; Zuo W, Li A, Li Z H and Lombardo U 2004 *Phys. Rev. C* **70** 055802
[13] Zhu Z Y, Li A, Hu J N and Sagawa H 2016 *Phys. Rev. C* **94** 045803
[14] Li A, Dong J M, Wang J B and Xu R X 2016 ApJS **223**, 16
[15] Li A and Dong J M, to be submitted
[16] Andersson N, Glampedakis K, Ho W C and Espinoza C M 2012 *Phys. Rev. Lett.* **109** 241103
[17] Chamel N 2013 *Phys. Rev. Lett.* **110** 011101
[18] Link B, Epstein R I and Lattimer J M 1999 *Phys. Rev. Lett.* **83** 3362
[19] Li A 2015 *Chin. Phys. Lett.* **32** 079701
[20] Morrison I A, Baumgarte T W, Shapiro S L and Pandharipande V R 2004 ApJ **617** L135
[21] Bejger M, Bulik T and Haensel P 2005 MNRAS **364**, 635
[22] Lattimer J M and Schutz B F 2005 ApJ **629** 979
[23] Urbanec M, Miller J C and Stuchlik Z 2013 MNRAS **433** 1903
[24] Rikovska S J, Guichon P A M, Matevosyan H H and Thomas A W 2007 *Nucl. Phys. A* **792** 341
[25] Li A, Xu R X and Lu J F 2010 MNRAS **402** 2715; Li A, Peng G X and Lu J F 2011 *Res. Astron. and Astrophys.* **11** 482
[26] Akmal A, Pandharipande V R and Ravenhall D G 1998 *Phys. Rev. C* **58** 1804

IOP Conf. Series: Journal of Physics: Conf. Series **861** (2017) 012015 doi:10.1088/1742-6596/861/1/012015

Effects of magnetic fields in white dwarfs

Bruno Franzon[1] **and Stefan Schramm**[2]
FIAS, Ruth-Moufang 1, 60438 Frankfurt am Main, Germany
E-mail: [1]franzon@fias.uni-frankfurt.de
E-mail: [2]schramm@fias.uni-frankfurt.de

Abstract. We perform calculations of white dwarfs endowed with strong magnetic fields. White dwarfs are the progenitors of supernova Type Ia explosions and they are widely used as candles to show that the Universe is expanding and accelerating. However, observations of ultraluminous supernovae have suggested that the progenitor of such an explosion should be a white dwarf with mass above the well-known Chandrasekhar limit $\sim 1.4\ M_{\odot}$. In corroboration with other works, but by using a fully general relativistic framework, we obtained also strongly magnetized white dwarfs with masses $M \sim 2.0\ M_{\odot}$.

1. Introduction

White dwarfs (WD) are formed in catastrophic astrophysical events such as supernova explosions. It is also in these events that other forms of compact objects, as neutron stars or black holes, can be created. It is understood that stars born with masses less than ~ 8 solar masses end up their evolutions as white dwarfs, see e.g. Ref. [1]. With a typical composition mostly made of carbon, oxygen, or helium, endowed with a background of electron-degenerate matter, white dwarfs are extremely dense objects, with a mass comparable to that of the Sun, which is distributed in a volume comparable to that of the Earth. As a result, the densities in WD's can reach values up to $\sim 10^{11}\ \mathsf{g/cm^3}$ at the stellar center.

The mass of white dwarfs is primarily given by the total mass of the nuclei, whereas the main contribution to the pressure comes from the electrons. In addition, in Ref. [2] was found that there is a limit in the stellar mass, beyond which degenerate white dwarfs are unstable. This critical mass is the so-called Chandrasekhar limit and is about $1.4\ M_{\odot}$.

White dwarfs are not only very dense objects, but they are associated also with fast rotation (see Ref. [3] and references therein) and also strongly magnetized. Observations have shown that the observed surface magnetic fields in white dwarfs can be so strong as 10^9 G [4, 5]. Although the internal magnetic fields of these stars cannot be directly constrained by observations, according to Virial theorem arguments, which give an upper estimate for the magnetic inside neutron stars, they can possess central magnetic fields as large as 10^{13} G. This value is obtained in Newtonian mechanics by equating the stellar magnetic energy stored within the star, $B^2/8\pi \times 4\pi/3R^3$, and the gravitational binding energy, $3GM^2/5R$. In doing do, the magnetic field strength satisfies $B \sim 1.4 \times 10^{14}\,(1.40\,M_{\odot}/M)(R/\,1000\,\mathrm{km})^{-2}$ G. On the other hand, analytic and numeric calculations, both in Newtonian theory as well as in General Relativity theory, showed that WD's may have internal magnetic fields as large as 10^{12-16} G [1, 6, 7].

IOP Conf. Series: Journal of Physics: Conf. Series **861** (2017) 012015 doi:10.1088/1742-6596/861/1/012015

The influence of magnetic fields on the structure of white dwarfs is an important problem, since super-massive magnetized WD's, whose existence is partially supported by magnetic forces, could simplify the explanation of observed ultra-luminous explosions of supernovae Type Ia (SNe Ia). SNe Ia occurs in binary systems, when a white dwarf accreting matter from a companion approaches the Chandrasekhar mass limit of 1.40 $\mathsf{M}_\odot$ and becomes unstable. As a result, thermonuclear explosion develops and an immense amount of energy is suddenly released. A feature of these events is that SNe Ia have spectra and light curves (evolution of the supernova brightness with time) amazingly uniform, indicating a common origin and a common intrinsic luminosity. However, recent observations of several supernovae appear to be more luminous than expected. This would indicate that such explosions occur when the white dwarf has a mass well above the Chandrasekhar mass limit.

Previous calculations showed that magnetic white dwarfs can have their masses increased up to 2.58 $\mathsf{M}_\odot$ for a magnetic field strength of 10^{18} G at the center of the star [8]. Nonetheless, such approach violates not only macro physics aspects, as for example, the breaking of spherical symmetry due to the magnetic field, but also microphysics considerations, which are relevant for a self-consistent calculation of the structure of these objects [9]. In addition, a self-consistent Newtonian structure calculation of strongly magnetized white dwarfs showed that these stars exceed the traditional Chandrasekhar mass limit significantly ($\mathsf{M} \sim 1.9\,\mathsf{M}_\odot$) for a maximum field strength of the order of 10^{14} G [6].

Here, we model static and rotating magnetized white dwarfs in a self-consistent way by solving the coupled Einstein-Maxwell equations by means of a pseudo-spectra method. The presence of such a strong magnetic field can locally affect the microphysics of the stellar matter, as for example, due to Landau quantization. However, in Ref. [6] was shown that Landau quantization does not affect the global properties of white dwarfs. Globally, the magnetic field can affect the structure of WD's, since it contributes to the Lorenz force, which acts against gravity. In addition, it contributes also to the structure of spacetime, since the magnetic field is now a source for the gravitational field through the Maxwell energy-momentum tensor. In the following, as we are interested in global effects that magnetic fields and rotation can induce in WD's, we simplify the discussion assuming white dwarfs that are predominately composed by $^{12}\mathsf{C}$ ($\mathsf{A/Z} = 2$) in an electron background.

2. Stellar structure

We follow the numerical technique as developed in Refs. [10, 11] to obtain magnetized white dwarf configurations in a fully general relativity way. Within this approach, the Einstein-Maxwell equations are solved numerically by means of a pseudo-spectra method for axisymmetric stellar configurations within the 3+1 formalism in General Relativity. The accuracy of solutions is controlled by a 2D general-relativistic virial theorem [10] and, for the stars presented here, it is typical of the order of 10^{-5}. Recently, this formalism was applied to neutron stars in Refs. [12, 13, 14] and to study magnetized white dwarfs in Ref. [7].

With the assumption of a stationary and axisymmetric spacetime, the line element in spherical-like coordinates (r, θ, ϕ) can be written as:

$$\begin{aligned} ds^2 = &- N^2 dt^2 + \Psi^2 r^2 \sin^2\theta (d\phi - \omega dt)^2 \\ &+ \lambda^2 (dr^2 + r^2 d\theta^2), \end{aligned} \tag{1}$$

with N, ω, Ψ and λ being functions of (r, θ). The gravitational field is deduced from the integration of a coupled system of four elliptic partial differential equations for the four metric functions, see e.g. Ref. [10]. The final system of gravitational equations can be cast in the form:

$$\Delta_2[(N\Psi - 1) r \sin\theta] = 8\pi N \lambda^2 \Psi r \sin\theta (S^r_r + S^\theta_\theta), \tag{2}$$

$$\Delta_2[\ln\lambda + \nu] = 8\pi\lambda^2 S^\phi_\phi + \frac{3\Psi^2 r^2 \sin^2\theta}{4N^2}\partial\omega\partial\omega - \partial\nu\partial\nu, \tag{3}$$

$$\Delta_3\nu = 4\pi\lambda^2(E+S) + \frac{\Psi^2 r^2 \sin^2\theta}{2N^2}\partial\omega\partial\omega - \partial\nu\partial(\nu + \ln\Psi), \tag{4}$$

and

$$\left[\Delta_3 - \frac{1}{r^2\sin^2\theta}\right](\omega r\sin\theta) = -16\pi\frac{N\lambda^2}{\Psi^2}\frac{J_\phi}{r\sin\theta} + r\sin\theta\partial\omega\partial(\nu - 3\ln\Psi), \tag{5}$$

where the short notation was introduced:

$$\begin{aligned}
\Delta_2 &= \frac{\partial^2}{\partial r^2} + \frac{1}{r}\frac{\partial}{\partial r} + \frac{1}{r^2}\frac{\partial^2}{\partial\theta^2} \\
\Delta_3 &= \frac{\partial^2}{\partial r^2} + \frac{2}{r}\frac{\partial}{\partial r} + \frac{1}{r^2}\frac{\partial^2}{\partial\theta^2} + \frac{1}{r^2\tan\theta}\frac{\partial}{\partial\theta} \\
\nu &= \ln N.
\end{aligned}$$

In addition, in the final gravitational field equations system, Eqs. 2-5, terms as $\partial\omega\partial\omega$ are defined as:

$$\partial\omega\partial\omega := \frac{\partial\omega}{\partial r}\frac{\partial\omega}{\partial r} + \frac{1}{r^2}\frac{\partial\omega}{\partial\theta}\frac{\partial\omega}{\partial\theta},$$

and the total energy density, momentum density and stress tensors of the system are:

$$E = \Gamma^2\,(e+p) - p + E^{EM} \tag{6}$$

$$J_\phi = (E+p)\,\lambda^2\Psi r\sin\theta U + J^{EM}_\phi \tag{7}$$

$$S^r_r = p + S^{EM\ r}_r \tag{8}$$

$$S^\theta_\theta = p + S^{EM\ \theta}_\theta, \tag{9}$$

$$S^\phi_\phi = p + (E+p)\,U^2 + S^{EM\ \phi}_\phi, \tag{10}$$

with e and p being the energy and the isotropic pressure of the fluid, U the fluid velocity, Γ represents the Lorentz factor which relates the Eulerian and the fluid comoving observers, and E^{EM}, J^{EM}_ϕ, $S^{EM\ r}_r$, $S^{EM\ \theta}_\theta$ and $S^{EM\ \phi}_\phi$ correspond to the electromagnetic contribution to the energy, momentum and stress tensor of the system, see e.g. Refs. [12, 11].

In the case of rigid rotation, the equation of motion ($\nabla_\mu T^{\mu\nu} = 0$) of a star endowed with magnetic fields yields:

$$H\,(r,\theta) + \nu\,(r,\theta) + M\,(r,\theta) - \Gamma(r,\theta) = const, \tag{11}$$

where $H(r,\theta)$ is the heat function defined in terms of the baryon number density n:

$$H = \int_0^n \frac{1}{e(n_1)+p(n_1)}\frac{dP}{dn}(n_1)dn_1, \tag{12}$$

IOP Conf. Series: Journal of Physics: Conf. Series **861** (2017) 012015 doi:10.1088/1742-6596/861/1/012015

and the magnetic potential $M(r,\theta)$ can be expressed as:

$$M(r,\theta) = M(A_\phi(r,\theta)) := -\int_{A_\phi(r,\theta)}^{0} f(x)\,\mathrm{d}x, \tag{13}$$

with A_ϕ being the magnetic vector potential and $f(x)$ an arbitrary function $f(x)$ that needs to be chosen [11]. In our case, we construct stellar models for constant values of $f(x) = k_0$. As already shown by [11], the current function can have a more complex structure, however our general conclusions should remain the same.

3. Results

In this section, we present the mass-radius diagram for static magnetized white dwarfs by solving self-consistently the Einstein-Maxwell equations, which take into account the anisotropy of the energy-momentum tensor due to magnetic field. Recently, studies of modified mass-radius relations of magnetic white dwarfs were proposed in Ref. [6]. As we also found in this work, these authors show that the mass of white dwarfs increases in the presence of magnetic fields.

In Fig. 1, we show the mass density distribution for a static star with central enthalpy of $\mathsf{H_c} = 0.0063\,\mathsf{c}^2$, with c being the speed of light. This value of the enthalpy results in the maximum gravitational mass of relativistic, static and magnetized white dwarfs achieved within the code, namely, 2.09 $\mathsf{M}_\odot$, which corresponds to a central mass density of $2.79 \times 10^{10}\,\mathsf{g/cm^3}$. As expected, the mass density is not spherically distributed and the maximum mass density is not at the center of the star. This is due to the fact that the Lorentz force exerted by the magnetic field breaks the spherical symmetry of white dwarfs considerably and acts as a centrifugal force that pushes the matter off-center. In this case, the central magnetic field reaches a value of 1.03×10^{15} G, whereas the surface magnetic field was found to be 2.02×10^{14} G.

For a better understanding of this aspect, we make use the dimensionless equation of motion for the static case, $\Gamma = 0$,

$$\mathsf{H}(\mathsf{r},\theta) + \nu(\mathsf{r},\theta) + \mathsf{M}(\mathsf{r},\theta) = \mathsf{C}, \tag{14}$$

and plot these quantities in the equatorial plane as shown in Fig. 2. The constant C can be calculated at every point in the star. We have chosen the center, since the central value of the magnetic potential $\mathsf{M}(\mathsf{r},\theta)$ is zero and the central enthalpy $\mathsf{H_c}$ is our input to construct solutions. The Lorentz force is the derivative of the magnetic potential $\mathsf{M}(\mathsf{r},\theta)$ in the equation Eq. (14) and reaches its maximum value off-center ($\mathsf{r_{eq}} \sim 350$ km, see Fig. 2). As already discussed in Ref. [15], the direction of the magnetic forces in the equatorial plane depends on the current distribution inside the star. In addition, the magnetic field changes its direction in the equatorial plane and, therefore, the Lorenz force reverses the direction inside the star. In our case, this can be seen from the qualitative change in behaviour of the function $\mathsf{M}(\mathsf{r},\theta)$ around $\mathsf{r_{eq}} \sim 350$ km (Fig. 2). It is worth to mention that we have studied also the effects of the Lorentz force due to high magnetic fields on the geometry of the neutron-star crust in Ref. [16], showing that the thickness of the crust can change (either increase or decrease), depending on the polar angle under the influence of the Lorentz force.

Fig. 3 depicts the relation between the mass and the circular equatorial radius for magnetized white dwarfs. The magnetic field is included in the calculation through the current function $\mathsf{k_0}$. The higher $\mathsf{k_0}$, the higher the magnetic field strength. The stellar sequence (different curves labeled by different $\mathsf{k_0'}$s) is obtained by changing the central enthalpy of the star (or the central density). The value $\mathsf{k_0} = 800$ roughly corresponds to the solution with maximum field configuration achieved within the code. For low magnetic fields, the sequences follow the mass-radius relation as for non-magnetic WD's, ultimately reaching the Chandrasekhar mass limit. As the magnetic field strength is increased, the deviation from the non-magnetic curve increases, resulting in configurations with masses well above the Chandrasekhar limit of 1.4 $\mathsf{M}_\odot$.

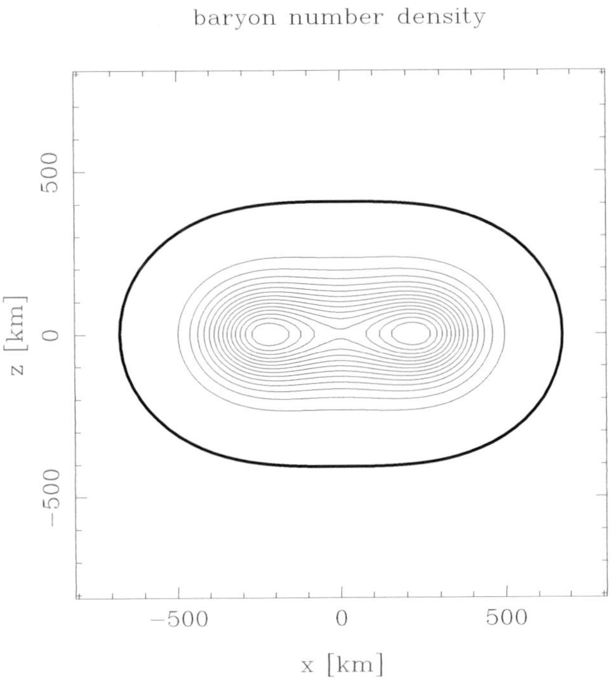

Figure 1. Isocontours of the baryon number density for a star with central magnetic field of 1.03×10^{15} G.

In this calculation, we found a maximum mass for a relativistic white dwarf of 2.09 $M_\odot$ for a magnetic field strength at the stellar center of B $\sim 10^{14}$ G (end of the yellow line in Fig. 3). Fig. 3 clearly demonstrates that WD masses larger than the standard Chandrasekhar limit for the non-magnetic case can be supported by strong magnetic fields.

Recently, collaborators and I improved the work presented above, taking into account possible

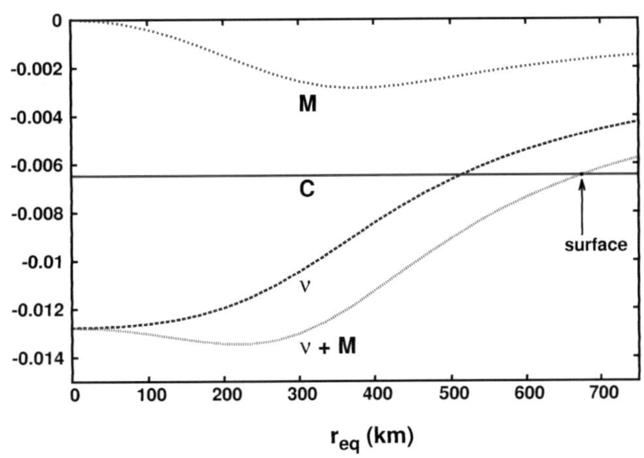

Figure 2. Behaviour of the different terms of the equation of motion as a function of the equatorial coordinate radius for the same star as shown in Fig. 1.

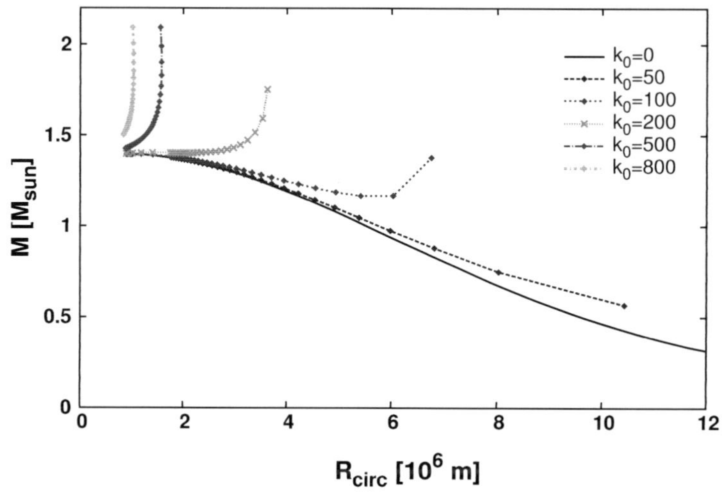

Figure 3. Mass-radius diagram for magnetized white dwarfs. Different curves represent different values of the current function k_0. We also compare the maximum white dwarf mass obtained in the Newtonian case (in black) and in the relativistic one (in red) for the maximum electric current value for which numerical convergence is achieved. This diagram is quite similar to the MR diagram calculated in [6]. However, those authors have evaluated only Newtonian white dwarfs. All curves in this figure were calculated for a ratio between the magnetic and matter pressure less than or quite close to 1 at the center of the star.

instabilities due to electron capture and nuclear fusion reactions in the cores of white dwarfs [17]. The stellar interior was composed of a regular crystal lattice made of carbon ions immersed in a degenerate relativistic electron gas. We found that magnetized white dwarfs violate the standard Chandrasekhar mass limit significantly, even when electron capture and pycnonuclear instabilities are present in the stellar interior. In addition, the maximum magnetic field found is an order of magnitude smaller than in Ref. [7]. This is because we modeled the stellar interior with a much more realistic equation of state than just a simple electron gas, and we considered the density threshold for nuclear fusion reactions, which restricts the central density of white dwarfs in $\sim 9.25 \times 10^9$ g/cm^3, limiting the stellar masses and, therefore, their radii, which for very massive and magnetized white dwarfs cannot be smaller than $\sim$ 1100 km.

4. Conclusions

We were able to calculate super-heavy white dwarfs in the presence of strong fields. This is an interesting and timely important problem, since such stars can exceed the Chandrasekhar mass limit and contribute to superluminous Type Ia supernovae. With this in mind, static equilibrium of stationary and axisymmetric white dwarf stars endowed with strong poloidal magnetic fields were carried out. Moreover, we presented a modication on the white dwarf mass-radius relation generated by the magnetic field. We found that a maximum white dwarf mass of about 2.00 $M_\odot$ may be supported if the interior field is as strong as approximately 10^{14} G. This mass is over 40 percent larger than the traditional Chandrasekhar limit.

References

[1] Shapiro S L and Teukolsky S A 2008 *Black holes, white dwarfs and neutron stars: the physics of compact objects*

IOP Conf. Series: Journal of Physics: Conf. Series **861** (2017) 012015 doi:10.1088/1742-6596/861/1/012015

[2] Chandrasekhar S, 1931 *The Astrophysical Journal* **74** 81
[3] Boshkayev K, Rueda J and Ruffini R 2014 *J. Korean Phys. Soc.* **65** 855
[4] Terada, Yukikatsu and Hayashi, Takayuki and Ishida 2008 *Publ. Astron. Soc. Jap.* **60** 387
[5] Reimers D, Jordan S, Koester D and Bade N 1996 *Astron. Astrophys.* **311** 572-578
[6] Bera, P and Bhattacharya D 2014 *Monthly Notices of the Royal Astronomical Society* **445** 3951-3958
[7] Franzon B and Schramm S 2015 *Phys. Rev. D* **82** 083006
[8] Das U and Mukhopadhyay B 2013 *Physical review letters* **110** 071102
[9] Chamel N, Fantina AF and Davis PJ 2013 *Physical Review D* **88** 081301
[10] Bonazzola S, Gourgoulhon E, Salgado M and Marck J 1993 *Astronomy and Astrophysics* **278** 421-443
[11] Bocquet M, Bonazzola S, Gourgoulhon E and Nova, J 1995 *Astronomy and Astrophysics* **301** 757
[12] Franzon B, Dexheimer V and Schramm S 2016 *Mon. Not. Roy. Astron. Soc.* **456** 2937-2945
[13] Franzon B, Dexheimer V and Schramm S 2016 *Phys. Rev. D* **94** 044018
[14] Franzon B, Gomes RO and Schramm S 2016 *Mon. Not. Roy. Astron. Soc.* **463** 571-579
[15] Cardall C, Prakash M and Lattimer J 2001 *The Astrophysical Journal* **554** 322
[16] Franzon B, Negreiros R and Schramm S. 2016 *arXiv* **1612.04670**
[17] Otoniel E, Franzon B, Malheiro M, Schramm S and Weber F 2016 *arXiv* **1609.05994**

IOP Conf. Series: Journal of Physics: Conf. Series **861** (2017) 012016 doi:10.1088/1742-6596/861/1/012016

The population of highly magnetized neutron stars

R O Gomes[1], V Dexheimer[2], B Franzon[3] and S Schramm[3]

[1]Department of Astronomy, Universidade Federal do Rio Grande do Sul, Porto Alegre, Brazil
[2]Department of Physics, Kent State University, Kent OH 44242 USA
[3]FIAS, Ruth-Moufang 1, 60438 Frankfurt am Main, Germany

E-mail: rosana.gomes@ufrgs.br

Abstract. In this work, we study the effects of strong magnetic field configurations on the population of neutron stars. The stellar matter is described within a relativistic mean field formalism which considers many-body force contributions in the scalar couplings. We choose the parametrization of the model that reproduces nuclear matter properties at saturation and also describes massive hyperon stars. Hadronic matter is modeled at zero temperature, in beta-equilibrium, charge neutral and populated by the baryonic octet, electrons and muons. Magnetic effects are taken into account in the structure of stars by the solution of the Einstein-Maxwell equations with the assumption of a poloidal magnetic field distribution. Our results show that magnetic neutron stars are populated essencialy by nucleons and leptons, due to the fact that strong magnetic fields decrease the central density of stars and, hence, supress the appearance of exotic particles.

1. Introduction

The topic of strong magnetic fields in neutron stars have been extensively studied in the literature. Magnetic field effects have been explored both on the equation of state [1, 2, 3, 4, 5, 6, 7, 8, 9, 10, 11, 12, 13, 14, 15, 16, 17] and on the strucuture of magnetic neutron stars [18, 19, 20, 21, 22], in order to identify their impact on the global properties of neutron stars, such as masses and radii. In particular, Refs. [21, 22] have shown that magnetic effects in the equation of state are not strongly significant for the determination of global properties of stars (i.e., mass and radius), although magnetic field dependence on microscopic processess as neutrino emission and consequently the cooling are still important [23].

Another relevant effect of strong magnetic fields in neutron stars is their equatorial radius enhancement and the consequent decrease of the central baryon densities [22]. Such effect can have a severe impact on the population of highly magnetized neutron stars. The decrease of the central chemical potential, related to the baryon density, can prevent the appearance of phase transitions [22, 24] as well as the existence of exotic degrees of freedom. In this work, we show for a particular parametrization of the many-body forces model (MBF model) that the hyperon population of neutron stars is completely supressed in the presence of central magnetic fields $B_c \sim 10^{18}$ G.

2. Hadronic Model

In order to describe hadronic matter inside magnetic stars, we make use of a many-body forces model (MBF model) [25], which is a relativistic mean field model that takes into account nonlinear meson-meson contribuitions in the scalar interactions. The lagrangian density of the MBF model reads [25]:

$$\begin{aligned}\mathcal{L} = &\sum_b \overline{\psi}_b \left[\gamma_\mu \left(i\partial^\mu - g_{\omega b}\omega^\mu - g_{\phi b}\phi^\mu - g_{\varrho b} I_{3b} \varrho_3^\mu\right) - m_{b\zeta}^*\right]\psi_b + \sum_l \overline{\psi}_l \gamma_\mu \left(i\partial^\mu - m_l\right)\psi_l \\ &+ \left(\frac{1}{2}\partial_\mu\sigma\partial^\mu\sigma - m_\sigma^2\sigma^2\right) + \frac{1}{2}\left(-\frac{1}{2}\omega_{\mu\nu}\omega^{\mu\nu} + m_\omega^2\omega_\mu\omega^\mu\right) + \frac{1}{2}\left(-\frac{1}{2}\phi_{\mu\nu}\phi^{\mu\nu} + m_\phi^2\phi_\mu\phi^\mu\right) \\ &+ \frac{1}{2}\left(-\frac{1}{2}\boldsymbol{\varrho}_{\mu\nu}.\boldsymbol{\varrho}^{\mu\nu} + m_\varrho^2\boldsymbol{\varrho}_\mu.\boldsymbol{\varrho}^\mu\right) + \left(\frac{1}{2}\partial_\mu\boldsymbol{\delta}.\partial^\mu\boldsymbol{\delta} - m_\delta^2\boldsymbol{\delta}^2\right),\end{aligned} \tag{1}$$

where the indices b and l denote the degrees of freedom of baryons (p^+, n, Λ, Σ^+, Σ^0, Σ^-, Ξ^0, Ξ^-) and leptons (e^-, μ^-), respectively. The first line represents the Dirac lagrangian for baryons and leptons, respectively. The second line presents the lagrangian densities of the scalar-isoscalar σ field and the vector-isoscalar ω and ϕ fields, which reproduce the attractive and repulsive features of nuclear interaction. The isovector fields δ (scalar) and ϱ (vector) are introduced in the third line, and are responsible for the description of isospin asymmetry present in neutron stars. The inclusion of the ϕ allows for a more accurate description of the hyperon-hyperon interaction. In this work we do not include the σ^* meson in order to focus on massive hyperon stars (for more results of the MBF model and the inclusion of the σ^* meson, see Ref. [25]).

The many-body forces contributions are introduced in the effective couplings of the scalar mesons:

$$g_{\sigma b}^* = \left(1 + \frac{g_{\sigma b}\sigma + g_{\delta b} I_{3b}\delta_3}{\zeta m_b}\right)^{-\zeta} g_{\sigma b}, \qquad g_{\delta b}^* = \left(1 + \frac{g_{\sigma b}\sigma + g_{\delta b} I_{3b}\delta_3}{\zeta m_b}\right)^{-\zeta} g_{\delta b}, \tag{2}$$

making them no longer constants, but dependent on the scalar fields. Consequently, the effective baryon mass is also affected by many-body contribuitions as: $m_{\zeta b}^* = m_b - g_{\sigma b}^*\sigma - g_{\delta b}^* I_{3b}\delta_3$. The parameter ζ regulates the intensity of the nonlinear contributions from the scalar fields, which are interpreted as higher order meson-meson interactions contributions.

In order to describe massive hyperon stars, we use the parametrization $\zeta = 0.040$ that reproduces a binding energy per nucleon $B/A = -15.75\,\text{MeV}$ and a saturation density $\rho_0 = 0.15\,\text{fm}^{-3}$. Nuclear matter properties values at saturation for this choice of parameters are: effective mass of the nucleon $m_n^* = 0.66 m_n$, compressibility $K_0 = 297$ (MeV), symmetry energy $J_0 = 32\,\text{MeV}$ and slope of the symmetry energy $L_0 = 97\,\text{MeV}$ [25]. Also, we use the spin-flavor SU(6) symmetry in order to define the hyperon couplings with the ω, ϕ, ϱ and δ fields [26, 27], and fix the hyperon potentials depths $U_\Lambda^N = -28\,\text{MeV}$, $U_\Sigma^N = +30\,\text{MeV}$ and $U_\Xi^N = -18\,\text{MeV}$ to obtain their coupling with the σ field.

3. Structure of magnetic stars

In order to describe the structure of magnetic neutron stars, we solve the Einstein-Maxwell equations system. In particular, we use the open source LORENE C++ library for numerical relativity developed by Bonazzola et al. [18, 19].

This formalism solves the Einstein's equations coupled to the Maxwell's equations for an axisymmetric metric using the 3+1 decomposition, which is a typical tool for numerical relativity. The system of equations is solved under the assumptions of meridional currents absence, infinite conductivity, magnetostatic equilibrium (from momentum and energy conservation) and for a poloidal magnetic field distribution. As the electromagnetic tensor fulfills the homogeneous Maxwell equations, only the Maxwell-Gauss and Maxwell-Ampère equations are left to be solved. For more details on the formalism, see Refs. [18, 19, 21, 22].

The formalism used in this work does not include the dynamics that would ultimately give rise to a magnetic field in the stars. For this reason, the magnetic field is generated by a current,

IOP Conf. Series: Journal of Physics: Conf. Series **861** (2017) 012016 doi:10.1088/1742-6596/861/1/012016

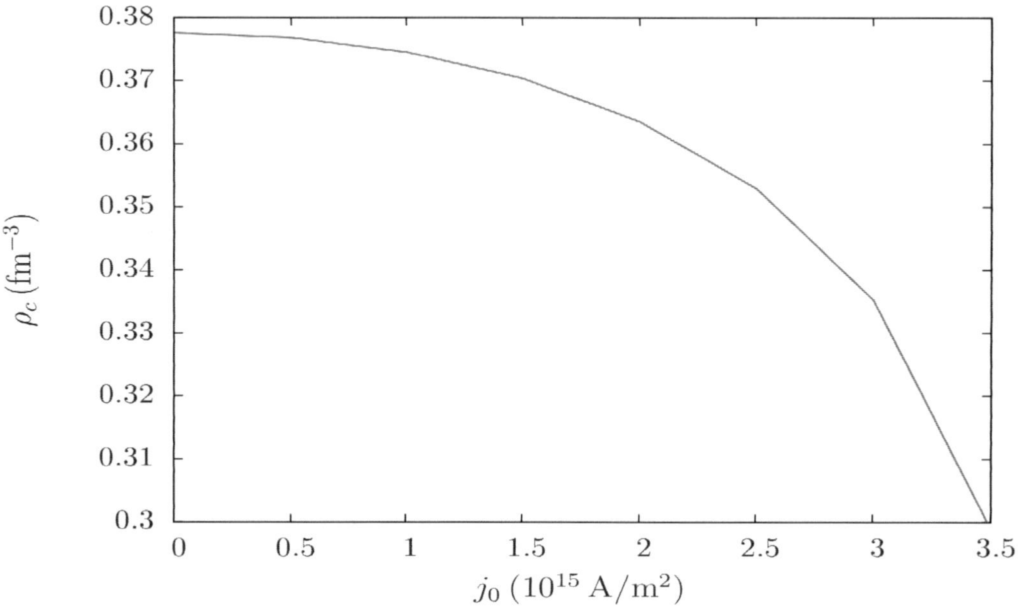

Figure 1. Central baryon density (vertical axis) of a $M_B = 2.2$ M$_\odot$ baryon mass star as a function of the current function (horizontal axis).

which is obtained from the EoS together with a current function j_0, which is introduced as a parameter. It has been shown in previous works that magnetic effects on the equation of state do not play a significant role on the determination of macroscopic properties of highly magnetized neutron stars [21, 22] and in this work, for this reason, we do not include such effects on the equation of state of the MBF model.

4. Maximum mass stars population

Strong magnetic fields have a very large impact on the population of neutron stars. First, the Lorentz force in magnetic stars acts agaist gravity, increasing the radius of stars and, consequently, reducing substantially their central density (and respective central baryon chemical potential). This can be seen in Fig. 1, in which the central density of a $M_B = 2.2$ M$_\odot$ baryon mass star is shown as function of different choices of the current function j_0, i.e. for different magnetic field distributions.

We can now turn our attention to the maximum gravitational mass stars generated with our EoS including or not magnetic fields. Fig. 2 shows the particle population as a function of the radius for the non magnetic star, which is spherical and has a gravitational mass $M_G = 2.15$ M$_\odot$ and central density $\rho_c = 0.86\,\text{fm}^{-3}$. The population is dominated by n, Λ and Ξ^- particles in the inner region, due to the high densities reached at the stellar core [25].

Fig. 3 shows the population for the margnetic star that has a gravitational mass $M_G = 2.22$ M$_\odot$ and the central density $\rho_c = 0.70\,\text{fm}^{-3}$. This is the maximum gravitational mass star for a magnetic configuration given by a current function $j_0 = 3.5 \times 10^{15}\,\text{A/m}^2$, which describes the strongest magnetic field distribution allowed by the code for the MBF model, with surface $B_s = 3.8 \times 10^{17}$ G and central $B_c = 1.1 \times 10^{18}$ G magnetic fields values. The particle population for the magnetic case is shown as a function of the polar radius of the star, but the results are qualitatively the same for the equatorial direction. It can be seen that, in case of central

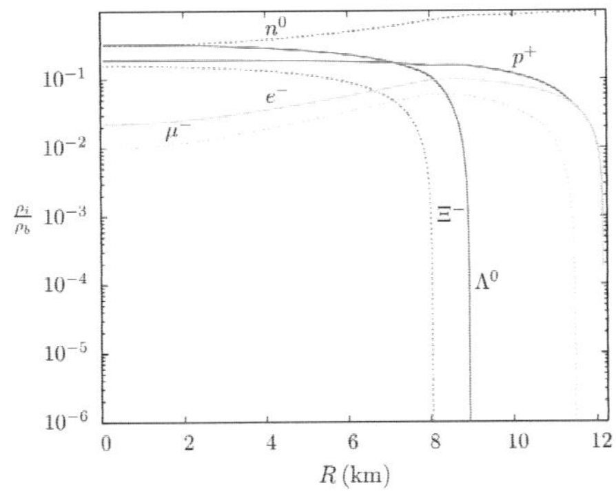

Figure 2. Particle stellar population for the maximum achieved gravitational mass star (M_G = 2.15 $M_\odot$) in the non magnetic case. The vertical axis shows the particles densities normalized by the baryon density and the horizontal axis shows the radius of the star.

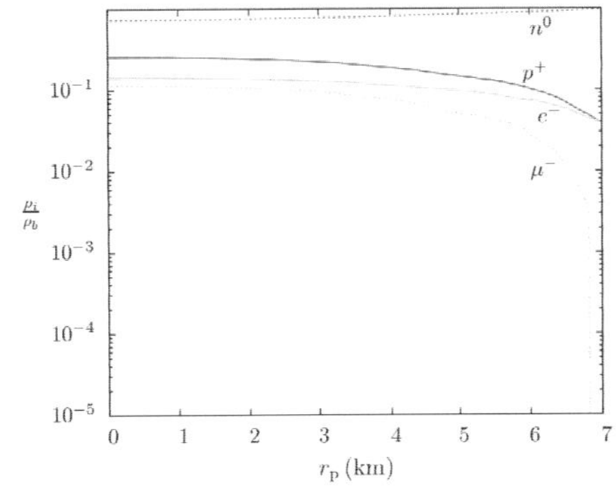

Figure 3. Particle stellar population for the maximum achieved gravitational mass star (M_G = 2.22 $M_\odot$) in the magnetic case ($j_0 = 3.5 \times 10^{15}$ A/m^2). The vertical axis shows the particles densities normalized by the baryon density and the horizontal axis shows the polar radius of the star.

magnetic fields of the order $\sim 10^{18}$ G, the hyperon population is completely supressed.

The decrease of strangeness at the core of magnetic stars is directly caused by the decrease of their central density. In such stars, the central region does not provide enough energy to reach the hyperons threshold of appearance. Similar results have already been discussed in the literature for highly magnetized hybrid stars, proving that strong magnetic fields can also prevent the occurance of phase transitions in such objects [22, 24]. Furthermore, although not investigated in this work, the microscopic effects of Landau quantization in the EoS also shifts the threshold of the appearance of hyperons to higher densities due to the extra magnetic contribuition to the particles energy levels [2, 7, 28].

5. Discussion

Strong magnetic fields in neutron stars have long been studied, usually focusing on their effects on the global properties of stars, such as masses, radii and deformation. In this work, we have applied a formalism that allows for describing consistently the structure of magnetic neutron stars, taking into account their deformation as well as solving the Einstein-Maxwell equations system for determining their strutucture (instead of using the Tolman-Oppenheimer-Volkoff spherical solutions). The hadronic matter interaction inside the stars was described by the MBF model, which considers many-body forces contributions in the effective coupling of scalar mesons. The matter was modeled to be at zero temperature, charge neutral, in beta-equilibrium and populated by nucleons, hyperons and leptons (for the non magnetic case).

We have shown that in the presence of strong magnetic field distributions, the central density of stars decreases due to the Lorentz force. This has the direct impact of decreasing the strangeness fraction at their core. Comparing the case of the maximum gravitational mass for the scenarios of a non magnetic star and for the highest magnetic field configuration allowed by the code, we find that central magnetic fields as strong as $\sim 10^{18}$ G supress completely the hyperon population of magnetic neutron stars, similarly to previous results for phase transitions in hybrid stars. These results imply that, as magnetic field strengths decay over time, a repopulation mechanism is driven inside stars, similarly to the one investigated for slowing down pulsars [29]. Such results are important for the detection of signals for exotic matter (hyperons, delta isobars, quark matter) inside magnetic neutron stars such as their cooling, tidal deformation and gravitational wave emisions. A similar work considering other parametrizations of the MBF model and performing a similar population investigation but for atars with the same baryonic mass is already in preparation.

References

[1] Chakrabarty S, Bandyopadhyay D and Pal S 1997 *Phys. Rev. Lett.* **78** 2898-2901
[2] Broderick A E, Prakash M and Lattimer J M 2000 *ApJ* **537** 351
[3] Broderick A E, Prakash M and Lattimer J M 2002 *Phys. Lett. B* **531** 167-74
[4] Sinha M, Mukhopadhyay B and Sedrakian A 2013 *Nucl. Phys. A* **898** 43-58
[5] Lopes L L and Menezes D P 2012 *Braz. J. Phys.* **42**
[6] Casali R H, Castro L B and Menezes D P 2014 *Phys. Rev. C* **89** 015805
[7] Gomes R O, Dexheimer V and Vasconcellos C A Z 2014 *Astron.Nachr.* **335** 666
[8] Gao Z F, Wang N, Xu Y and Li X D 2015 *Astron. Nachr.* **336** 866
[9] Martinez A P, Rojas H P and Mosquera Cuesta H 2008 *Int. J. Mod. Phys. D* **17** 2107-23
[10] Orsaria M, Ranea-Sandoval I F and Vucetich H 2011 *ApJ* **734** 41
[11] Dexheimer V, Menezes D P and Strickland M 2014 *J. Phys. G* **41** 015203
[12] Denke R Z and Pinto M B 2013 *Phys. Rev. D* **88** 056008
[13] Isayev A A 2015 *J. Phys.: Conf. Series* **607** 012013
[14] Gonzales Felipe R, Manreza Paret D and Perez Martinez A 2011 *Eur. Phys. J. A* **47** 1
[15] Paulucci L, Ferrer E J, de la Incera V and Horvath J E 2011 *Phys. Rev. D* **83** 043009
[16] Rabhi A, Pais H, Panda P K and Providencia C 2009 *J. Phys. G* **36** 115204
[17] Dexheimer V, Negreiros R, Schramm S and Hempel M 2013 *AIP: Conf. Proc.* **1520** 264-9
[18] Bonazzola S, Gourgoulhon E, Salgado M and Marck J A 1993 *Astron. Astrophys.* **278** 421-43
[19] Bocquet M, Bonazzola S, Gourgoulhon E and Novak J 1995 *Astron. Astrophys.* **301** 757
[20] Cardall C Y, Prakash M and Lattimer J M 2001 *ApJ* **554** 322-39
[21] Chatterjee D, Elghozi T, Novak J and Oertel M 2015 *Mon. Not. Roy. Astron. Soc.* **447** 3785
[22] Franzon B, Dexheimer V and Schramm S 2015 *Mon. Not. Roy. Astron. Soc.* **456** 2937-45
[23] Dexheimer V, Negreiros R and Schramm S 2012 *Eur. Phys. J. A* **48** 189
[24] Franzon B, Gomes R O and Schramm S 2016 *Mon. Not. Roy. Astron. Soc.* **463** 571-9
[25] Gomes R O, Dexheimer V, Schramm S and Vasconcellos C A Z 2015 *ApJ* **808** 8
[26] Dover C B and Gal A 1985 *Prog. Part. Nucle. Phys.* **12** 171-239
[27] Schaffner J, Dover C B, Gal A, Greiner C, Millener D J and Stoecker H 1994 *Annals Phys.* **235** 35-76
[28] Strickland M, Dexheimer V and Menezes D P 2012 *Phys. Rev. D* **86** 125032
[29] Negreiros R, Schramm S and Weber F 2013 *Phys. Lett. B* **718** 1176

IOP Conf. Series: Journal of Physics: Conf. Series **861** (2017) 012017 doi:10.1088/1742-6596/861/1/012017

Meson physics in asymmetric matter

Andrea Mammarella[1] **and Massimo Mannarelli**

Laboratori Nazionali del Gran Sasso, via G. Acitelli 22, Assergi (AQ), Italy

E-mail: [1]andrea.mammarella@lngs.infn.it

Abstract. This paper describes dynamic and thermodynamic (at $T = 0$) properties of mesons in asymmetric matter in the framework of Chiral Perturbation Theory. We consider the effect of nonzero isospin and strangeness chemical potentials on a mesonic system and report on the corresponding phase diagram. We also study meson masses and mixing in the resulting normal phase, pion condensation phase and kaon condensation phase. We find differences with previous papers regarding meson masses and mixing in the condensed phases; the results presented here are supported by theory group analysis and direct calculations. Pressure, density and equation of state of the system at $T = 0$ and nonzero μ_I are calculated, finding remarkable agreement with analogue studies performed by lattice calculations.

1. Introduction

Understanding the QCD phase diagram is important for many different phenomena, for example the astrophysics of compact stars or heavy-ion collisions. It is known that depending on the value of the isospin chemical potential, μ_I, and on the value of the strangeness chemical potential, μ_S, three different phases can be realized: the normal phase, the pion condensed (πc) phase and the kaon condensed (Kc) phase [1, 2, 3]. The realization of a mesonic condensate can drastically change the low energy properties of matter, including the mass spectrum and the lifetime of mesons.

Previous analysis of the meson condensed phases by QCD-like theories were developed in [4, 5]. Pion condensation in two-flavor quark matter was studied in [2, 6] and in three-flavor quark matter in [3]. In particular, the phase diagram as a function of μ_I and μ_S was presented in [3]. Finite temperature effects in $SU(2)_L \times SU(2)_R$ chiral perturbation theory (χPT) have been studied in [7, 8, 9]. One remarkable property of quark matter with nonvanishing isospin chemical potential is that it is characterized by a real measure, thus the lattice realization can be performed with standard numerical algorithms [10, 11]. The πc and the Kc phases have been studied by NJL models in [12, 13, 14] and by random matrix models in [15]. All these models find results in qualitative and quantitative agreement; in particular, the phase diagram of matter has been firmly established. We have studied the masses and mixing of mesonic states and we have found results that disagree with [3] that, up to our knowledge, is the only paper in which they are studied.
The thermodynamic properties of the condensed phases have been studied by LQCD simulations in [16, 17]. Previously, various results on the πc phase were derived in [9] by an NJL model. In particular, the equation of state (EoS) of the NJL model was presented. Recently in [18, 19, 20] a perturbative analysis of QCD at large isospin density has been presented. Those pQCD results

are consistent with LQCD for $\mu_I \gtrsim 3m_\pi$ [20], where m_π is the pion mass. At smaller values of μ_I it seems that pQCD underestimates the energy density and it is not able to capture the condensation mechanism. However, for small values of μ_I (and μ_S), χPT can be used.

In this article we briefly review how to include chemical potentials in χPT [21, 22, 23, 24, 25], then we describe some phenomenological properties related to this inclusion, like the existence of the different phases already listed and the changes in meson masses and mixing. We will also show how the inclusion of chemical potentials affects meson properties [26]. Finally we will study the effect of nonzero isospin and strangeness chemical potentials on the pressure, density and equation of state of the system [27].

The paper is organized as follows. In Sec. 2 we describe the model and how it predicts different phases. In Sec. 3 we show how group theory tools can be used to calculate the mass eigenstates and the meson mixing in the condensed phases and we list and discuss the results obtained. In sec 4 we study the termodynamic properties of the system. Finally, in Sec. 5, we summarize the results.

2. Model

2.1. Lagrangian and definitions

In this section we briefly review the model that we are going to use in the following. It is the one described in [3]. It is a chiral effective Lagrangian at the lowest order in the momenta. The general $\mathcal{O}(p^2)$ Lorentz invariant Lagrangian density describing the pseudoscalar mesons can be written as [21]

$$\mathcal{L} = \frac{F_0^2}{4}\mathrm{Tr}(D_\nu \Sigma D^\nu \Sigma^\dagger) + \frac{F_0^2}{4}\mathrm{Tr}(X\Sigma^\dagger + \Sigma X^\dagger)\,, \tag{1}$$

where Σ describes the meson fields, $X = 2B_0(s + ip)$ represents scalar and pseudoscalar external fields and the covariant derivative is defined as

$$D_\mu \Sigma = \partial_\mu \Sigma - \frac{i}{2}[v_\mu, \Sigma] - \frac{i}{2}\{a_\mu, \Sigma\}\,, \tag{2}$$

where v_μ and a_μ are vectorial and axial external currents, respectively. The Lagrangian has two free parameters F_0 and B_0. Because this is an effective theory, they have to be fixed by mesons phenomenology: they are related to the pion decay and to the quark-antiquark condensate, respectively. See for example [21, 22, 23, 24, 25].

If the meson field transforms as

$$\Sigma \to R\Sigma L^\dagger\,, \tag{3}$$

the Lagrangian density possess the symmetry $SU(N_f)_L \times SU(N_f)_R$ and the chiral symmetry breaking corresponds to the spontaneous global symmetry breaking $SU(N_f)_L \times SU(N_f)_R \to SU(N_f)_{L+R}$. In standard χPT, the mass eigenstates are charge eigenstates as well. This ensures that mesons are particles with a well defined mass and charge. If the system is in a medium the ground state can be tilted to different vacua that can also be charged with respect to some of the $SU(3)$ generators. This would imply that the eigenstates will not be eigenstates of the broken charges. The effects of a medium can be taken into account by considering appropriate external currents in Eq. (1).

At vanishing temperature the ground state is determined by maximizing the Lagrangian density with respect to the external currents. The pseudoscalar mesons are then described as oscillations around the vacuum. We use the same nonlinear representation of [3] corresponding to

$$\Sigma = u\bar{\Sigma}u \qquad \text{with} \qquad u = e^{iT\cdot\phi/2}\,, \tag{4}$$

where T_a are the $SU(N_f)$ generators and $\bar{\Sigma}$ is a generic $SU(N_f)$ matrix to be determined by maximizing the static Lagrangian. The reasoning behind the above expression is that under

$SU(N_f)_L \times SU(N_f)_R$ mesons can be identified as the fluctuations of the vacuum as in Eq. (3) with $\theta^R_a = -\theta^L_a = \phi_a$.

In the following we will assume that $a_\mu = 0$, $p = 0$, $X = 2GM$, where M is the $N_f \times N_f$ diagonal quark mass matrix and G is a constant, that with these conventions is equal to B_0. Moreover, we will assume that

$$v^\nu = 2\mu\delta^{\nu 0} \tag{5}$$

meaning that the vectorial current consists of the quark chemical potential, with μ a $SU(N_f)$ matrix in flavor space. Its explicit expression in our case of interest, ie $SU(3)$, is:

$$\mu = \text{diag}\,(\mu_u, \mu_d, \mu_s) = \text{diag}\left(\frac{1}{3}\mu_B + \frac{1}{2}\mu_I, \frac{1}{3}\mu_B - \frac{1}{2}\mu_I, \frac{1}{3}\mu_B - \mu_S\right) = \frac{\mu_B - \mu_S}{3} I + \frac{\mu_I}{2}\lambda_3 + \frac{\mu_S}{\sqrt{3}}\lambda_8\,, \tag{6}$$

It is important to remark that this model only holds for $|\mu_B| \lesssim 940$ MeV, $|\mu_I| \lesssim 770$ MeV and $|\mu_S| \lesssim 550$ MeV [3].

2.2. Ground state and different phases

To find the ground state we have to substitute (4) in the Lagrangian (1). It is not necessary to use a complete SU(3) parametrization for $\bar{\Sigma}$, but it is sufficient:

$$\bar{\Sigma} = \begin{pmatrix} 1 & 0 & 0 \\ 0 & \cos\beta & -\sin\beta \\ 0 & \sin\beta & \cos\beta \end{pmatrix} \begin{pmatrix} \cos\alpha & \sin\alpha & 0 \\ -\sin\alpha & \cos\alpha & 0 \\ 0 & 0 & 1 \end{pmatrix} \begin{pmatrix} 1 & 0 & 0 \\ 0 & \cos\beta & \sin\beta \\ 0 & -\sin\beta & \cos\beta \end{pmatrix}, \tag{7}$$

because $\bar{\Sigma}$ has to be orthogonal to the chemical potential in the SU(3) generator space [26]. Substituting (7) in (1) and maximizing, we find three different vacua, thus there are three ground states, each one describing a different phase [3]:

- Normal phase:

$$\mu_I < m_\pi \qquad \mu_S < m_K - \frac{1}{2}\mu_I\,, \tag{8}$$

  characterized by

$$\alpha_N = 0, \quad \beta_N \in (0, \pi), \quad \bar{\Sigma}_N = \text{diag}(1, 1, 1)\,. \tag{9}$$

- Pion condensation phase:

$$\mu_I > m_\pi \qquad \mu_S < \frac{-m_\pi^2 + \sqrt{(m_\pi^2 - \mu_I^2)^2 + 4m_K^2\mu_I^2}}{2\mu_I}\,, \tag{10}$$

  characterized by $\cos\alpha_\pi = \left(\frac{m_\pi}{\mu_I}\right)^2$, $\beta_\pi = 0$,

$$\bar{\Sigma}_\pi = \begin{pmatrix} \cos\alpha_\pi & \sin\alpha_\pi & 0 \\ -\sin\alpha_\pi & \cos\alpha_\pi & 0 \\ 0 & 0 & 1 \end{pmatrix} = \frac{1 + 2\cos\alpha_\pi}{3} I + i\lambda_2 \sin\alpha_\pi + \frac{\cos\alpha_\pi - 1}{\sqrt{3}}\lambda_8\,.$$

- Kaon condensation phase:

$$\mu_S > m_K - \frac{1}{2}\mu_I \qquad \mu_S > \frac{-m_\pi^2 + \sqrt{(m_\pi^2 - \mu_I^2)^2 + 4m_K^2\mu_I^2}}{2\mu_I}\,, \tag{11}$$

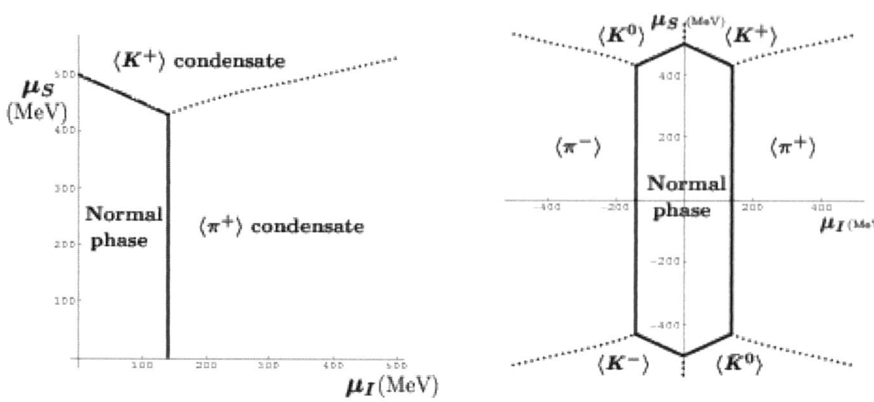

Figure 1. Phase diagram of our model in presence of nonzero isospin and strangeness chemical potentials and at $T = 0$ [3].

characterized by $\cos\alpha_K = \left(\frac{m_K}{\frac{1}{2}\mu_I+\mu_S}\right)^2, \beta_K = \pi/2,$

$$\bar{\Sigma}_K = \begin{pmatrix} \cos\alpha & 0 & \sin\alpha \\ 0 & 1 & 0 \\ -\sin\alpha & 0 & \cos\alpha \end{pmatrix} = \frac{1+2\cos\alpha_K}{3} I + \frac{\cos\alpha_K - 1}{2\sqrt{3}}\left(\sqrt{3}\lambda_3 - \lambda_8\right) + i\lambda_5 \sin\alpha_K \,.$$

Note that the kaon condensation can only happen for

$$\mu_S > \bar{\mu}_S = m_K - \frac{m_\pi}{2} \,. \tag{12}$$

These results are summarized in figure 1 [3].

3. Meson Masses and Mixing

3.1. Mixing

As we have seen in Sec. 2.2 the ground state in the condensed phases is not diagonal and thus has SU(3) charges that cause symmetry breaking that will lead to mix mesons. Studying the breaking pattern can guide us to identify the allowed mixing in the condensed phases.

The starting Lagrangian has an $SU(3)_L \times SU(3)_R$ symmetry, broken to $SU(3)_V$ by the quark masses. The introduction of chemical potentials via the external current (5) further reduces this symmetry to $U(1)_{L+R} \times U(1)_{L+R}$, meaning that the λ_3 and λ_8 terms in (5) break isospin and hypercharge conservation. When the system enters one of the condensed phases the vacuum acquires a charge and thus there is no symmetry left.

In all of these cases we can use the quantum numbers of the $SU(2)$ subgroups of $SU(3)$ to label the states and to find out which ones can mix. We only need two of them, because they are not independent. The states that can mix and the related quantum numbers are shown in table 1

Unfortunately, this is not sufficient to determine if π_0 and η can mix, because they do not have a well defined U and V spin, so we have to study deeply how the ground state affects them. Let us first consider the normal phase. In the normal phase there is no operator that can induce the mixing of the mesonic states, thus they remain unchanged but the Q_3 and Q_8 charges will induce Zeeman-like mass splittings.

In any of the condensed phases, there is an additional charge that is spontaneously induced, and the corresponding operator will lead to mixing.

Mixing states	(T, U)
π_+, π_-	$(1, 1/2)$
K_+, K_-	$(1/2, 1/2)$
$K_0, \bar{K}_0$	$(1/2, 1)$

Table 1. Mixing mesons with the corresponding T-spin and U-spin quantum numbers. These quantum numbers label the $SU(3)$ subspace spanned by the corresponding mesonic states. The π_0 and the η do not appear because they are not U-spin eigenstates.

Let us first focus on isospin (or T-spin). We have to consider two cases. Suppose that the vacuum has a charge that commutes with T^2, as in the πc phase, say the charge corresponding to $T_2 = i(T_- - T_+)$, see Eq. (11). The $T_\pm$ operators can induce mixing among the charged pions and among the kaons. On the other hand, T-spin conservation does not allow the $|\pi_0\rangle = |T = 1, T_3 = 0\rangle$ to mix with the $|\eta\rangle = |T = 0, T_3 = 0\rangle$.

Now suppose instead that the vacuum has a charge that does not commute with T^2 as in the Kc phase. Any operator that does not commute with isospin will commute with U-spin or with V-spin. In the Kc phase $Q_5|0\rangle \neq 0$, then the vacuum is not invariant under this charge. However, since $[T_5, U] = 0$ it follows that U-spin is conserved. The lowering and raising operators inducing the mixing will be $U_\pm$. Regarding the π_0 and the η, in this case we have that $|U = 1, U_3 = 0\rangle$ and $|U = 0, U_3 = 0\rangle$ do not mix. Since $|U = 1, U_3 = 0\rangle = \frac{|\pi_0\rangle + \sqrt{3}|\eta\rangle}{2}$ and $|U = 0, U_3 = 0\rangle = \frac{\sqrt{3}|\pi_0\rangle - |\eta\rangle}{2}$, these will be the mass eigenstates.

3.2. Masses

The mass eigenstates are found diagonalizing the Lagrangian in the different phases. They present the mixing predicted by the group theory analysis of the previous subsection. Masses and mixings have been calculated [26]. The procedure is straightforward: we have to diagonalize the quadratic part of the Lagrangian by solving the related secular equation, obtaining the masses and subsequently the eigenstates. The resulting expressions are quite involved so we only show plots of the masses μ_I dependence for different values of μ_S in figure 2.

4. Thermodynamic

The pressure of the system can be found substituting $\bar{\Sigma} \to \Sigma$ in 1. In the different phases we get:

$$p^N_{\rm LO} = 0 \qquad p^{\pi c}_{\rm LO} = \frac{F_0^2 \mu_I^2}{2}\left(1 - \frac{m_\pi^2}{\mu_I^2}\right)^2 \qquad p^{Kc}_{\rm LO} = \frac{F_0^2 \mu_K^2}{2}\left(1 - \frac{m_K^2}{\mu_K^2}\right)^2$$

with $\mu_K = \mu_I/2 + \mu_S$. It is important to observe that the expressions for the pressure in the condensed phases are exactly the same if we change $\mu_I \to \mu_K$ and $m_\pi \to m_K$. Number density and equation of state can be derived using:

$$n_{i,\rm LO} = \frac{\partial p_{\rm LO}}{\partial \mu_i} \qquad \text{and} \qquad \epsilon = \mu_I n_I + \mu_S n_S - p\,, \tag{13}$$

where the index i represents I=isospin or K=kaon. In the πC phase we obtain:

$$n^{\pi C}_{I,\rm LO} = f_\pi^2 \mu_I \left(1 - \frac{m_\pi^4}{\mu_I^4}\right) \qquad \text{and} \qquad \epsilon^{\pi c}_{\rm LO} = \frac{f_\pi^2 \mu_I^2}{2}\left(1 + 2\frac{m_\pi^2}{\mu_I^2} - 3\frac{m_\pi^4}{\mu_I^4}\right) \tag{14}$$

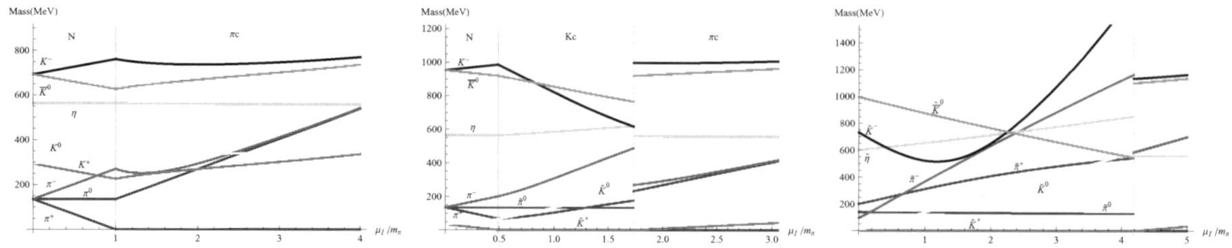

Figure 2. Mass spectrum of the pseudoscalar mesonic octet. Left panel: results obtained for $\mu_S = 200$ MeV. The vertical solid line represents the second order phase transition between the normal phase and the pion condensation phase. In this case the strange quark chemical potential is below the threshold value for kaon condensation, 425 MeV, thus the kaon condensed phase does not take place for any value of μ_I. Central panel: results obtained for $\mu_S = 460$ MeV. The vertical solid line represents the second order phase transition from the normal phase to the kaon condensation phase. The dashed line corresponds to the first order phase transition between the kaon condensed phase to the pion condensed phase. Right panel: results obtained for $\mu_S = 550$ MeV, corresponding to the largest value of μ_S.

while the analogous results in the kaon condensation phase can be obtained substituting $m_\pi \to m_K$ and $\mu_I \to \mu_K$, see [17, 27]. To compare the χPT energy density with the results obtained by LQCD simulations in [16] and by pQCD in [20], we divide it by the Stefan-Boltzmann limit, which has been defined in [16] as $\epsilon_{SB} = 9\mu_I^4/(4\pi^2)$ and describes a cold, degenerate gas of weakly interacting quarks which is the limit of our system if there was no interactions. In Fig. 3 we report our ratio $\epsilon_{\rm LO}^{\pi c}/\epsilon_{\rm SB}$ and the results of [16] and [20]. We immediately notice that the χPT curve perfectly captures the peak structure at low μ_I, while it begins to depart from the LQCD results after $\mu_I \sim 2m_\pi$, indicating the breakdown of the LO approximation. An interesting result is that within our framework we obtain an analytic expression for the position of the peak, which for the πc phase is given by $\mu_I^{\rm peak} = \left(\sqrt{13}-2\right)^{1/2} m_\pi \simeq 1.276\, m_\pi$ and is independent of f_π. This result is very close to the LQCD result obtained in [16], where the values $\mu_I^{\rm peak} = \{1.20, 1.25, 1.275\} m_\pi$ have been obtained considering different spatial volumes L^3 with side $L = \{16, 20, 24\}$, respectively. The continuum-linearly-extrapolated value for the peak position is $\mu_I^{\rm peak} = 1.30\ m_\pi$.

It is not clear if the peak point is related to some undergoing physics. In our model the pion condensate has reached a significant value at that point, almost approaching its asymptotic limit B, see Fig. 4, so we tend to exclude that the peak appearance is related to the occurrence of the condensation.
More specifically, the pion condensate can be evaluated by introducing external source terms in the chiral Lagrangian and by differentiating with respect to them. After the phase transition it grows as $\langle \pi^+ \rangle = B\sqrt{1 - m_\pi^4/\mu_I^4}$ [3, 5], so that at $\mu_I^{\rm peak}$ it reaches already $\sim 79\%$ of its asymptotic value. We are therefore more inclined to interpret this peak structure as a consequence of the filling of the condensate, rather than of the onset of its formation, which at zero-temperature should occur at $\mu_I = m_\pi$. We also obtain an analytic expression for the ratio at the peak

$$\left.\frac{\epsilon}{\epsilon_{SB}}\right|_{\rm peak} = \frac{4(\sqrt{13}-5)\pi^2}{9(\sqrt{13}-2)}\frac{f_\pi^2}{m_\pi^2}\,, \tag{15}$$

which would give information on the f_π/m_π scaling if precise LQCD data were available. It is worth mentioning that the above results hold in the Kc phase if one considers $\epsilon_{SB} = 9\mu_K^4/(4\pi^2)$.

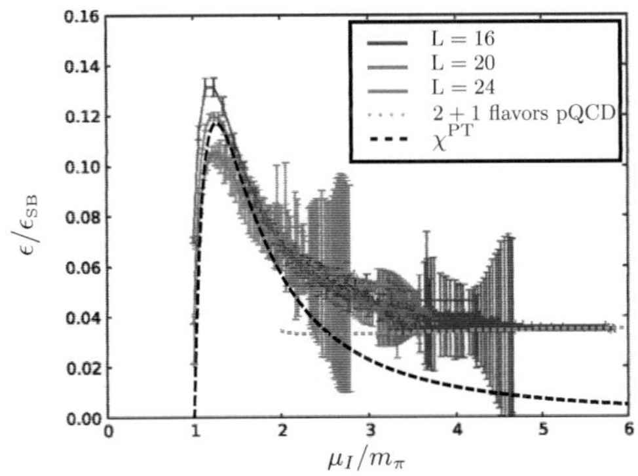

Figure 3. Energy density over the Stefan-Boltzmann limit. The lattice points have been obtained at $T = 20$ MeV; the colors correspond to the different lattice volumes considered in [16]. The orange dotted line corresponds to the pQCD results of [20]. The dashed black line corresponds to our results.

Then, we obtain $\mu_K^{\text{peak}} = (\sqrt{13} - 2)^{1/2} m_K$ and an expression analogous to Eq.(15), with $m_\pi \to m_K$.

From Fig. 3 it seems that for $\mu_I > 2m_\pi$ the LO χPT breaks down, resulting in an underestimate of the energy density. The basic reason is that for large μ_I we have that $\epsilon_{\text{LO}}^{\pi c} \propto f_\pi^2 \mu_I^2$ from Eq. (13), while pQCD correctly predicts $\epsilon^{\pi c} \propto \mu_I^4$. At the next-to-leading order (NLO), it is possible to show that the χPT energy density includes a term proportional to $(2l_1 + 2l_2 + l_3)\mu_I^4$, where l_1, l_2 and l_3 are NLO low energy constants [24]. Comparing our results with the pQCD energy density we obtain

$$2l_1 + 2l_2 + l_3 = \frac{3}{8\pi^2}\left(\frac{\epsilon_{\text{pQCD}}}{\epsilon_{\text{SB}}} - \frac{2 f_\pi^2 \pi^2}{9\mu_I^2} + \mathcal{O}(\mu_I^{-4})\right), \tag{16}$$

leading to $2l_1 + 2l_2 + l_3 \simeq 0.6 \times 10^{-3}$ for $\mu_I = 3m_\pi$, consistent with the empirical values of the low energy costants [24].

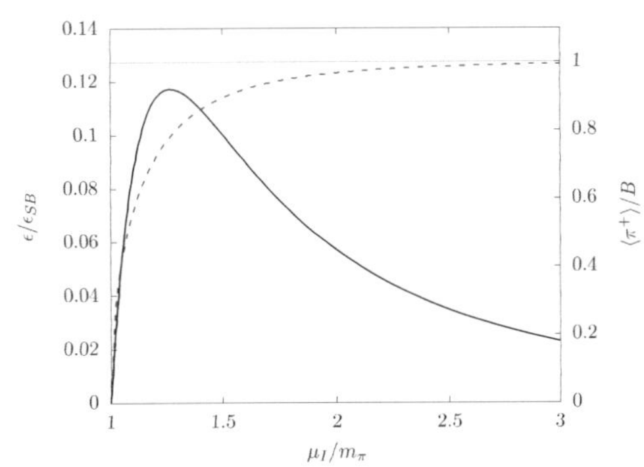

Figure 4. χPT results for the ratio ϵ/ϵ_{SB} (solid line) and the pion condensate normalized to its maximum value B (dashed blue line).

5. Conclusion

We have shown how meson physics in presence of chemical potentials can be described using Chiral Perturbation Theory. There are three different phases: a normal phase, a pion

IOP Conf. Series: Journal of Physics: Conf. Series **861** (2017) 012017 doi:10.1088/1742-6596/861/1/012017

condensation phase and a kaon condensation phase. The condensed phases ground states have a charge and thus can generate mixing among mesons.
In this context we have illustrated how the mixing is influenced by model symmetries and that group theory constrains the mixing possibilities. These results, obtained by group theory alone, are expected to hold in any theory describing meson states. We have then described mesons masses in the three phases, calculating them from the Lagrangian (1). These masses are in perfect agreement with the group theory reasoning.
Finally we have derived the pressure and equation of state of the system at $T = 0$, comparing them with the equivalent results obtained in lattice simulations and obtaining a very good agreement on a non trivial feature like a peak structure in the meson energy density over the Stefan Boltzmann one.
These results can be applied for example in the physics of compact stars, the study of cosmic rays and the study of in medium nuclear decays.

References

[1] Migdal A B, Saperstein E, Troitsky M and Voskresensky D 1990 *Phys.Rept.* **192** 179–437
[2] Son D and Stephanov M A 2001 *Phys.Rev.Lett.* **86** 592–595 (*Preprint* hep-ph/0005225)
[3] Kogut J and Toublan D 2001 *Phys.Rev.* **D64** 034007 (*Preprint* hep-ph/0103271)
[4] Kogut J, Stephanov M A and Toublan D 1999 *Phys.Lett.* **B464** 183–191 (*Preprint* hep-ph/9906346)
[5] Kogut J, Stephanov M A, Toublan D, Verbaarschot J and Zhitnitsky A 2000 *Nucl.Phys.* **B582** 477–513 (*Preprint* hep-ph/0001171)
[6] Son D and Stephanov M A 2001 *Phys.Atom.Nucl.* **64** 834–842 (*Preprint* hep-ph/0011365)
[7] Loewe M and Villavicencio C 2003 *Phys.Rev.* **D67** 074034 (*Preprint* hep-ph/0212275)
[8] Loewe M and Villavicencio C 2004 *Phys.Rev.* **D70** 074005 (*Preprint* hep-ph/0404232)
[9] He L y, Jin M and Zhuang P f 2005 *Phys.Rev.* **D71** 116001 (*Preprint* hep-ph/0503272)
[10] Alford M G, Kapustin A and Wilczek F 1999 *Phys.Rev.* **D59** 054502 (*Preprint* hep-lat/9807039)
[11] Kogut J and Sinclair D 2002 *Phys.Rev.* **D66** 034505 (*Preprint* hep-lat/0202028)
[12] Toublan D and Kogut J 2003 *Phys.Lett.* **B564** 212–216 (*Preprint* hep-ph/0301183)
[13] Barducci A, Casalbuoni R, Pettini G and Ravagli L 2004 *Phys.Rev.* **D69** 096004 (*Preprint* hep-ph/0402104)
[14] Barducci A, Casalbuoni R, Pettini G and Ravagli L 2005 *Phys.Rev.* **D71** 016011 (*Preprint* hep-ph/0410250)
[15] Klein B, Toublan D and Verbaarschot J 2005 *Phys.Rev.* **D72** 015007 (*Preprint* hep-ph/0405180)
[16] Detmold W, Orginos K and Shi Z 2012 *Phys. Rev.* **D86** 054507 (*Preprint* 1205.4224)
[17] Detmold W, Orginos K, Savage M J and Walker-Loud A 2008 *Phys. Rev.* **D78** 054514 (*Preprint* 0807.1856)
[18] Andersen J O, Haque N, Mustafa M G and Strickland M 2016 *Phys. Rev.* **D93** 054045 (*Preprint* 1511.04660)
[19] Cohen T D and Sen S 2015 *Nucl. Phys.* **A942** 39–53 (*Preprint* 1503.00006)
[20] Graf T, Schaffner-Bielich J and Fraga E S 2016 *Phys. Rev.* **D93** 085030 (*Preprint* 1511.09457)
[21] Gasser J and Leutwyler H 1984 *Annals Phys.* **158** 142
[22] Leutwyler H 1994 *Annals Phys.* **235** 165–203 (*Preprint* hep-ph/9311274)
[23] Ecker G 1995 *Prog.Part.Nucl.Phys.* **35** 1–80 (*Preprint* hep-ph/9501357)
[24] Scherer S 2003 *Adv.Nucl.Phys.* **27** 277 (*Preprint* hep-ph/0210398)
[25] Scherer S and Schindler M R 2005 (*Preprint* hep-ph/0505265)
[26] Mammarella A and Mannarelli M 2015 *Phys. Rev.* **D92** 085025 (*Preprint* 1507.02934)
[27] Carignano S, Mammarella A and Mannarelli M 2016 *Phys. Rev.* **D93** 051503 (*Preprint* 1602.01317)

IOP Conf. Series: Journal of Physics: Conf. Series **861** (2017) 012018 doi:10.1088/1742-6596/861/1/012018

Crystalline chiral condensates in compact stars

Stefano Carignano
INFN, Laboratori Nazionali del Gran Sasso, Via G. Acitelli, 22, I-67100 Assergi (AQ), Italy
E-mail: carignano@lngs.infn.it

Abstract. I discuss the phenomenon of inhomogeneous chiral symmetry breaking in dense quark matter, with a particular emphasis on its relevance for the physics of compact stars. After briefly reviewing the formalism employed for the study of crystalline chiral condensates within effective models, I focus on their effects on the equation of state and the possible consequences for mass-radius relations. Finally, I discuss how model extensions which provide a more realistic description of matter inside compact stellar objects affect the formation of these inhomogeneous condensates.

1. Introduction

The study of the properties of quantum chromodynamics (QCD) at finite density is one of the most challenging tasks in contemporary nuclear physics. In spite of our little knowledge, the phase diagram of strong interaction is nevertheless expected to be extremely rich. A particularly interesting phenomenon which might be realized in dense quark matter is the formation of spatially inhomogeneous condensates formed by pairs of fermions carrying a finite momentum. This translates into an explicit spatial dependence of these condensates and the formation of crystalline structures. Prominent examples of inhomogeneous phases in dense quark matter are related to the phenomena of color-superconductivity and chiral symmetry breaking. Inhomogeneous color-superconductors might appear in isospin imbalanced quark matter, where the separation between the quark Fermi surfaces makes the formation of diquark Cooper pairs with finite momentum the energetically favored channel (see [1] for a recent review). Inhomogeneous chiral condensates on the other hand arise in the region between the low-density phase where chiral symmetry is spontaneously broken and the high-density restored one. In the usual picture limited to homogeneous chiral condensates made of quark-antiquark pairs, the presence of a chemical potential induces a stress on the formation of these pairs, eventually leading to chiral restoration. On the other hand, at finite densities the formation of pairs made of quarks and holes at the Fermi surface becomes a competitive condensation channel. Since pairs with large relative momenta are disfavored, the preferred pairing mechanism leads to a condensate with a non-vanishing net total momentum, i.e. a spatially inhomogeneous one (for a more detailed discussion, see e.g. [2]).

Explicit model calculations indeed support this hypothesis by suggesting that an inhomogeneous "island" where the favored structure for the chiral condensate is a spatially modulated one appears at low temperatures and intermediate densities (for a recent review, see [3]). A typical phase diagram resulting from this type of calculations is shown in Fig. 1. The actual extension of this window is still uncertain: some NJL and Dyson-Schwinger studies suggest that, at least at low temperatures, inhomogeneous chiral symmetry breaking might occur

IOP Conf. Series: Journal of Physics: Conf. Series **861** (2017) 012018 doi:10.1088/1742-6596/861/1/012018

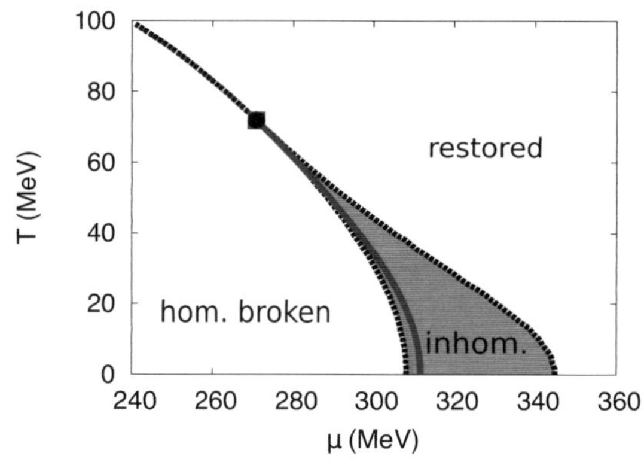

Figure 1. Typical phase diagram in the μ-T plane when allowing for inhomogeneous phases in model calculations. The shaded region indicates the inhomogenous phase, covering the first-order phase boundary (blue solid line) between the homogeneous chirally broken and restored phases.

in a very wide range of densities [4, 5], while other models predict a smaller size. In this sense, Fig. 1 could be seen as a rather conservative estimate.

At any rate, the realization of inhomogeneous chiral symmetry breaking appears to be a rather robust prediction for the region of the phase diagram which is expected to be relevant for the description of compact stars, which provide the only known realization of ultra-dense matter in nature. The formation of crystalline chiral condensates might therefore have significant consequences on the physical properties of these objects, possibly leading to new signatures.

2. Model description of inhomogeneous chiral symmetry breaking

The Nambu–Jona-Lasino (NJL) model provides one of the most commonly employed frameworks for the study of chiral symmetry breaking in dense quark matter. Starting point is the Lagrangian

$$\mathcal{L}_{\rm NJL} = \bar\psi \left(i\gamma^\mu \partial_\mu - m\right)\psi + G\left(\left(\bar\psi\psi\right)^2 + \left(\bar\psi i\gamma^5\tau^a\psi\right)^2\right), \tag{1}$$

where ψ denotes a quark field with bare mass m (which in this work will be taken for simplicity to zero, i.e. the chiral limit), interacting by local four-point vertices proportional to the dimensionful coupling constant G.

In order to analyze the thermodynamic properties at temperature T and quark chemical potential μ one typically calculates the grand potential Ω of the system within a mean-field approximation in the presence of static scalar and pseudoscalar condensates (here for simplicity the possibility of charged pion condensation is neglected):

$$S(\mathbf{x}) = \langle\bar\psi\psi\rangle \quad \text{and} \quad P(\mathbf{x}) = \langle\bar\psi i\gamma^5\tau^3\psi\rangle, \tag{2}$$

If the condensates are allowed to retain an explicit spatial dependence, the derivation of the mean-field thermodynamic potential becomes significantly more difficult than for space-independent condensates. As discussed in detail in [3], one finds

$$\Omega_{\rm MF}(T,\mu;S,P) = \Omega_{kin} + \Omega_{cond}\,, \tag{3}$$

with the condensate part

$$\Omega_{cond} = \frac{1}{V}\int_V d^3x\, G\left(S^2(\mathbf{x}) + P^2(\mathbf{x})\right) \tag{4}$$

IOP Conf. Series: Journal of Physics: Conf. Series **861** (2017) 012018 doi:10.1088/1742-6596/861/1/012018

and the kinetic part

$$\Omega_{kin} = -\frac{1}{V}\sum_{\lambda}\left[\frac{E_\lambda - \mu}{2} + T\log\left(1 + e^{-\frac{E_\lambda-\mu}{T}}\right)\right], \tag{5}$$

where E_λ are the eigenvalues of the effective Dirac Hamiltonian

$$H(\mathbf{x}) = \gamma^0\left[-i\gamma^i\partial_i + m - 2G(S(\mathbf{x}) + i\gamma^5\tau^3P(\mathbf{x}))\right], \tag{6}$$

In contrast to the homogeneous case, the diagonalization of H for arbitrary space-dependent condensate functions is rather difficult and must in general be performed numerically. After that, $\Omega_{\rm MF}$ must be minimized with respect to the functions $S(\mathbf{x})$ and $P(\mathbf{x})$ in order to find the energetically favored ground state. Since this problem has not yet been solved in 3+1 dimensions, one considers simple ansätze for the condensate functions. This reduces the problem to finding the minimum of $\Omega_{\rm MF}$ with respect to a limited number of variational parameters[1].

In this contribution I will focus on two popular one-dimensional ansätze which even allow for an analytical diagonalization of H. Choosing the modulation to be along the z-direction and introducing the complex mass function

$$M(z) = -2G\big(S(z) + iP(z)\big), \tag{7}$$

the most simple (and therefore most popular) one is the so-called (dual) chiral density wave (CDW) [6, 7]

$$M(z) = \Delta e^{iqz}, \tag{8}$$

i.e. a single plane wave with amplitude Δ and wave number q, corresponding to a spatial period of $L = 2\pi/q$.

A more sophisticated ansatz is the "real kink crystal" (RKC), corresponding to an array of domain-wall solitons [8, 9]. The mass function reads

$$M(z) = \Delta\sqrt{\nu}\,\mathrm{sn}(\Delta z|\nu) \tag{9}$$

and depends on two parameters, Δ and ν. Here $\mathrm{sn}(\alpha|\nu)$ denotes a Jacobi elliptic function with elliptic modulus $\nu \in [0,1]$, interpolating between $\tanh(\alpha)$ for $\nu = 1$ and $\sin(\alpha)$ for $\nu = 0$. The amplitude of the modulation is then given by $\Delta\sqrt{\nu}$ and its spatial period by $L = 4\mathbf{K}(\nu)/\Delta$, where $\mathbf{K}(\nu)$ is a complete elliptic integral of the first kind. In contrast to the CDW, the RKC is real (hence the name), i.e. the pseudoscalar condensate vanishes identically.

The resulting phase diagrams for the two modulations look qualitatively similar to Fig. 1, although one finds that the transition from the homogeneous broken to the inhomogeneous phase is first-order for the CDW while it is second-order for the RKC [3, 10]. Furthermore, the CDW is found to be disfavored against the RKC [9].

It is possible to calculate spatially averaged quark-number densities $\bar{n} = -\partial\Omega/\partial\mu$ for these two ansätze (see [11] for a more detailed discussion). For the RKC ansatz, the inhomogeneous phase starts essentially at $\bar{n} = 0$, corresponding to infinitely separated domain-wall solitons [12, 13]. For the CDW solution, on the other hand, the minimum density is around 1.4 n_0, where $n_0 \approx 0.5$ fm^{-3} is the quark number density in symmetric nuclear matter at saturation. In both cases the upper density of the inhomogeneous island reaches about 2.2 n_0. Keeping in mind that these numbers are model and parameter dependent and could also be higher, this lies in the ballpark of densities expected in a possible quark phase in compact stars.

[1] Most parametrizations used in literature correspond to one-dimensional spatial modulations. Two- and higher-dimensional modulations have been studied as well, but were found to be disfavored against one-dimensional ones, at least within the mean-field approximation [14, 15]

IOP Conf. Series: Journal of Physics: Conf. Series **861** (2017) 012018 doi:10.1088/1742-6596/861/1/012018

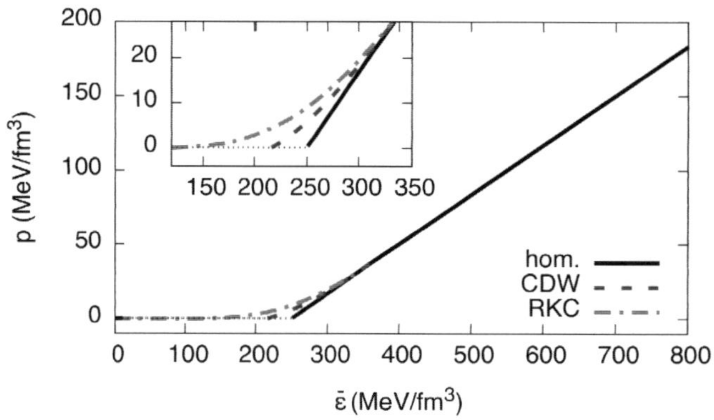

Figure 2. Equation of state (pressure p vs. spatially averaged energy density $\bar{\varepsilon}$) at $T = 0$ for homogeneous matter (solid) and in the inhomogeneous phase with CDW (dashed) or RKC (dash-dotted) modulations. All equations of state begin at $p = \bar{\varepsilon} = 0$ in the homogeneously broken phase. The dotted horizontal lines indicate discontinuities related to first-order transitions to the restored phase or to the CDW solution. The inset shows an enlarged detail of the transition region.

Knowing the density and the pressure for these solutions, it is possible to calculate the corresponding equations of state at $T = 0$, i.e. the pressure as a function of the spatially averaged energy density $\bar{\varepsilon} = -p + \mu\bar{n}$. They are shown in Fig. 2. In general, the inhomogeneous equations of state and in particular the RKC one are stiffer than the homogeneous one, although the differences are rather small. One can therefore anticipate that the effect of inhomogeneous phases on the mass-radius relation of compact stars will be negligible. This has been checked by solving the Tolman-Oppenheimer-Volkoff (TOV) equation using these equations of state of the system as an input [11]. The results of this calculation are shown in Fig. 3. In this context it is important to recall that the aim here is not to build a realistic $M(R)$ sequence which can reach the $2M_\odot$ ($M_\odot$ being the solar mass) observed values[2] [16, 17], but rather to examine the influence of inhomogeneous chiral condensation on the mass-radius relations. For this, only results for pure non self-bound quark stars with different crystalline structures are presented, neglecting effects like charge neutrality or the presence of magnetic fields. As expected, the presence of inhomogeneous condensates in the core of the stars makes very little difference on the resulting mass-radius sequences (except for the low-mass region which should become irrelevant once a more realistic description of the hadronic phase is implemented). It looks therefore that it would be impossible to infer the existence of an inhomogeneous phase on the basis of a mass-radius measurement, even if it was very precise.

3. Effects of model extensions on inhomogeneous chiral symmetry breaking

In order to provide a realistic description of matter inside compact stars, several extensions can be implemented on top of the simple NJL model considered in the previous section. Their influence on inhomogeneous chiral symmetry breaking are briefly described in the following. For simplicity, only the two light quark flavors will be considered.

[2] Such an analysis has been performed e.g. in [18].

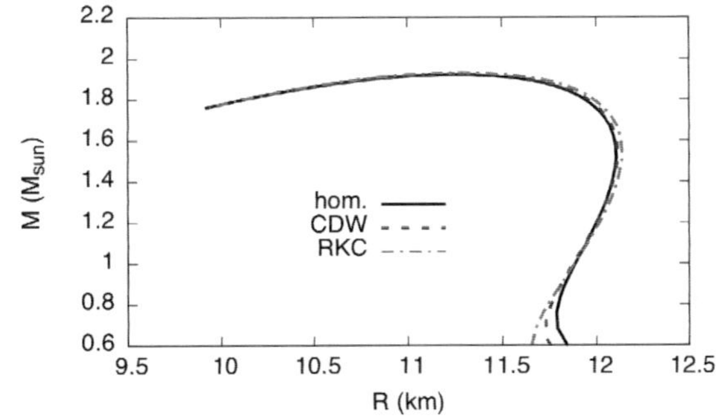

Figure 3. Mass-radius sequences for a pure quark star, neglecting electric neutrality and magnetic fields, for a RKC and CDW modulations as well as for homogeneous matter. The differences between the three curves turn out to be minimal.

3.1. Electric neutrality and isospin asymmetry

First of all, when describing compact stars it is necessary to take into account weak decays and ensure global electric neutrality. In order to describe this situation, electrons are included in the model[3] and the system is then characterized by two conserved global charges: the net quark number and the total electric charge. The thermodynamic potential Ω then depends on two independent chemical potentials, the quark number chemical potential μ and the electric charge chemical potential μ_Q. The chemical potentials of the individual particle species are given as linear combinations of μ and μ_Q, according to their quantum numbers, i.e.

$$\mu_u = \mu + \frac{2}{3}\mu_Q\,, \qquad \mu_d = \mu_s = \mu - \frac{1}{3}\mu_Q\,, \qquad \mu_\ell = -\mu_Q\,. \tag{10}$$

The condition for (spatially averaged) electric neutrality is then given by $\bar{n}_Q = -\partial\Omega/\partial\mu_Q{=}0$. In the quark sector, μ_Q is equivalent to an isospin chemical potential $\mu_I \equiv \mu_u - \mu_d = \mu_Q$. The treatment of isospin-asymmetric matter within the NJL model is not entirely trivial, as it turns out that the exact isospin structure of the interaction becomes relevant in this case. The $\mathcal{L}_{\rm NJL}$ of Eq. (1) can be viewed as a special case of the more general Lagrangian [19, 20]

$$\begin{aligned}\mathcal{L} = \;& \bar\psi(i\gamma^\mu\partial_\mu - \hat m)\psi \\ &+ G_1\Big((\bar\psi\psi)^2 + (\bar\psi\tau^a\psi)^2 + (\bar\psi i\gamma_5\psi)^2 + (\bar\psi i\gamma_5\tau^a\psi)^2\Big) \\ &+ G_2\Big((\bar\psi\psi)^2 - (\bar\psi\tau^a\psi)^2 - (\bar\psi i\gamma_5\psi)^2 + (\bar\psi i\gamma_5\tau^a\psi)^2\Big)\,,\end{aligned} \tag{11}$$

with the current quark matrix $\hat m = \mathrm{diag}_f(m_u, m_d)$ and two $SU(2)_L \times SU(2)_R \times U_V(1)$ symmetric interaction terms, proportional to the coupling constants G_1 and G_2, respectively. The first term respects an additional $U_A(1)$ symmetry, which is explicitly broken by the second. A detailed study of inhomogeneous phases in this framework is in progress and will be presented soon [21].

Finally, instead of considering a homogeneous leptonic background and imposing the neutrality condition for the spatially averaged charge density, one could in principle do better and allow for an inhomogeneous lepton density as well. While the density gradient would increase the kinetic energy of the electrons, it could lower the local Coulomb energy, and the optimal shape would result from balancing these two contributions. This possibility will also be analyzed in the near future [22].

[3] Neutrinos can freely leave the star, so that it is sufficient to take into account charged leptons.

3.2. Magnetic fields

Another very important element characterizing compact stars is the presence of strong magnetic fields. For magnetars, surface fields of up to $10^{15}G$ have been measured, and even higher values are expected to be reached in their cores. Aside from the well known effects on the usual homogeneous chiral symmetry breaking mechanism, such as the phenomenon of magnetic catalysis (see [23, 24] for recent reviews), the presence of strong magnetic fields has a significant influence on the formation and the properties of inhomogeneous phases. For a CDW-type ansatz (Eq. (8)) in the presence of a static background magnetic field H pointing in the z direction, the quark energies for each flavor have been calculated in [25] and are given by

$$E_n^f = \pm \begin{cases} \sqrt{\Delta^2 + p_z^2} + q, & n = 0 \\ \sqrt{\left(\pm\sqrt{\Delta^2 + p_z^2} + q\right)^2 + 2|e_f H| n} & n > 0 \,. \end{cases} \tag{12}$$

By inspecting this expression one can see that the quark momenta transverse with respect to the direction of H become quantized into $p_\perp^2 \sim 2n|e_f H|$, with the integer n labelling the so-called Landau levels. In turn, this implies that at the lowest Landau level (LLL), for which $n = 0$, the problem becomes effectively 1+1-dimensional, a configuration in which inhomogeneous chiral condensation is always favored [26, 27]. In particular, as a consequence of the spectral asymmetry at the LLL, at low temperatures the CDW becomes the thermodynamically favored state over homogeneous matter for any $\mu > 0$. Furthermore, it has been argued that due to the presence of magnetic fields new types of inhomogeneous condensates characterized by complex order parameters can become favored as well [28].

In summary, the presence of a magnetic field qualitatively alters the model phase structure at finite density, significantly enhancing the size of the inhomogeneous phase. Of course, these effects will be relatively small until the strength of the magnetic field approaches the scales given by the quark chemical potential (see e.g. [25, 11] for a discussion), which are typically around $H \gtrsim 10^{17}$G.

3.3. Vector interactions

Finally, in compact stars, where matter is expected to reach several times nuclear-matter density, vector-interaction effects could be particularly important. The effect of vector interactions on inhomogeneous matter has been investigated in [12]. To this end the NJL Lagrangian (Eq. (1)) was extended by the term

$$\mathcal{L}_V = -G_V(\bar{\psi}\gamma^\mu\psi)^2 \,, \tag{13}$$

where G_V is the vector coupling constant.

Within the mean-field approximation, only the temporal part $V^0 = \langle\bar{\psi}\gamma^0\psi\rangle$ of the vector condensate, which is identical to the quark number density n, was retained. The effect of this field can then be absorbed in a shift of the chemical potential[4]: $\mu \to \tilde{\mu} = \mu - 2G_V n$.

For $G_V = 0$, as seen in Fig. 1, the inhomogeneous phase covers the first-order phase boundary between the homogeneous broken and restored phases, and ends at the critical point [9, 29]. Since vector interactions weaken or even remove this homogeneous first-order phase transition, it was expected that the inhomogeneous phase becomes smaller or disappears as well when G_V is increased. It turned out, however, that the inhomogeneous phase becomes larger instead, keeping its extension in temperature and enhancing its size in the chemical potential direction [12].

[4] When dealing with inhomogeneous condensates, the quark density is in general spatially modulated, implying that the shifted chemical potential also depends on the position. For a CDW this is not the case, as the density is constant, whereas it is for the RKC modulation. Nevertheless, in [12] the shift in the chemical potential was implemented only through the spatial average of the vector condensate, an approximation which is expected to be rather inaccurate close to the onset of the inhomogeneous island, but otherwise reasonably valid, particularly when approaching the restored phase.

4. Conclusions

I discussed some properties of the phenomenon of inhomogeneous chiral symmetry breaking, focusing on some of its implications for the physics of compact stars. Several models suggest that spatially modulated structures of the chiral order parameters become energetically favored in a window of the phase diagram which is relevant for cold and dense stellar matter. Furthermore, model extensions providing a more realistic description of matter inside compact stars, such as the inclusion of background magnetic fields or vector interactions have been shown to enlarge the inhomogeneous window significantly, suggesting that if quark matter is present at all inside these objects, it is likely in a crystalline phase.

The interplay of these crystalline chiral condensates with color superconductivity, on the other hand, is not yet completely clear. Preliminary NJL model studies have shown that inhomogeneous chiral condensates may coexist with a homogeneous diquark condensate if the diquark coupling is small, but become suppressed for larger couplings [30, 31]. This could change, however, if electric neutrality is imposed, since it strongly disfavors the BCS pairing of up and down quarks. As a possible result, the diquark condensates could become spatially modulated as well or be suppressed completely. In particular, the realization of a coexistence phase where both color-superconducting and chiral condensates are inhomogeneous is a fascinating possibility, which might be worth investigating.

The determination of clear signatures relating macroscopic observables with the microscopic properties of matter within stellar objects is obviously a very challenging task. While the influence of inhomogeneous condensates on mass-radius relations turned out to be marginal, more promising signatures can be expected from observables related to transport properties, since they are directly influenced by the crystalline structure of the matter and the corresponding low-energy excitations (phonons). Some of these signatures, like gravitational waves and glitches, have been investigated for crystalline color superconductors (see e.g. [1]), but not yet for the specific case of inhomogeneous chiral condensates[5].

References

[1] R. Anglani et al., Rev. Mod. Phys. **86**, 509 (2014), arXiv:1302.4264
[2] T. Kojo, Y. Hidaka, L. McLerran, R. Pisarski, Nucl. Phys. A **843**, 37 (2010), arXiv:0912.3800
[3] M. Buballa, S. Carignano, Prog. Part. Nucl. Phys. **81**, 39 (2015), arXiv:1406.1367
[4] S. Carignano, M. Buballa, Acta Phys. Polon. Supp. **5**, 641 (2012), arXiv:1111.4400
[5] D. Müller, M. Buballa, J. Wambach, Phys. Lett. B **727**, 240 (2013), arXiv:1308.4303
[6] W. Broniowski, A. Kotlorz, M. Kutschera, Acta Phys. Polon. B **22**, 145 (1991)
[7] E. Nakano, T. Tatsumi, Phys. Rev. D **71**, 114006 (2005)
[8] O. Schnetz, M. Thies, K. Urlichs, Annals Phys. **314**, 425 (2004)
[9] D. Nickel, Phys. Rev. D **80**, 074025 (2009), arXiv:0906.5295
[10] S. Carignano, Proceedings TNPI 2016, arXiv:1702.03547
[11] M. Buballa, S. Carignano, Eur.Phys.J. A **52**, 57 (2016) arXiv:1508.04361
[12] S. Carignano, D. Nickel, M. Buballa, Phys. Rev. D **82**, 054009 (2010), arXiv:1007.1397
[13] M. Buballa, S. Carignano, Phys. Rev. D **87**, 054004 (2013), arXiv:1210.7155
[14] H. Abuki, D. Ishibashi, K. Suzuki, Phys.Rev. D **85**, 074002 (2012), arXiv:1109.1615
[15] S. Carignano, M. Buballa, Phys. Rev. D **86**, 074018 (2012), arXiv:1203.5343
[16] P. Demorest, T. Pennucci, S. Ransom, M. Roberts, J. Hessels, Nature **467**, 1081 (2010), arXiv:1010.5788
[17] J. Antoniadis et al., Science **340**, 6131 (2013), arXiv:1304.6875
[18] S. Carignano, E.J. Ferrer, V. de la Incera, L. Paulucci, Phys.Rev. D **92**, 105018 (2015), arXiv:1505.05094
[19] M. Asakawa, K. Yazaki, Nucl. Phys. A **504**, 668 (1989)
[20] M. Frank, M. Buballa, M. Oertel, Phys. Lett. B **562**, 221 (2003), arXiv:hep-ph/0303109
[21] D. Nowakowski, M. Buballa, S. Carignano, In preparation
[22] S. Carignano, M. Mannarelli, In preparation
[23] V.A. Miransky, I.A. Shovkovy, Phys. Rept. **576**, 1 (2015), arXiv:1503.00732
[24] J.O. Andersen, W.R. Naylor, A. Tranberg (2014), arXiv:1411.7176

[5] An exception is the neutrino emissivity, which has been worked out for a CDW in [32].

[25] I. Frolov, V. Zhukovsky, K. Klimenko, Phys.Rev. D **82**, 076002 (2010), arXiv:1007.2984
[26] V. Schön, M. Thies, Phys. Rev. D **62**, 096002 (2000), arXiv:hep-th/0003195
[27] G. Basar, G.V. Dunne, M. Thies, Phys. Rev. D **79**, 105012 (2009)
[28] K. Nishiyama, S. Karasawa, T. Tatsumi, Phys. Rev. D **92**, 036008 (2015), arXiv:1505.01928
[29] D. Nickel, Phys. Rev. Lett. **103**, 072301 (2009), arXiv:0902.1778
[30] M. Sadzikowski, Phys. Lett. B **553**, 45 (2003), arXiv:hep-ph/0210065
[31] D. Nowakowski, S. Carignano, Proceedings MPCS2015, arXiv:1602.04798
[32] T. Tatsumi, T. Muto, Phys. Rev. D **89**(10), 103005 (2014), arXiv:1403.1927

IOP Conf. Series: Journal of Physics: Conf. Series **861** (2017) 012019 doi:10.1088/1742-6596/861/1/012019

Anomalous Transport Properties of Dense QCD in a Magnetic Field

Vivian de la Incera
Dept. of Eng. Sci. and Physics, College of Staten Island, CUNY, and CUNY-Graduate Center, New York 10314, USA
E-mail: vivian.incera@csi.cuny.edu

Abstract. Despite recent advancements in the study and understanding of the phase diagram of strongly interacting matter, the region of high baryonic densities and low temperatures has remained difficult to reach in the lab. Things are expected to change with the planned HIC experiments at FAIR in Germany and NICA in Russia, which will open a window to the high-density-low-temperature segment of the QCD phase map, providing a unique opportunity to test the validity of model calculations that have predicted the formation of spatially inhomogeneous phases with broken chiral symmetry at intermediate-to-high densities. Such a density region is also especially relevant for the physics of neutron stars, as they have cores that can have several times the nuclear saturation density. On the other hand, strong magnetic fields, whose presence is fairly common in HIC and in neutron stars, can affect the properties of these exotic phases and lead to signatures potentially observable in these two settings. In this paper, I examine the anomalous transport properties produced by the spectral asymmetry of the lowest Landau level (LLL) in a QCD-inspired NJL model with a background magnetic field that exhibits chiral symmetry breaking at high density via the formation of a Dual Chiral Density Wave (DCDW) condensate. It turns out that in this model the electromagnetic interactions are described by the axion electrodynamics equations and there is a dissipationless Hall current.

1. Introduction

A growing body of works indicates that with increasing density, the chirally broken phase of quark matter is not necessarily replaced by an homogeneous, chirally restored phase, but instead, at least for an intermediate region of densities, the system may favor the formation of inhomogeneous phases. To gain insight on why this occurs notice that with increasing density the homogeneous chiral condensate becomes disfavored due to the high-energy cost of exciting the antiquarks from the Dirac sea to the Fermi surface where the pairs form. At the same time, with higher densities, co-moving quarks and holes at the Fermi surface may pair with minimal energy cost through a mechanism analogous to Overhauser's [1], giving rise to a spatially modulated chiral condensate [2]. Spatially modulated chiral condensates in QCD have been discussed in the context of quarkyonic matter [3], where they appear in the form of quarkyonic chiral spirals [4] at zero magnetic field, or double quarkyonic chiral spirals [5] in the presence of a magnetic field. Inhomogeneous chiral condensates have been also studied in the context of NJL models (for a review see [6]) that share the chiral symmetries of QCD and are then useful to study the chiral phase transition.

In the present paper, we explore the properties of a QCD-inspired NJL model with a background magnetic field that exhibits chiral symmetry breaking at high density via

IOP Conf. Series: Journal of Physics: Conf. Series **861** (2017) 012019 doi:10.1088/1742-6596/861/1/012019

the formation of a Dual Chiral Density Wave (DCDW) condensate. As will be shown below, the electromagnetic interactions in this model are described by the equations of axion electrodynamics, exhibiting a dissipationless Hall current perpendicular to the magnetic field, and an anomalous electric charge directly proportional to the magnetic field strength and the modulation of the condensate.

2. DCDW Phase in a Magnetic Field

Let us describe cold and dense quark matter in a magnetic field with the help of a QCD-inspired the three-color, two-flavor Nambu-Jona-Lasinio (NJL) Lagrangian density that includes electromagnetism,

$$\mathcal{L} = -\frac{1}{4}F_{\mu\nu}F^{\mu\nu} + \bar{\psi}[i\gamma^\mu(\partial_\mu + iQA_\mu) + \gamma_0\mu]\psi + G[(\bar{\psi}\psi)^2 + (\bar{\psi}i\tau\gamma_5\psi)^2], \quad (1)$$

Here $\psi^T = (u, d)$ is the fermion doublet in flavor space with electric charge matrix $Q = \mathrm{diag}(e_u, e_d) = \mathrm{diag}(\frac{2}{3}e, -\frac{1}{3}e)$; μ is the baryon chemical potential; G is the four-fermion coupling; and τ_a are the isospin Pauli matrices. A constant and uniform magnetic field along the x^3-direction is introduced by the background field $\bar{A}_\mu = (0, 0, Bx, 0)$. Hence, the electromagnetic potential A_μ is the sum of the fluctuation field plus the background field $\bar{A}_\mu$. Notice that the flavor symmetry $SU(2)_L \times SU(2)_R$ of the NJL model is reduced to the subgroup $U(1)_L \times U(1)_R$ here due to the electromagnetic coupling.

As shown in [7, 8], this system favors the formation of the inhomogeneous condensate,

$$\langle\bar{\psi}\psi\rangle = \Delta\cos(\mathbf{q}\cdot\mathbf{x}), \quad \langle\bar{\psi}i\tau_3\gamma_5\psi\rangle = \Delta\sin(\mathbf{q}\cdot\mathbf{x}), \quad (2)$$

which represents a density wave with magnitude Δ and modulation vector $q_i = (0, 0, q)$ along the field direction. This type of density wave condensate is known in the literature as the DCDW condensate [9].

The corresponding mean-field Lagrangian is

$$\mathcal{L}_{MF} = \bar{\psi}[i\gamma^\mu(\partial_\mu + iQA_\mu) + \gamma_0\mu]\psi - m\bar{\psi}e^{i\tau_3\gamma_5 qz}\psi - \frac{1}{4}F_{\mu\nu}F^{\mu\nu} - \frac{m^2}{4G}, \quad (3)$$

with $m = -2G\Delta$. The spatially dependent mass term can be eliminated with the help of the local chiral transformation

$$\psi \to U_A\psi \qquad \bar{\psi} \to \bar{\psi}\bar{U}_A \quad (4)$$

$$\text{with } U_A = e^{-i\tau_3\gamma_5\theta} \qquad \bar{U}_A = \gamma_0 U^\dagger\gamma_0 = e^{-i\tau_3\gamma_5\theta} \quad (5)$$

with $\theta(t, \mathbf{x}) = \frac{1}{2}q_\mu x^\mu = \frac{qz}{2}$.

The mean-field Lagrangian then becomes

$$\mathcal{L}_{MF} = \bar{\psi}[i\gamma^\mu(\partial_\mu + iQA_\mu - i\tau_3\gamma_5\partial_\mu\theta) + \gamma_0\mu - m]\psi - \frac{1}{4}F_{\mu\nu}F^{\mu\nu} - \frac{m^2}{4G} \quad (6)$$

The energy spectrum of the quarks in (6) separates into the LLL ($l = 0$) modes

$$E_k^{LLL} = \epsilon\sqrt{\Delta^2 + k_3^2} + q/2, \quad \epsilon = \pm, \quad (7)$$

and higher ($l > 0$) Landau level modes

$$E_k^{l>0} = \epsilon\sqrt{(\xi\sqrt{\Delta^2 + k_3^2} + q/2)^2 + 2|e_f B|l}, \quad \epsilon = \pm, \xi = \pm, l = 1, 2, 3, \ldots \quad (8)$$

Notice that the LLL spectrum is not symmetric about the zero energy level [7]-[8]. An asymmetric spectrum is a sign of a nontrivial topology. Indeed, it is known that such a theory is characterized by the topological quantity $\eta = \sum_k \text{sgn}(E_k)$, known as the Atiyah-Patodi-Singer (APS) invariant [10]. Henceforth, we use the notation η_B for the APS invariant and keep in mind that due to the magnetic field B, the sum over the energy modes E_k is equivalent to a sum over all the Landau levels and an integral over momenta. Notice that η_B is divergent. In order to ensure that the energies with equal magnitude and opposite signs cancel out in the sum, it needs proper regularization. Using the regularization discussed in [11], it can be shown that the contributions of all the $l > 0$ modes cancel out because they are symmetric about zero, so only the asymmetric modes contribute to η_B, as was expected, given the connection between the nontrivial topology and the LLL spectral asymmetry. The regularized index $\eta_B = \lim_{s\to 0} \sum_k \text{sgn}(E_k)|E_k|^{-s}$ gives rise to an anomalous baryon (quark) number density ρ_B^A [8] given by

$$\rho_B^A = -N_c\eta_B/2 = N_c \sum_f \frac{|e_f|}{4\pi^2}\mathbf{B}\cdot\nabla(\mathbf{q}\cdot\mathbf{x}) = 3\frac{|e|}{4\pi^2}qB \tag{9}$$

when $q < 2m$. The use of a different regularization procedure that allows to extract the anomalous part of the thermodynamic potential, led to the same anomalous quark number density, obtained in this case not from the index η_B, but as the derivative of the thermodynamic potential with respect to the baryon chemical potential μ [7]. The extension of this calculation to the isospin asymmetric case was done in [12]. When $q > 2m$, the quark density acquires an additional, non-topological contribution [8] and becomes

$$\rho_B^A = -N_c\eta_B/2 = -N_c|eB|\left[-q + \sqrt{q^2 - 4m^2}\right]/4\pi^2 \tag{10}$$

As discussed in [13], the lack of invariance of the fermion measure in the path integral under the U_A transformation produces an additional contribution to the Lagrangian density (6) given by

$$(\det U_A)^{-2} = e^{-2\text{Tr}\log U} = e^{i\int d^4x\theta(x)d(x)} \tag{11}$$

where $d(x) = -2\delta^{(4)}(0)\text{tr}\tau_3\gamma_5$ and $\delta^{(4)}(0) = \langle x \mid x\rangle = \int \frac{d^4p}{(2\pi)^4}e^{ip(x-y)}|_{x=y}$. In the above equation, Tr means functional and matrix trace, while tr means matrix trace. This expression is ill-defined since it is the product of infinite ($\delta^{(4)}(0)$) times zero ($\text{tr}\tau_3\gamma_5 = 0$). It hence needs regularization and this should be done in a way that preserves the gauge-invariance of the theory. With that aim, we introduce an arbitrary, smooth regularizing function $f(t)$, such that $f(0) = 1$, $f(\infty) = 0$, $tf'(t) = 0$ at $t = 0$ and $t = \infty$. We then regularize the exponent following Fujikawa's approach [14] to obtain

$$\int_R d^4x\theta(x)d(x) = -2\lim_{\Lambda\to\infty}\text{Tr}\left[\theta(x)\tau_3\gamma_5 f((i\not{D}/\Lambda)^2)\right] = \int d^4x\frac{\kappa}{4}\theta F_{\mu\nu}\widetilde{F}^{\mu\nu} \tag{12}$$

with $D_\mu = \partial_\mu + iQA_\mu - i\tau_3\gamma_5\frac{q}{2}$ the corresponding Dirac operator, and $\kappa = \frac{N_c}{2\pi^2}[e_u^2 - e_d^2] = \frac{e^2}{2\pi^2}$.

The effective Lagrangian of the model is then

$$\begin{aligned}\mathcal{L}_{eff} &= \bar\psi[i\gamma^\mu(\partial_\mu + iQA_\mu - i\tau_3\gamma_5\frac{q}{2}) + \gamma_0\mu - m]\psi - \frac{m^2}{4G} \\ &- \frac{1}{4}F_{\mu\nu}F^{\mu\nu} + \frac{\kappa}{4}\theta F_{\mu\nu}\widetilde{F}^{\mu\nu},\end{aligned} \tag{13}$$

Thus, the measure contribution generated an axion term $\frac{\kappa}{4}\theta F_{\mu\nu}\widetilde{F}^{\mu\nu}$ which, as will be seen below, leads to important new physics [13].

IOP Conf. Series: Journal of Physics: Conf. Series **861** (2017) 012019 doi:10.1088/1742-6596/861/1/012019

3. Spectral Asymmetry and Index of the Dirac Operator

The origin of the axion term in (13) can be traced back to the spectral asymmetry of the LLL. This can be better understood with the help of the index of the Dirac operator in Euclidean space

$$\text{index}(i\not{D}_E) = n^+ - n^-, \tag{14}$$

which is defined as the difference between the number of zero modes of $i\not{D}_E$ with positive n^+ and negative n^- chirality. To see the connection, let us consider

$$\int d^4x d(x) = -2 \lim_{\Lambda\to\infty} \text{Tr}\left[\tau_3\gamma_5 f((i\not{D}_E/\Lambda)^2)\right] = \int d^4x \frac{\kappa}{4} F_{\mu\nu}\widetilde{F}^{\mu\nu} \tag{15}$$

which can be seen from (12) after going to Euclidean space and taking $\theta(x) = 1$. To keep the notation simple, we are not attaching an E index to the tensors and integral variables, but they are all in Euclidean.

Since τ_3 and $i\not{D}_E$ are diagonal in flavor space, we can do our analysis separately for the u and the d quark and add the two results at the end. In Euclidean space the operator $i\not{D}_E$ is hermitian, hence its eigenvalues λ_k are real. Consider a set of orthogonal and complete eigenfunctions ϕ_k of $i\not{D}_E$ with eigenvalues λ_k, $i\not{D}_E\phi_k = \lambda_k\phi_k$. Then, for each ϕ_k, we have $i\not{D}_E\gamma_5\phi_k = -\lambda_k\gamma_5\phi_k$, so $\gamma_5\phi_k$ is also an eigenfunction of $i\not{D}_E$ but with eigenvalue $-\lambda_k$. This implies that for $\lambda_k \neq 0$ the eigenfunctions ϕ_k and $\gamma_5\phi_k$ are orthogonal and with opposite eigenvalues. It also follows that for $\lambda_k \neq 0$, we can construct two combinations of ϕ_k and $\gamma_5\phi_k$,

$$\phi_{k,\pm} = \frac{1 \pm \gamma_5}{2}\phi_k \tag{16}$$

that are eigenfunctions of γ_5 and of $(i\not{D}_E)^2$, but not of $i\not{D}_E$.

We then have that if $\lambda_k \neq 0$ the functions $\phi_{k,\pm}$ satisfy

$$\gamma_5\phi_{k,\pm} = \pm\phi_{k,\pm}, \qquad (i\not{D}_E)^2\phi_{k,\pm} = \lambda_k^2\phi_{k,\pm} \tag{17}$$

meaning that for $\lambda_k \neq 0$ the eigenfunctions of $(i\not{D}_E)^2$ come in pairs with equal energy square eigenvalue, but opposite chirality.

On the other hand, for $\lambda_k = 0$, both eigenfunctions ϕ_k and $\gamma_5\phi_k$ have the same zero eigenvalue under the operator $i\not{D}_E$. In the subspace of these zero-mode-eigenfunctions, which we denote by $\phi_l^{(0)}$, γ_5 can be diagonalized, thus the $\phi_k^{(0)}$ can be separated into $\phi_k^{(0+)}$, with plus γ_5-eigenvalue (positive chirality), and $\phi_k^{(0-)}$, with minus γ_5-eigenvalue (negative chirality). However, there is no reason to claim that the numbers of eigenfunctions with positive and negative chiralities in the zero-mode-subspace have to be equal. In other words, the eigenfunctions $\phi_k^{(0\pm)}$ do not have to come in pairs.

Let us go back now to Eq. (15), and consider the contribution coming from the u-quark. Taking the Tr operator in the space of the eigenfunctions $\phi^\pm$ of $(i\not{D}_E)^2$ we find

$$\begin{aligned}
& -\frac{1}{2}\int d^4x d^{(u)}(x) = \lim_{\Lambda\to\infty} \text{Tr}\left[\gamma_5 f((i\not{D}_E/\Lambda)^2)\right] \\
&= \lim_{\Lambda\to\infty}\sum_k \langle\phi_k \mid \gamma_5 f((i\not{D}_E/\Lambda)^2) \mid \phi_k\rangle = \lim_{\Lambda\to\infty}\sum_k f(\lambda_k^2/\Lambda^2)\langle\phi_k \mid \gamma_5 \mid \phi_k\rangle \\
&= \lim_{\Lambda\to\infty}\sum_{l=1}^{n^u+} f(0)\langle\phi_l^{(0+)} \mid \phi_l^{(0+)}\rangle - \lim_{\Lambda\to\infty}\sum_{l=1}^{n^u_-} f(0)\langle\phi_l^{(0-)} \mid \phi_l^{(0-)}\rangle \\
&= \sum_{l=1}^{n^u+}\langle\phi_l^{(0+)} \mid \phi_l^{(0+)}\rangle - \sum_{l=1}^{n^u_-}\langle\phi_l^{(0-)} \mid \phi_l^{(0-)}\rangle = n^u_+ - n^u_-
\end{aligned} \tag{18}$$

IOP Conf. Series: Journal of Physics: Conf. Series **861** (2017) 012019 doi:10.1088/1742-6596/861/1/012019

where we used that $f(0) = 1$. In going from line 2 to 3 in Eq.(18), we took into account that the eigenfunctions with nonzero λ_k come in pairs of opposite chirality, so their contributions cancel out in the sum. This means that only the zero-energy modes contribute.

Doing the same calculation for the d quark and combining the two results we obtain

$$-\frac{1}{2}\int d^4x d(x) = n_+ - n_- = \text{index}(i\not{D}_E) = -\int d^4x \frac{\kappa}{8} F_{\mu\nu}\widetilde{F}^{\mu\nu} \tag{19}$$

so the integral of d(x) is proportional to the index of the Dirac operator and we can interpret d(x) as the index "density." The above relation remains valid also if the first and last terms are written back in Minkowski space.

Now, we can see from the expressions (7)-(8) in the massless limit case $\Delta = 0$, that in the presence of a background magnetic field the higher Landau levels ($l > 0$) lead to no zero modes. Moreover, from the two asymmetric modes in the LLL, only one of them can be a zero mode. Therefore, the spectral asymmetry of the LLL gives rise to a nonzero index($i\not{D}_E$). We conclude that the index "density" d(x) is connected to the spectral asymmetry of the LLL, and so is the axion term in (13).

4. Axial Electrodynamics in Dense QCD Matter in a Magnetic Field

The electromagnetic effective action, found from the partition function after integrating in the fermion fields and expanding in powers of A_μ, takes the form

$$\begin{aligned}\Gamma(A) = \;& -V\Omega + \int d^4x \left[-\tfrac{1}{4}F_{\mu\nu}F^{\mu\nu} + \tfrac{\kappa}{4}\theta F_{\mu\nu}\widetilde{F}^{\mu\nu}\right] \\ & - \int d^4x A^\mu(x) J_\mu(x) + \cdots ,\end{aligned} \tag{20}$$

with V the four-volume, $\Omega = \Omega(\mu, B)$ the thermodynamic potential [7], and $J_\mu(x) = (J_0, \mathbf{J})$ the ordinary (nonaxion) electric four-current determined by the sum of the tadpole diagrams of each flavor.

The linear equations of motion derived from (20) are

$$\nabla \cdot \mathbf{E} = J_0 + \tfrac{e^2}{4\pi^2} qB, \tag{21}$$

$$\nabla \times \mathbf{B} - \tfrac{\partial \mathbf{E}}{\partial t} = \mathbf{J} - \tfrac{e^2}{4\pi^2}\mathbf{q} \times \mathbf{E}, \tag{22}$$

$$\nabla \cdot \mathbf{B} = 0, \quad \nabla \times \mathbf{E} + \tfrac{\partial \mathbf{B}}{\partial t} = 0, \tag{23}$$

These are the equations of axion electrodynamics. In the above equations, the terms J_0 and $\mathbf{J}$ represent the ordinary charge and current density contributions obtained from the tadpole diagrams, while $J_0^{anom} = \frac{e^2}{4\pi^2}qB$ and $\mathbf{J}^{anom} = -\frac{e^2}{4\pi^2}\mathbf{q} \times \mathbf{E}$ are anomalous charge and current densities generated by the axion term in (20). As shown in [13], the tadpoles contributions do not cancel the anomalous ones.

As already discussed, the origin of the axion term in the effective action and thus the anomalous terms in Eqs. (21-22), can be traced back to the asymmetry of the LLL. This relation can also be seen from the connection between the anomalous electric charge and the anomalous quark number density associated with the Atiyah-Patodi-Singer index. Notice that the same expression for the anomalous electric charge density is found if one takes the anomalous quark number density of flavor f, $\rho^A_{Bf} = N_c \frac{|e_f|}{4\pi^2} qB$, multiplies it by the flavor's electric charge e_f, and sums in flavor. We highlight that the anomalous current is a dissipationless Hall current, perpendicular to both, the magnetic and the electric field. This could have important consequences for the transport properties of the system. Notice that no CME current is generated here because the axion field θ is time-independent.

5. Anomalous transport

The magnetic DCDW phase exhibits quite interesting properties. The most important is the existence of the dissipationless anomalous Hall current, $\mathbf{J}^{anom} = -\frac{e^2}{4\pi^2}\mathbf{q} \times \mathbf{E}$. Since it is perpendicular to $\mathbf{E}$ and to the condensate modulation $\mathbf{q}$, which in turn is parallel to $\mathbf{B}$, the Hall current is produced as long as $\mathbf{E}$ and $\mathbf{B}$ are not parallel.

These results can be relevant for neutron stars. Consider a neutron star with a core of DCDW matter threaded by a poloidal magnetic field. Any electric field present in the core, whether due to the anomalous electric charge or not, and as long as it is not parallel to the magnetic field, will lead to dissipationless Hall currents in the plane perpendicular to the magnetic field. Could these currents serve to resolve the issue of the stability [16] of the magnetic field strength in magnetars? How will the electric transport be affected by the anomalous Hall conductance? These and other questions highlight the importance to explore which observable signatures could be identified and then used as telltales of the presence of the DCDW phase in the core. Notice that the condition of electrical neutrality does not need to be satisfied locally for compact hybrid stars [17], which could have a core in the magnetic DCDW phase with an anomalous charge contribution and Hall currents circulating inside and at the surface.

The anomalous Hall current could be also produced in future HIC like those planned at the Nuclotron-based Ion Collider Facility (NICA) at Dubna, Russia [18] and at the Facility for Antiproton and Ion Research (FAIR) at Darmstadt, Germany [19], which will explore the high density, cold region of the QCD phase map, and where event-by-event off-central collisions will likely generate perpendicular electric and magnetic fields [20]. It will be interesting to carry out a detailed quantitative analysis of how these currents could lead to observable signatures, even after taking into account that there the QGP distributes itself more as an ellipsoid than as an sphere about the center of the collision. The Hall currents will tend to deviate the quarks from the natural outward direction from the collision center and one would expect a different geometry of the particle flow in the DCDW phase compared to other dense phases that have no anomalous electric current. The realization of the DCDW phase in the QGP of future HIC experiments is likely viable because the inhomogeneity of the phase is characterized by a length $\Delta x = \hbar/q \sim 0.6fm$ for $q \sim \mu = 300$ MeV [7], much smaller than the characteristic scale $L \sim 10fm$ of the QGP at RHIC, NICA, and FAIR, while the time scale for this phase will be the same as for the QGP.

Other interesting effects might emerge by considering the fluctuations $\delta\theta$ of the axion field. If one goes beyond the mean-field approximation, there will be mass and kinetic terms of the axion field fluctuation. Besides, due to the background magnetic field, the axion fluctuation couples linearly to the electric field via the term $\kappa\delta\theta\mathbf{E}\cdot\mathbf{B}$, so the field equations of the axion fluctuation and the electromagnetic field will be mixed, giving rise to a quasiparticle mode known as the axion polariton mode [21]. The axion polariton mode is gapped with a gap proportional to the background magnetic field. This implies that electromagnetic waves of certain frequencies will be attenuated by the DCDW matter, since in the DCDW medium they propagate as polaritons. The axion polariton could help to probe the possible realization of the magnetic DCDW in future HIC experiments at high baryon densities, due to its effect in the attenuation of certain light frequencies.

6. Conclusions

Cold and dense QCD in the magnetic DCDW phase exhibits a richness of macroscopically observable quantum topological effects that can be relevant for the HIC physics and neutron stars. It is rather fortunate that we will not have to wait too long to test the realization of these topological phases in the experiment. Future HIC experiments will certainly generate strong magnetic and electric fields in their off-central collisions and will open a much more sensitive window to look into a very challenging region of QCD. For example, the Compressed

Baryonic Matter (CBM) at FAIR [19] have been designed to run at unprecedented interaction rates to provide high-precision measures of observables in the high baryon density region. That is why it is so timing and relevant to carry out detailed theoretical investigations of all potential observables of the magnetic DCDW phase.

We hope that the findings presented here will serve to stimulate quantitative studies to identify signatures of the anomalous effects that will allow to probe the realization of this dense QCD phase both in neutron stars and in HIC experiments.

Acknowledgments

The results presented in this paper were based on work done in collaboration with Efrain J Ferrer.

References

[1] Overhauser A W 1978 *Adv. Phys.* **27** 343.
[2] Deryagin D V, Grigoriev D Y and Rubakov V A 1992 *Int. J. Mod. Phys. A* **7** 659; Shuster E and Son D T 2000 *Nucl. Phys. B* **573** 434; Park B-Y et. al 2000 *Phys. Rev. D* **62** 034015; Rapp R, Shuryak E and Zahed I 2001 *Phys. Rev. D* **63** 034008.
[3] McLerran L and Pisarski R D 2007 textitNucl. Phys A **796** 83.
[4] Kojo T, Hidaka T Y, McLerran L and Pisarski R D 2010 *Nucl. Phys. A* **843** 37; Kojo T, et al. 2012 *Nucl.Phys. A 875* 94; Kojo T, Pisarski R D and Tsvelik A M 2010 *Phys. Rev. D* **82** 074015; Kojo T 2012 *Nucl. Phys A* **877** 70.
[5] Ferrer E J, de la Incera V and Sanchez A 2012 *Acta Phys. Polon. Supp.* **5** 679.
[6] Buballa M and Carignano S 2015 *Prog. Part. Nucl. Phys.* **81** 39.
[7] Frolov I E Zhukovsky V Ch and Klimenko K G 2010 *Phys. Rev. D* **82** 076002.
[8] Tatsumi T, Nishiyama K and Karasawa S 2015 *Phys. Lett. B* **743** 66.
[9] Nakano E and Tatsumi T 2005 *Phys. Rev. D* **71** 114006.
[10] Atiyah M, Patodi V and Singer I 1975 *Proc. Cambridge Phils. Soc.* **77** 42; 1975 **78** 405; 1976 **79** 71.
[11] Niemi A J and Semenoff G W 1986 *Phys. Reports* **135** 99.
[12] Carignano S, Ferrer E J, de la Incera V and Paulucci L 2015 *Phys. Rev. D* **92** 105018.
[13] Ferrer E J, de la Incera V 2017 *Phys. Lett. B* **769** 208; arXiv: 1512.03972.
[14] Fujikawa K 1980 *Phys. Rev. D* **21** 2848.
[15] Bermudez A, Mazza L, Rizzi M, Goldman N, Lewenstein M and Martin-Delgado M A 2010 *Phys. Rev. Lett.* **105** 190404.
[16] Harding A K and Lai D 2006 *Rept. Prog. Phys.* **69** 2631; Spruit H C 2008 *AIP Conf. Proc. 983* 391.
[17] Glendenning N K 2000 *Compact Stars, Nuclear Physics, Particle Physics and General Relativity*, Springer-Verlag, NY
[18] Deng W-T and Huang X-G 2012 *Phys. Rev. C* **85** 044907; Toneev V, Rogachevsky O and Voronyuk V 2016 *Eur. Phys. J. A* **52** 264.
[19] Ablyazimov T *et. al.* arXiv:1607.01487.
[20] Ferrer E J and de la Incera V 2016 *Eur. Phys. J. A* **52** 266.
[21] Li R, Wang J, Qi X-L and Zhang S-C 2010 *Nature Phys.* **6** 284.

IOP Conf. Series: Journal of Physics: Conf. Series **861** (2017) 012020 doi:10.1088/1742-6596/861/1/012020

Equations of State of different phases of dense quark matter

E J Ferrer

Dept. of Eng. Sci. and Physics, College of Staten Island, CUNY, and CUNY-Graduate Center, New York 10314, USA

E-mail: efrain.ferrer@csi.cuny.edu

Abstract. Compact stars with significant high densities in their interiors can give rise to quark deconfined phases that can open a window for the study of strongly interacting dense nuclear matter. Recent observations on the mass of two pulsars, PSR J1614-2230 and PSR J0348+0432, have posed a great restriction on their composition, since their equations of state must be hard enough to support masses of about at least two solar masses. The onset of quarks tends to soften the equation of state, but due to their strong interactions, different phases can be realized with new parameters that affect the corresponding equations of state and ultimately the mass-radius relationships. In this paper I will review how the equation of state of dense quark matter is affected by the physical characteristics of the phases that can take place at different baryonic densities with and without the presence of a magnetic field, as well as their connection with the corresponding mass-radius relationship, which can be derived by using different models.

1. Introduction

The composition of neutron stars is still an open question. Propositions ranging from nuclear matter (possibly with hyperons and superfluid nucleons) to deconfined quark matter (in any of its possible phases as color superconducting or with inhomogeneous particle-hole pairing, etc.) are under exam. New data on masses and radii, transport properties, as well as the modeling of other phenomena as glitches, bursts episodes, etc., help to constraint the equation of state (EoS) corresponding to different matter phases that can prevail in their interior, but no firm conclusion has been drawn yet. In both, nuclear and quark matter descriptions, there still exist a few parameters to be adjusted, which should be further constrained by nuclear matter and heavy-ion collision data, so to be able to discriminate among the different star constituents, but their region of reliability at low temperature and large chemical potential poses a big challenge.

The proposition that quark stars could be a stable configuration for compact stellar objects has been around for almost 50 years [1, 2, 3, 4]. The idea is based on the fact that matter composed of up, down, and strange quarks, the so called strange quark matter (SQM), may have a lower energy per baryon number than the nucleon, thus being absolutely stable. More recent developments brought the idea that color superconductivity should be the favored state of deconfined quark matter, since the Cooper pair condensation lower the total energy per baryon number of the system [5, 6, 7, 8]. On this basis, many works have been done to study the characteristics of the EoS of the different phases of quark matter.

In this regard, recent data [9, 10, 11] have determined masses and/or radii for some compact objects with high precision (although some of these remain to be confirmed [12]), rendering some information about the possible composition of these objects. In this scenario, the recent observation for the pulsar J1614-2230, a binary system for which the mass of the neutron star was measured rather accurately through the Shapiro delay yielding M = 1.97±0.04 $M_\odot$ [11], posed an important question about the existence of SQM. Of course the answer to this is directly related to the prevailing matter phase that can produce an EoS stiff enough to reach such large mass value.

In this paper, I am reviewing several results obtained in a set of papers that will be conveniently referred, where different approaches in describing the quark matter EoS were used. A main purpose will be to discuss some effects that can affect the stiffness of the EoS, so to determine if it is possible to reach the reported mass value M = 1.97±0.04 $M_\odot$. I will finish discussing the magnetic field effect on the EoS of quark matter in the color superconducting color-flavor-locked (CFL) phase, pointing out the new challenges that the presence of a magnetic field poses on the calculation of the M-R relationship.

2. Bag vs NJL model in the CFL phase

In the study of the EoS of the possible matter phases that could form neutron stars, two frameworks have been mainly used: the MIT bag model and the Nambu-Jona-Lasinio (NJL) model. Both models are inspired in some of the fundamental properties of strongly interacting systems, although some aspects could be lacking.

The MIT bag model is a phenomenological model that was proposed to explain hadrons [13]. In this model, the quarks are asymptotically free and confinement is provided through the bag constant B, which artificially constrains the quarks inside a finite region in space. In the case of bulk quark matter, the asymptotically-free phase of quarks will form a perturbative vacuum (inside a "bag"), which is immersed in the nonperturbative vacuum. The creation of the 'bag" costs free energy. Then, in the energy density, the energy difference between the perturbative vacuum and the true one should be added. Essentially, that is the bag constant characterizing a constant energy per unit volume associated to the region where the quarks live. From the point of view of the pressure, B can be interpreted as an inward pressure needed to confine the quarks into the bag.

Let's consider now the difference between these two approaches in the case of SQM. For the free quark system, the thermodynamic potential is given in the MIT framework by

$$\Omega = \sum_i \Omega_i + B, \tag{1}$$

where

$$\Omega_i = \frac{\mu_i^4}{4\pi^2}, \tag{2}$$

with i running for quarks u, d, s and electrons, and μ_i being the corresponding chemical potential for each particle i.

When the color superconducting pairing is considered in the CFL phase, which is the most stable phase at high densities [14], the thermodynamical potential is assumed to be the sum of the one corresponding to the unpaired state (1), plus a gap (Δ) depending term [15, 16],

$$\Omega_{CFL}^{MIT} = \sum_i \Omega_i - \frac{3}{\pi^2}\mu^2\Delta_{CFL}^2 + B \tag{3}$$

The term $\sum_i \Omega_i$ represents a fictitious non-paired state proportional to the volume of the Fermi sphere in which all quarks have a common Fermi momentum and the extra Δ-depending term represents a surface contribution of the Fermi sphere that is associated with the Cooper pair biding energy. Here, as in (1), B represents the bag constant contribution.

In the context of the NJL model, the thermodynamic potential of the CFL phase is given by [17],

$$\Omega^{NJL}_{CFL} = -\frac{1}{4\pi^2}\int_0^\infty dpp^2 e^{-p^2/\Lambda^2}(16|\epsilon| + 16|\overline{\epsilon}|) + \\ -\frac{1}{4\pi^2}\int_0^\infty dpp^2 e^{-p^2/\Lambda^2}(2|\epsilon'| + 2|\overline{\epsilon'}|) + \frac{3\Delta^2_{CFL}}{G} + B, \quad (4)$$

where

$$\varepsilon = \pm\sqrt{(p-\mu)^2 + \Delta^2_{CFL}}, \quad \overline{\varepsilon} = \pm\sqrt{(p+\mu)^2 + \Delta^2_{CFL}}, \\ \varepsilon' = \pm\sqrt{(p-\mu)^2 + 4\Delta^2_{CFL}}, \quad \overline{\varepsilon'} = \pm\sqrt{(p+\mu)^2 + 4\Delta^2_{CFL}}, \quad (5)$$

are the quasiparticles dispersion relations. See that in (5) the particle masses are neglected at the high chemical potentials where the CFL phase is realized. In (4) a smooth cutoff depending on the effective-theory energy scale Λ was introduced to guarantee continuous thermodynamical quantities.

The system EoS is then obtained from the thermodynamic potential as the relation between the energy density and the pressure, which for an isotropic system can be respectively obtained as

$$\epsilon = \Omega - \mu\frac{\partial\Omega}{\partial\mu}, \quad (6)$$

$$P = -\Omega \quad (7)$$

On the other hand, the mass-radius (M-R) relationship is obtained by integrating the relativistic equations for stellar structure, that is, the well-known Tolman-Oppenheimer-Volkoff (TOV) and mass continuity equations, which in natural units, $c = G = 1$ are given by

$$\frac{dM}{dR} = 4\pi R^2\epsilon \quad (8)$$

$$\frac{dP}{dR} = -\frac{\epsilon M}{R^2}\Big(1+\frac{P}{\epsilon}\Big)\Big(1+\frac{4\pi R^3 P}{M}\Big)\Big(1-\frac{2M}{R}\Big)^{-1} \quad (9)$$

with ϵ and P taken from (6) and (7) respectively.

A fundamental difference between models (3) and (4) is related to the way the pairing term Δ is implemented. In the MIT bag model the value for the gap parameter is fixed by hand, hence it does not explicitly depend on changes in other parameters characterizing the system. While in the NJL approach the pairing gap is self-consistently obtained through the gap equation

$$\frac{\partial\Omega_{CFL}}{\partial\Delta_{CFL}} = 0. \quad (10)$$

In this way, as it was shown in Ref. [18], Δ depends on the density and diquark coupling constant G, as can be seen in Fig 1. Thus, changing the coupling constant in this case from G = 4.32 to 7.10 GeV^{-2}, the corresponding gap parameters obtained at $\mu = 500$ MeV change from $\Delta = 10$ to 100 MeV, respectively. We point out that the value $G = 7.10 GeV^{-2}$ is used with the only purpose of comparing the NJL results with the MIT ones, where the use of $\Delta = 100$ MeV is a common practice.

The EoS corresponding to NJL and MIT models are represented in the region of interest for SQM in Fig. 2 for two different values of the gap parameter. It can be observed that the splitting between the EoS for different Δ's is more significant for the values introduced in the

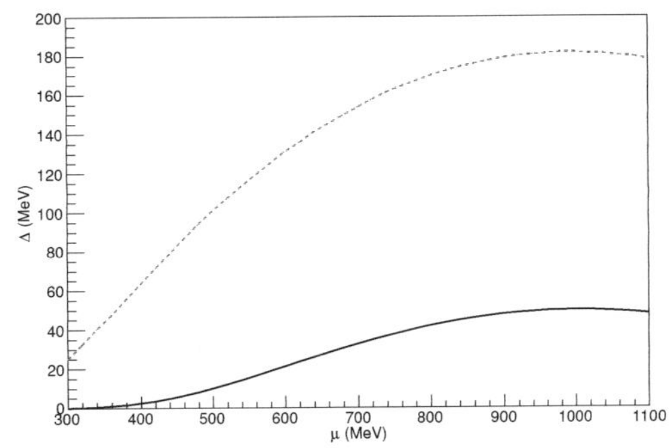

Figure 1. Gap parameter, Δ, vs baryonic chemical potential, μ, for NJL model in the CFL phase for two coupling constant values G=4.32 GeV^{-2} (full line) and G=7.10 GeV^{-2} (dashed line).

MIT model than in those obtained in the NJL model, although they are really significant only for high gap values. As it was found in [18], in the MIT case, the higher the gap, the stiffer the EoS. Hence, as the mass supported by a given star configuration is related to the stiffness of the EoS, a higher value of the gap renders higher maximum masses for stable strange stars [8]. However, this is *not* the case for the NJL calculations. When a higher value of G is used, although the corresponding gap parameter increases for a fixed value of μ, the EoS does not change considerably and actually weakly softens in the region of interest for neutron star interior. Therefore, in the NJL approach it is not possible to increase the maximum star mass of strange stars even when one uses unphysically large values of the coupling constant as can be seen from Fig. 3. The origin of the softening of the EoS in the CFL-NJL approach is due to the term $3\Delta^2/G$ that enters with a negative sign in the pressure.

As discussed in [18], from a physical point of view, the increase of G beyond a certain value in the CFL-NJL model implies a softening of the EoS because for large enough G's the system begins to crossover from BCS to BEC [19]-[22]. The crossover is reflected in the decay of the system pressure, which is due to an increment in the number of diquarks that become Bose molecules and hence cannot contribute to the Pauli pressure of the system. As shown in Refs. [22, 23], if the diquark coupling were high enough to produce a complete crossover from the BCS to the BEC regime, the pressure of the bosonic system at zero temperature would become zero implying a collapse of the stellar system.

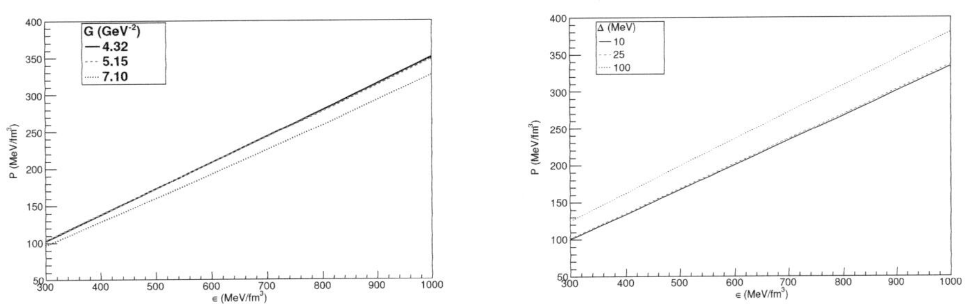

Figure 2. The EoS for CFL matter within the NJL theory (left panel) for different values of G at $B = 0$. The same for the MIT bag model for different values of Δ in the right panel.

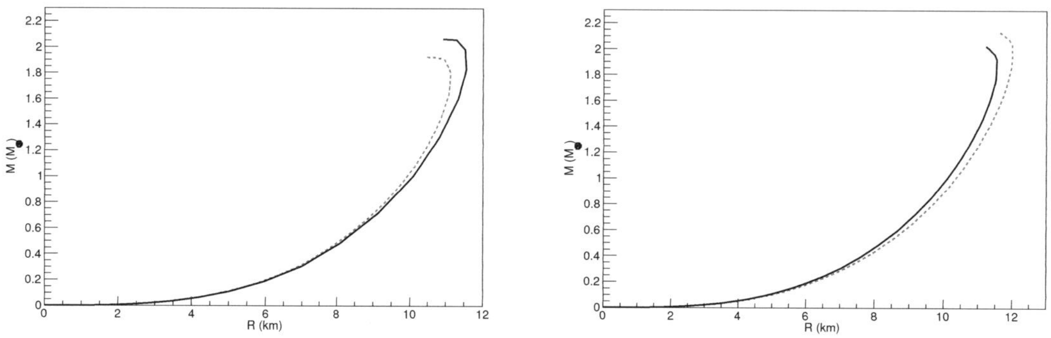

Figure 3. Mass-radius relation for CFL matter with B=58 MeV/fm^3. Results obtained with NJL theory in the left panel: full line for G=4.32 GeV^{-2} and dashed line for G=7.10 GeV^{-2}. Results obtained with MIT bag model in the right panel with Δ = 10 (full line) and 100 MeV (dashed line).

To conclude this section, using Eq. (3) one can indeed increase the maximum allowed mass for CFL strange stars by using an arbitrarily high gap value to match the recent reported data in [9]-[11]. Nevertheless, this effect is not possible within the more consistent NJL approach (4), as can be seen in Fig. 3, since to increase the gap the coupling constant should be increased and it tends to decrease the mass for a given radius.

3. EoS of the CFL-gluonic phase

In the NJL approach, gluons degrees of freedom are usually disregarded as having a negligible effect at zero temperature. Nevertheless, as it was shown in Ref. [24], gluons can affect the EoS of the CFL phase. The reason is that, as shown in [25], the polarization effects of quarks in the CFL background embed all the gluons with Debye (m_D) and Meissner (m_M) masses that depend on the baryon chemical potential,

$$m_D^2 = \frac{21 - 8\ln 2}{18} m_g^2, \qquad m_M^2 = \frac{21 - 8\ln 2}{54} m_g^2, \qquad m_g^2 = g^2\mu^2 N_f/6\pi^2. \tag{11}$$

Here, N_f is the number of flavors of massless quarks, and g the quark-gluon gauge coupling constant. As a consequence, the gluons in the CFL phase have nonzero rest energy that produces a positive contribution to the system's energy density and consequently a negative contribution to the pressure.

To consider the gluon effect into the system EoS we started in [24] from the three-flavor gauged-NJL Lagrangian at finite baryon density, which is invariant under $SU_c(3) \times SU_L(3) \times SU_R(3) \times U_V(1)$ symmetry

$$\begin{aligned}\mathcal{L} = &-\bar{\psi}(\gamma^\mu D_\mu + \mu\gamma^0)\psi - G_V(\bar{\psi}\gamma_\mu\psi)^2 + G_S\sum_{k=0}^{8}\left[(\bar{\psi}\lambda_k\psi)^2 + (\bar{\psi}i\gamma_5\lambda_k\psi)^2\right] \\ &-K\left[det_f\left(\bar{\psi}(1+\gamma_5)\psi\right) + det_f\left(\bar{\psi}(1+\gamma_5)\psi\right)\right] + \frac{G_D}{4}\sum_{\eta}(\bar{\psi}P_\eta\bar{\psi}^T)(\psi^T P_\eta\psi) + \mathcal{L}_G,\end{aligned} \tag{12}$$

In (12), the quark fields ψ_i^a have flavor ($i = u, d, s$) and color ($a = r, g, b$) indexes. Here, the coupling G_S is associated with the quark-antiquark channel, the coupling K with the 't Hooft determinant term that excludes the $U_A(1)$ symmetry [26], the coupling G_D with the diquark channel $P_\eta = C\gamma_5\epsilon^{ab\eta}\epsilon_{ij\eta}$ and the coupling G_V with the repulsive vector channel. This last interaction term naturally appears after a Fierz transformation of a point-like four-fermion interaction with the Lorentz symmetry broken by the finite density [27]. We are neglecting the current quark masses m_{0i} since our main domain of interest will be the CFL phase with $\mu \gg m_{0i}$. The gluon Lagrangian density $\mathcal{L}_G$ is

$$\mathcal{L}_G = -\frac{1}{4} G^A_{\mu\nu} G^{\mu\nu}_A + \mathcal{L}_{gauge} + \mathcal{L}_{ghost}, \tag{13}$$

with $G^A_{\mu\nu} = \partial_\mu G^A_\nu - \partial_\nu G^A_\mu + g f^{ABC} G^B_\mu G^C_\nu$, the gluon strength tensor; $\mathcal{L}_{gauge}$, a gauge-fixing term; and $\mathcal{L}_{ghost} = -\eta^{A\dagger}\partial^\mu(\partial_\mu \eta^A + g f^{ABC} G^B_\mu \eta^C)$, the ghost-field Lagrangian. The coupling between gluons and quarks occurs through the covariant derivative

$$D_\mu = \partial_\mu - igT^A G^A_\mu, \tag{14}$$

where $T^A = \lambda^A/2$, $A = 1 - 8$, are the generators of the $SU_c(3)$ group in the fundamental representation (λ^A are the Gell-Mann matrices). The model (12) is nonrenormalizable due to the four-fermion interaction terms, so a cutoff Λ needs to be introduced in all the calculations to regularize the theory in the ultraviolet region. The parameter Λ defines the energy scale below which this effective theory is valid.

We want to consider the region of densities large enough to have a CFL phase, where the chiral condensate has already vanished and hence only the expectation values for the diquark condensate $\Delta_\eta = \langle \psi^T P_\eta \psi \rangle$ and the baryon charge density $\rho = \langle \bar{\psi}\gamma_0\psi\rangle$ should be considered. One can now bosonize the four-fermion interaction via a Hubbard-Stratonovich transformation and then take the mean-field approximation to obtain the zero-temperature thermodynamic potential [24],

$$\Omega = \Omega_q + \Omega_g - \Omega_{vac}, \tag{15}$$

where the quark, gluon and vacuum contributions are given respectively by

$$\Omega_q = -\frac{1}{4\pi^2}\int_0^\Lambda dpp^2(16|\epsilon| + 16|\bar{\epsilon}|) - \frac{1}{4\pi^2}\int_0^\Lambda dpp^2(2|\epsilon'| + 2|\bar{\epsilon'}|) + \frac{3\Delta^2}{G_D} - G_V\rho^2, \tag{16}$$

$$\Omega_g = \frac{2}{\pi^2}\int_0^\Lambda dpp^2\left(\sqrt{p^2 + \tilde{m}_D^2\theta(\Delta - p) + 3\tilde{m}_g^2\theta(\tilde{\mu} - p)\theta(p-\Delta)} + 3\sqrt{p^2 + \tilde{m}_M^2\theta(\Delta - p)}\right), \tag{17}$$

$$\Omega_{vac} \equiv \Omega(\mu = 0, \Delta = 0), \tag{18}$$

with energy spectra

$$\varepsilon = \pm\sqrt{(p-\tilde{\mu})^2 + \Delta^2}, \quad \bar{\varepsilon} = \pm\sqrt{(p+\tilde{\mu})^2 + \Delta^2}, \tag{19}$$

$$\varepsilon' = \pm\sqrt{(p-\tilde{\mu})^2 + 4\Delta^2}, \quad \bar{\varepsilon}' = \pm\sqrt{(p+\tilde{\mu})^2 + 4\Delta^2}, \tag{20}$$

and dynamical quantities, Δ and ρ, found from the equations,

$$\frac{\partial\Omega}{\partial\Delta} = 0, \quad \rho = -\frac{\partial\Omega_q}{\partial\tilde{\mu}} \tag{21}$$

We have that, as the mean value $\langle\bar{\psi}\gamma_0\psi\rangle$ enters in the covariant derivative as a shift to the baryon chemical potential, the effective chemical potential for the baryon charge is now $\tilde{\mu} = \mu - 2G_V\rho$ instead of μ. Moreover, while the solution of the gap equation (first equation

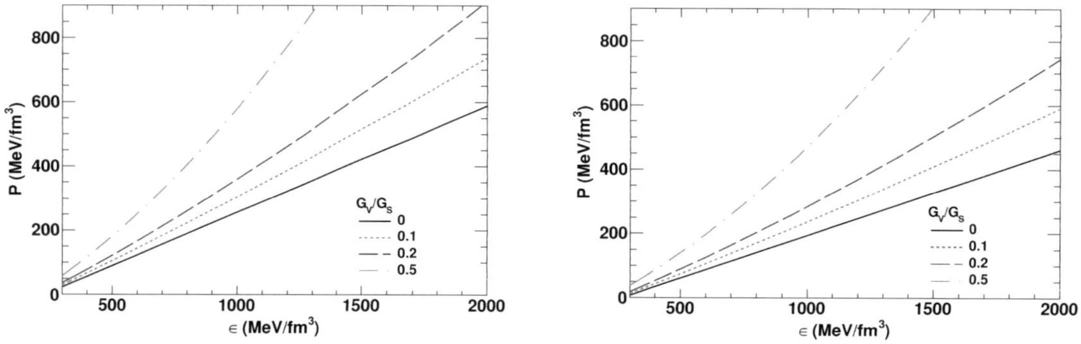

Figure 4. Equation of state for CFL matter with gluon contribution (right panel) and without it (left panel), for different values of the vector coupling G_V.

in (21)) is a minimum of the thermodynamic potential, the solution of the second equation is a maximum [28], since it defines, as usual in statistics, the particle number density ρ.

To find the EoS of this system we need the corresponding pressure P and energy density ϵ, that can be found from the thermodynamic potential (15) as

$$P = -(\Omega_q + \Omega_g - \Omega_{vac}) + (B - B_0), \quad \epsilon = \Omega_q + \Omega_g - \Omega_{vac} + \tilde{\mu}\rho - (B - B_0) \tag{22}$$

Notice that the chemical potential that multiplies the particle number density ρ in the energy density is $\tilde{\mu}$ instead of μ. This result can be derived following the same calculations of Ref. [29] to find the quantum-statistical average of the energy-momentum tensor component τ_{00}.

In (22), we added, as usual, the bag constant B (B_0 is introduced to ensure that $\epsilon = P = 0$ in vacuum). As we already pointed out, in the MIT bag model, B was introduced as a phenomenological input parameter to account for the free energy cost of quark matter relative to the confined vacuum [13]. Nevertheless, in the NJL model, the bag constant can be calculated in the mean-field approximation as a dynamical quantity related to the spontaneous breaking of chiral symmetry [30]. Following this dynamical approach, we showed in [24] that the bag constant in this case is given by $B = 0$ and the only remaining bag parameter in (22) is B_0, which is written as

$$B_0 = \frac{9}{\pi^2}\left[\int_0^\Lambda p^2 dp \left(\sqrt{m^2+p^2} - \sqrt{p^2 + \frac{2G_S m}{\sqrt{m^2+p^2}}}\right)\right] - 4K\left(\frac{3}{\pi^2}\right)^3 \left[\int_0^\Lambda dp p^2 \frac{m}{\sqrt{m^2+p^2}}\right]^3, \tag{23}$$

with the same vacuum dynamical mass for the three quarks found from

$$1 = 4G_S \frac{3}{\pi^2}\int_0^\Lambda p^2 dp \frac{1}{\sqrt{m^2+p^2}} + 2K\frac{9}{\pi^4}\left[\int_0^\Lambda p^2 dp \frac{m}{\sqrt{m^2+p^2}}\right]^2 \tag{24}$$

For the parameter set under consideration one obtains $B_0 = 57.3$ MeV/fm^3 [30], which is a value very close to what we were using in Section 2. Nevertheless, we should underline that in this calculation no contribution associated with the confinement/deconfinement phase transition has been taken into account.

Then, the EOS for a range of G_V with and without the gluon contribution to the thermodynamic potential was found in [24] and is plotted in Fig. 4. As can be seen by comparing the two panels, the inclusion of gluons' degrees of freedom soften the EOS. Only for relatively

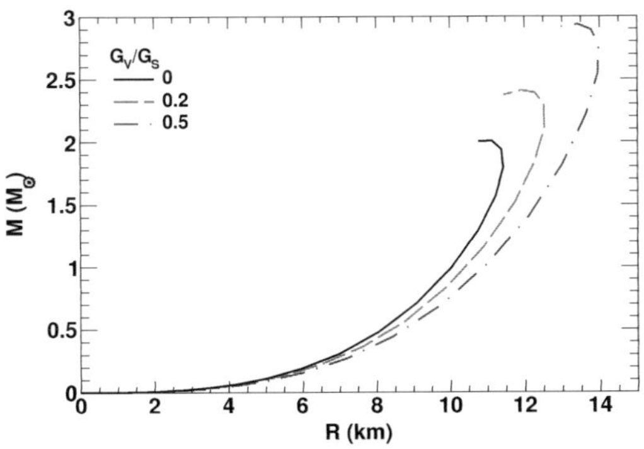

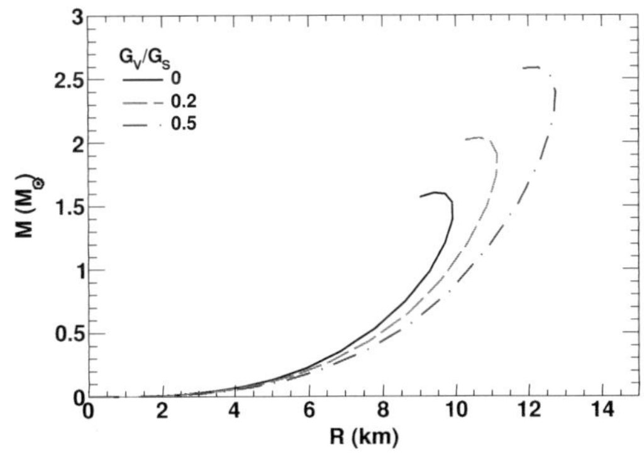

Figure 5. Mass-radius relationship of strange stars made of CFL matter with (right panel) and without (left panel) gluon contribution. The maximum mass values for vector interaction in the range $G_V/G_S < 0.2$ remains below $2M_\odot$.

high values of G_V the EOS of the CFL matter with gluons becomes stiffer than the EOS of the regular CFL matter with no gluons and no vector coupling, at least in the energy density region relevant for compact stars. Moreover, if we compare the curve for the CFL matter at certain value of G_V with the corresponding one including gluons and the same coupling value, the former will be stiffer than the second one, independent of the G_V value.

We can also note that with the increase of the energy density the shift in pressure due to a change in G_V for CFL matter with gluons increases for the same value of energy density. On the other hand, while at energy densities smaller than 500 MeV/fm^3 there is a negligible change even for $G_V/G_S = 0.5$, at $\epsilon = 1500$ MeV/fm^3, the jump in pressure is really noticeable. This is a natural result, since the effect of the vector interaction is proportional to the baryon density and should be more significant in the high density region.

Now it can be calculated the M-R relationship for strange stars made of CFL matter with gluon contribution by using Eqs. (8) and (9). The corresponding M-R sequences found in [24] are shown in Fig. 5 for the EoS of quark matter with (right panel) and without (left panel) the gluon contribution. Comparing them, it is evident that the gluons decrease the maximum mass for each sequence up to 20%, an effect that turns more prominent for lower values of the vector interaction. Sequences including gluons cannot reach $2M_\odot$ if $G_V/G_S < 0.2$. Here, it is important to point out that several results actually suggest that a low vector coupling between quarks may be favored at high-densities [31]. Even more, as discussed in [32], the vector interaction makes the chiral phase transition weaker in the low T and high μ region, with the possibility that the transition becomes a crossover in the region when the interaction is strong enough [28, 33], what is not expected from Lattice calculations. Then, the absence of the vector interaction would be preferable in the high density region. The same conclusion was posed by studying the phase transition from the hadronic phase to the quark phase in Ref. [34]. There, it was found that the hadron-quark phase transition may take place only at small G_V/G_S ratio. For larger ratio, the repulsive vector interaction makes the NJL phase too stiff to cross to the hadronic phase. However, taking a zero vector interaction for quark matter and further considering effects which soften the quark matter EOS, like diquark condensation, there could be a phase transition.

Then, if it is needed to consider a very weak or even zero vector interaction for high-dense quark matter, gluon effects would rule out stars composed entirely of CFL matter. More precisely, stars composed entirely of CFL quark matter with gluons, within the formalism presented here, should be disregarded if G_V is sufficiently low, as they cannot explain the mass

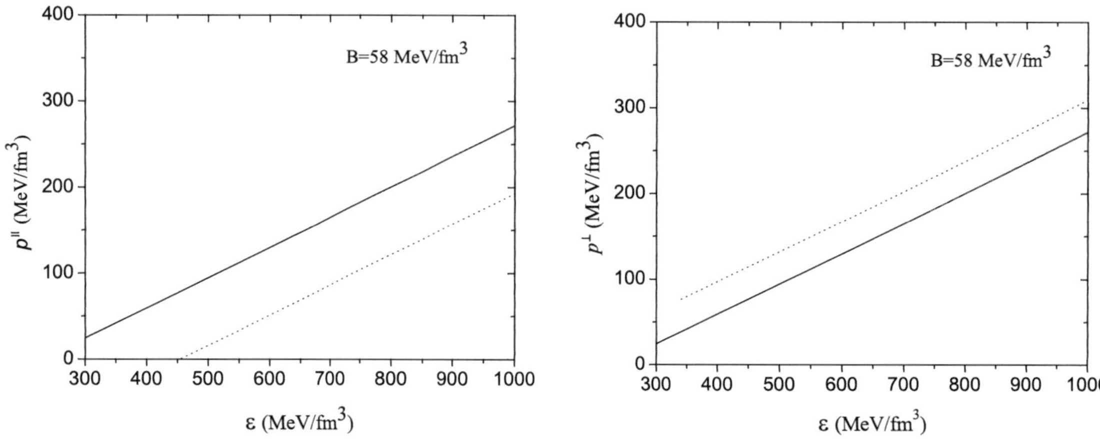

Figure 6. EoS for MCFL matter considering parallel (left panel) and perpendicular (right panel) pressures for different values of $\tilde{H}$: zero field (solid line), and $5 \times 10^{18} G$ (dotted line). The value of the bag constant was fixed to B=58 MeV/fm^3.

measurements of the mentioned compact stars PSR J1614-2230 and PSR J0348+0432.

4. EoS of Magnetized Quark Matter

In the presence of a uniform magnetic field H, the EoS becomes anisotropic [29]. In this case, the pressure of the system develops a splitting in the directions parallel and perpendicular to the applied field. This splitting has to be taken into account in the EoS, which is then modified from the usual form into

$$P_{\parallel} = -\Omega - \frac{H^2}{2}, \tag{25}$$

$$P_{\perp} = -\Omega - H\mathcal{M} + \frac{H^2}{2}, \tag{26}$$

$$\varepsilon = \Omega + \mu\rho + \frac{H^2}{2}, \tag{27}$$

where Ω is the thermodynamical potential evaluated at the physical minimum, $\rho = -\partial\Omega/\partial\mu$ is the particle density, ε the energy density and $\mathcal{M} = -\partial\Omega/\partial H$ the system magnetization.

In the case of SQM in the color superconducting phase, the magnetic field gives rise to a new phase that is called the magnetic CFL (MCFL) phase (see for review [35] and references therein). In this case the thermodynamic potential is given by [17]

$$\Omega_{MCFL} = \Omega_C + \Omega_N \tag{28}$$

with

$$\Omega_C = -\frac{\tilde{e}\tilde{H}}{4\pi^2}\sum_{n=0}^{\infty}(1-\frac{\delta_{n0}}{2})\int_0^{\infty} dp_3 e^{-(p_3^2+2\tilde{e}\tilde{H}n)/\Lambda^2}[8|\varepsilon^{(c)}| + 8|\overline{\varepsilon}^{(c)}|], \tag{29}$$

$$\Omega_N = -\frac{1}{4\pi^2}\int_0^{\infty} dpp^2 e^{-p^2/\Lambda^2}[6|\varepsilon^{(0)}|+6|\overline{\varepsilon}^{(0)}|] - \frac{1}{4\pi^2}\int_0^{\infty} dpp^2 e^{-p^2/\Lambda^2}\sum_{j=1}^{2}[2|\varepsilon_j^{(0)}|+2|\overline{\varepsilon}_j^{(0)}|] + \frac{\Delta^2}{G} + \frac{2\Delta_H^2}{G}, \tag{30}$$

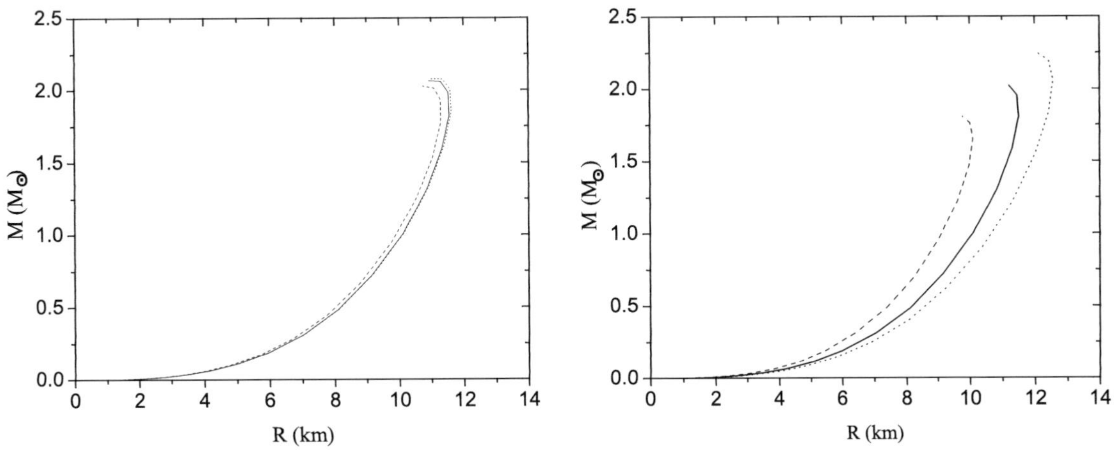

Figure 7. Mass-radius relation for strange stars in the MCFL phase with bag constant $B = 58$ MeV/fm^3. The full line indicates the M-R relation for zero magnetic field, whereas the dashed and dotted lines corresponds to MR relation calculated with parallel and perpendicular pressures, respectively, for $\tilde{H} = 1.7 \times 10^{17}$ G (left panel) and $\tilde{H} = 3 \times 10^{18}$ G (right panel).

and

$$\varepsilon^{(c)} = \pm\sqrt{(\sqrt{p_3^2 + 2\tilde{e}\widetilde{H}n} - \mu)^2 + \Delta_H^2},$$
$$\overline{\varepsilon}^{(c)} = \pm\sqrt{(\sqrt{p_3^2 + 2\tilde{e}\widetilde{H}n} + \mu)^2 + \Delta_H^2}, \tag{31}$$

$$\varepsilon^{(0)} = \pm\sqrt{(p-\mu)^2 + \Delta^2}, \qquad \overline{\varepsilon}^{(0)} = \pm\sqrt{(p+\mu)^2 + \Delta^2},$$
$$\varepsilon_1^{(0)} = \pm\sqrt{(p-\mu)^2 + \Delta_a^2}, \qquad \overline{\varepsilon}_1^{(0)} = \pm\sqrt{(p+\mu)^2 + \Delta_a^2},$$
$$\varepsilon_2^{(0)} = \pm\sqrt{(p-\mu)^2 + \Delta_b^2}, \qquad \overline{\varepsilon}_2^{(0)} = \pm\sqrt{(p+\mu)^2 + \Delta_b^2}, \tag{32}$$

being the dispersion relations of the charged (c) and neutral (0) quarks. In the above we used the notation $\Delta_{a/b}^2 = \frac{1}{4}(\Delta \pm \sqrt{\Delta^2 + 8\Delta_H^2})^2$. In this formulation we are neglecting again the quark masses since they are too much smaller than the chemical potential at the densities under consideration. Also we are not considering the possible effect of the magnetic field interaction with the quark anomalous magnetic moment, since as proved in [36] it has a negligible effect. Even the third gap appearing in the MCFL phase, which is associated to the Cooper condensate magnetic moment [37] is not taken into consideration because it is only relevant at extremely high magnetic fields when the particles are confined to the lowest Landau level.

The MCFL gaps Δ and Δ_H correspond respectively to the (d, s) pairing gap, which takes place only between neutral quarks, and to the (u, s) and (u, d) pairing gaps, which receive contribution from pairs of charged and neutral quarks. The separation of the gap in two different parameters in the MCFL case, as compared to the CFL, where there is only one gap, reflects the symmetry difference between these two phases [38, 39]. Here again, Λ-dependent smooth cutoffs were introduced. The notation, $\widetilde{H}$, is used for the in-medium rotated magnetic field that due to his long-range nature can penetrate the color superconductor [40].

The EoS, corresponding to Eqs. (25)-(27) evaluated in the solutions of the gap equations

$$\frac{\partial\Omega_{MCFL}}{\partial\Delta}=0, \qquad \frac{\partial\Omega_{MCFL}}{\partial\Delta_H}=0, \tag{33}$$

are plotted in Fig. 6. There, we can see that in the presence of a magnetic field the EoS becomes highly anisotropic. While for the parallel pressure the magnetic field shifts the EoS curve above the zero-field one, for the parallel pressure the magnetic field effect is the opposite. Moreover, the magnetic-field effect is much significant for the $\epsilon - p^{\|}$ relationship than for the $\epsilon - p^{\perp}$ one. At $\widetilde{H} \sim 10^{18}$ G, the shift in the energy density with respect to the zero-field value is ~ 200 MeV/fm^3 for the same pressure in the parallel-pressure case, while this field effect is too much smaller for the same range of field values in the perpendicular-pressure case.

Now we can use the obtained EoS to obtain the M-R relationship through the TOV equations (8)-(9) for the parallel and perpendicular pressures independently. The results are plotted in Fig. 7. There we can see that the effect of the parallel pressure is to increase the star mass for a given radius, while the perpendicular pressure tends to decrease the mass for the same radius. Hence, the anisotropy of the magnetized EoS opens an interesting question in regards to the calculation of the star M-R relationship, since each pressure has an opposite effect. To solve this paradox it will be needed to find an alternative formulation that includes the anisotropy from where to obtain the M-R relationship, since the TOV equations (8)-(9) were obtained under the supposition that the system has an isotropic configuration.

5. Summary

In this paper it was reviewed the EoS's of quark matter under several conditions and following different approaches with the goal to see under what conditions the observed mass value M = 1.97±0.04 $M_\odot$ can be reached.

First, the results for the EoS in the MIT bag model was compared to those of the NJL model. It was shown that while in the MIT model the introduction by hand of a large gap produces a stiffer EoS, when in the NJL model it is increased the gap self consistently by strengthening the diquark-diquark coupling the EoS becomes softer. This last effect is associated with the fact that a strong coupling can produce a BCS-BEC crossover with a decrease of pressure in the BEC phase. This result implies that to use the MIT bag model with large gap values to predict stellar masses can be misleading. Also, it is worth to notice that the maximum value of the mass that can be reached in both models is sensitive to the used bag constant value.

Then, it was consider the effect of gluons in the EoS of dense quark matter in the CFL phase at zero temperature. Since in this phase the gluons acquire Debye and Meissner masses they can contribute to the thermodynamic potential calculated in the framework of the gauged-NJL model. The gluon masses increase the system rest energy at zero temperature and hence decrease the pressure. As a consequence, the gluon effect softer the EoS in such a way that only for values of the coupling constant associated with the vector-vector interaction, G_V, larger than a certain critical value, the system can reach the observed stellar mass value M = 1.97±0.04 $M_\odot$.

Also, the effect of a strong magnetic field on the EoS and M-R relation for CFL matter was discussed. The main result is that a sufficiently high magnetic field produces a big anisotropy in the EoS that prevents the use of the TOV equations to study the M-R relationship in those conditions, since the TOV equations were derived for an isotropic medium with spherical symmetry. Moreover, it is known that at sufficiently high $\widetilde{H}$ new phases of dense quark matter can take place [41], as for example, a paramagnetic phase with a ground state of gluon vortices that are formed at rotated magnetic fields higher than the gluon Meissner mass square [42]-[43]. How this magnetic phase can affect the EoS of dense quark matter is an open question.

Finally, to extrapolate the results presented here to hybrid stars, other quark phases that are more likely to be realized at the moderated-high densities that can prevails in the core close to the transition from the existing nuclear matter in the envelope, should be considered. Other phases of quark matter, as 2SC at strong coupling with [44] or without [45] magnetic field, or the magnetized neutral DCDW phase [46], are good candidates in this context.

Acknowledgments
I would like to thank my collaborators V. de la Incera, J. E. Horvath and L. Paulucci who contributed to most of the results reviewed here.

References

[1] Bodmer A 1971 *Phys. Rev. D* **4** 1601
[2] Chin S A and Kerman A K 1979 *Phys. Rev. Lett.* **43** 1292
[3] Terazawa H 1979 *Tokyo U. Report* INS-336
[4] Witten E 1984 *Phys. Rev. D* **30** 272
[5] Alford M, Rajagopal K and Wilczek F 1999 *Nucl. Phys. B* **537** 433
[6] Rapp R, Schaefer T, Shuryak E V and Velkovsky M 2000 *Ann. Phys. (N.Y.)* **280** 35
[7] Alford M, Rajagopal K, Reddy S and Wilczek F 2001 *Phys. Rev. D* **64** 074017
[8] Lugones G and Horvath J E 2002 *Phys. Rev. D* **66** 074017
[9] Ozel F and Psaltis D 2009 *Phys. Rev. D* **80** 103003
[10] Ozel F, Baym G and Guver T 2010 *Phys. Rev. D* **82** 101301
[11] Demorest P B et al. 2010 *Nature* **467** 1081
[12] Steiner A W, Lattimer J M and Brown E F 2010 *Astrophys. J.* **722** 33
[13] Chodos A, Jaffe R L, Johnson K, Thorn C B and Weisskopf V F 1974 *Phys. Rev. D* **9** 3471
[14] Alford M, Kouvaris C and Rajagopal K 2004 *Phys. Rev. D* **92** 222001
[15] Rajagopal K and Wilczek F 2001 *Phys. Rev. Lett.* **86** 3492
[16] Alford M, Rajagopal K, Reddy S and Wilczek F 2001 *Phys. Rev. D* **64** 074017
[17] Paulucci L, Ferrer E J, de la Incera V and Horvath J E 2011 *Phys. Rev. D* **83** 043009
[18] Paulucci L, Ferrer E J, Horvath J E and de la Incera V 2013 *J. Phys. G* **40** 125202
[19] Nishida Y and Abuki H 2005 *Phys. Rev. D* **72** 096004
[20] Sun G-F, He L and Zhuang P 2007 *Phys. Rev. D* **75** 096004
[21] He L and Zhuang P 2007 *Phys. Rev. D* **75** 096003
[22] Ferrer E J, de la Incera V, Keith J P and Portillo I 2015 *Nucl. Phys. A.* **933** 229
[23] Ferrer E J and Keith J P 2012 *Phys. Rev. C* **86** 035205
[24] Ferrer E J, de la Incera V and Paulucci L 2015 *Phys. Rev. D* **92** 043010
[25] Rischke D H 2000 P*hys. Rev. D* **62** 054017
[26] Hooft 't 1976 *Phys. Rev. Lett.* **37** 8
[27] Buballa M 2005 *Phys. Rep.* **407** 205
[28] Kitazawa M, Koide T, Kunihiro T and Nemoto Y 2002 *Prog. Theor. Phys.* **108** 929
[29] Ferrer E J, de la Incera V, Keith J P, Portillo I and Springsteen P L 2010 *Phys. Rev. C* **82** 065802
[30] Buballa M and Oertel M 1999 *Phys. Lett. B* **457** 261
[31] Steinheimer J and Schramm S 2014 *Phys. Lett. B* **736** 241
[32] Kashiwa K, Kouno H, Sakaguchi T, Matsuzaki M and Yahiro M 2007 *Phys. Lett. B* **647** 446
[33] Abuki H, Gatto R and Ruggieri M 2009 *Phys. Rev. D* **80** 074019
[34] Bentz W, Horikawa T, Ishii N and Thomas A W 2003 *Nucl. Phys. A* **720** 95
[35] Ferrer E J and de la Incera V 2013 *Magnetism in Dense Quark Matter* (Lect. Notes in Phys. vol. 871) ed. by Kharzeev D, Landsteiner K, Schmitt A andYee H.-U. (Springer Berlin Heidelberg) pp. 399-432
[36] Ferrer E J, de la Incera V, Manreza Paret D, Perez Martinez A and Sanchez A 2015 *Phys. Rev. D* **91** 085041
[37] Feng B, Ferrer E J and de la Incera V 2011 *Nucl . Phys. B* **853** 213
[38] Ferrer E J, de la Incera V and Manuel C 2005 *Phys. Rev. Lett.* **95** 152002
[39] Ferrer E J, de la Incera V and Manuel C 2006 *Nucl. Phys. B* **747** 88
[40] Alford M, Berges J and Rajagopal K 2000 *Nucl. Phys. B* **571** 269
[41] Ferrer E J and de la Incera V 2007 *Phys. Rev. D* **76** 045011
[42] Ferrer E J and de la Incera V 2006 *Phys. Rev. Lett* **97** 122301
[43] Ferrer E J and de la Incera V 2007 *J. Phys. A* **40** 6913
[44] Mandal T and Jaikumar P 2016 *Phys.Rev. D* **94** 074016
[45] Rüster S B, Werth V, Buballa M, Shovkovy I A and Rischke D H 2005 *Phys. Rev. D* **72** 034004
[46] Carignano S, Ferrer E J, de la Incera V and Paulucci L 2015 *Phys. Rev. D* **92** 105018

IOP Conf. Series: Journal of Physics: Conf. Series **861** (2017) 012021 doi:10.1088/1742-6596/861/1/012021

Studies of the neutron star crust

S Schramm[1] and R Nandi

Frankfurt Institute for Advanced Studies, Ruth-Moufang-Str. 1. D-60438 Frankfurt, Germany

E-mail: [1]schramm@fias.uni-frankfurt.de

Abstract. The physics of the neutron star crust is highly relevant for many observable phenomena like the temperature evolution of accreting neutron stars or the evolution of supernovae. In order to investigate the structure and dynamics of the crust we perform extensive molecular-dynamics-type simulations of the nucleons in the crust including the neutralizing effect of the electron background. Simulations for a range of densities, temperatures and proton-to-neutron ratios have been performed. Results for the isospin dependence of the energy per particle and nucleonic correlation functions as obtained from the simulation runs are discussed.

1. Introduction

The structure of neutron stars is highly complex, encompassing a high-density homogeneous core of leptons, nucleons and potentially more exotic particles like hyperons and quarks in the center, followed by a solid crust of exotic nuclear structures and nuclei, finally surrounded by an atmosphere of light nuclei. The crust itself can be divided into an outer region with nuclei arranged in a crystal structure embedded in an electron plasma and the inner crust beyond the neutron drip, which includes a gas of free neutrons. The structure and properties of the crust region are important for many aspects in neutron star and supernova physics like heat transport and neutrino opacities. In order to study the inner crust we developed an efficient code that can be used for large-scale quantum molecular dynamics simulations of the system. In particular we study the development of the pasta phase in the transition region between crust and core. We investigate the sensitivity of the specific pasta structures and the density range of the transition on the isospin dependence of the nuclear interaction. In addition we compute correlation functions that can be used to derive basic transport coefficients of the crustal matter.

The actual composition of the outer crust of the neutron star is still quite uncertain (see [1]). It depends very strongly on the isospin dependence of the nuclear interactions, which is still not very well understood. As a consequence the position of the neutron drip line, beyond which additional neutrons are no longer bound, is quite unclear, in particular in the case of heavier elements. This has been demonstrated, e.g. in [2], where it was shown that the drip point for the example of uranium isotopes varies by more than 70 neutrons, looking at a range of standard Skyrme and relativistic mean field (RMF) approaches. Fig. 1 shows the least bound neutron states in uranium in a specific calculation. It demonstrates a distinct shell closure for a neutron number of 258, stabilizing isotopes up to this value, compared to other calculations that yield a drip point of N = 184. This uncertainty, related to the strength of the various shell closures, continues in the inner crust, where nuclei are immersed in a gas of dripped neutrons and where

IOP Conf. Series: Journal of Physics: Conf. Series **861** (2017) 012021 doi:10.1088/1742-6596/861/1/012021

the system becomes more and more isospin asymmetric. A modeling approach of this region is discussed in the following section.

2. Modeling Approach

The model description of the nuclear crust used in our simulations is based on the so-called Quantum Molecular Dynamics (QMD) - approach that was pioneered a long time ago in [3]. Within this approach the single-nucleon wave function is described by a Gaussian function

$$\psi(\mathbf{r_i}) = \frac{1}{(2\pi C_W)^{3/4}} \exp\left[-\frac{(\mathbf{r_i}-\mathbf{R_i})^2}{4C_W} + i\,\mathbf{r_i}\cdot\mathbf{P_i}\right], \tag{1}$$

with center-of-mass position R_i and momentum P_i. The main (very significant) approximations employed are that the width parameter C_W is chosen to be constant in time and the total wave function of the system is taken as product of the single-particle wave functions. This in the end leads to classical equations of motion of the particle positions and momenta R_i, P_i, where the densities involved in determining the forces are given by the overlap densities of the Gaussian wave functions. For an extensive discussion of the formalism see [4]. The dynamics of the system is governed by a Skyrme-type Hamiltonian. The overall structure of the Hamiltonian is given by

$$H = T + V_{\mathrm{Pauli}} + V_{\mathrm{Skyrme}} + V_{\mathrm{sym}} + V_{\mathrm{MD}} + V_{\mathrm{Coul}} \quad . \tag{2}$$

T is the kinetic energy and V_{Pauli} is a phenomenological Pauli potential which approximates the effect of the fully antisymmetrized many-body wavefunction [5]. V_{Skyrme} is the nucleon-nucleon Skyrme-like interaction with a momentum-dependent part V_{MD} and V_{Coul} contains the Coulomb interaction. V_{sym} is the isospin-dependent part of the potential. In this work we follow closely the formulation developed in [6], extending this treatment by non-linear isospin-dependent interactions in order to study isospin effects in more detail. All terms are explicitly discussed in [7].

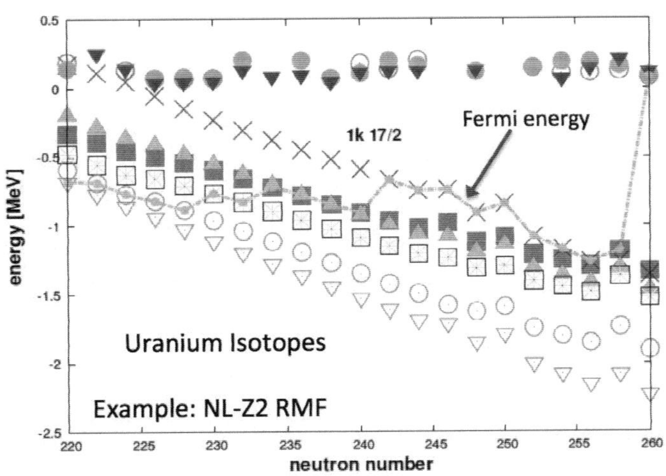

Figure 1. Single-particle energies of the least bound neutron states in the uranium isotope chain. The results are obtained in an RMF calculation using the NL-Z2 parameter set. The line shows the Fermi energy. A clear shell closure for N=258 can be observed.

IOP Conf. Series: Journal of Physics: Conf. Series **861** (2017) 012021 doi:10.1088/1742-6596/861/1/012021

3. Results of the Simulations

Comprehensive numerical studies of the inner crust of neutron stars require extensive long-time simulations for a variety of conditions, including scans of relevant densities, proton-to-neutron ratios, and temperatures, as well as repeating such scans for different interaction parameters.

After establishing that the properties of saturated nuclear matter and nuclear binding energies have reasonable values (see [7]), a closer study of isospin-asymmetric matter follows. In order to test this aspect we calculate the isospin dependence of the energy density around nuclear saturation $\rho_0 \approx 0.165\text{fm}^{-3}$. In the numerical runs this is simply done by changing the relative number of protons and neutrons in the simulation box. The results are shown in Fig. 2. In the left panel one can clearly observe significant deviations from a simple parametrization of the density-dependent energy per particle ϵ as quadratic function like

$$e(\rho, \delta) = e_0(\rho) + e_{\text{sym}}(\rho)\delta^2 \tag{3}$$

where δ is defined as $\delta = (\rho_n - \rho_p)/(\rho_n + \rho_p)$ and is related to the proton over nucleon ratio Y_p via $\delta = 1 - 2Y_p$. Adding a further term to the parametrization as $e_{\text{sym},4}(\rho)\delta^4$ [8] leads to a very good fit of the numerical results as shown in the right panel of the figure. This demonstrates that one has to take into account higher order terms in describing the isospin dependence of nuclear matter energies if one wants to describe a larger range of isospin values beyond symmetric nuclear matter.

In a range of densities of the inner crust close to the transition to the core, the transition from the regular lattice structure to the homogeneous core might give rise to unusual shapes of nuclei as was pointed out originally in [9, 10], generating the so-called nuclear phasta phase. The typical sequence of structures expected, starting from spherical clusters/nuclei, are rod-like nuclei and then slabs, followed by the inverted configurations as rod-like and spherical holes in nuclear matter. In order to quantitatively assess what types of structures are present at some given density in the simulated nucleonic configurations, one can calculate the so-called Minkowski functionals that can be used to characterize the shape of a 3D object [11].

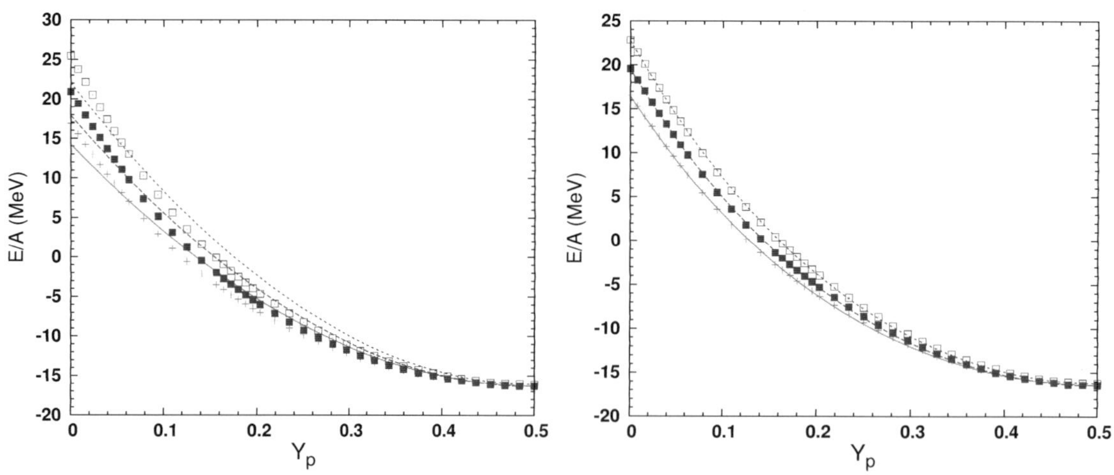

Figure 2. Energy per particle as function of Y_p. The simulation results are shown as symbols for three different densities $0.9\,\rho_0$, ρ_0, and $1.1\,\rho_0$, respectively. The left panel displays the result of a quadratic fit to the numerical values, whereas the right panel shows the corresponding result for a fit up to quartic terms in the asymmetry.

The four independent functionals are the volume and surface of the object as well as the mean integral curvature H and Euler characteristic χ defined as

$$H = \frac{1}{2}\int_{\partial K}(\kappa_1 + \kappa_2)dA \qquad \chi = \frac{1}{2\pi}\int_{\partial K}\kappa_1 \cdot \kappa_2\, dA\,. \tag{4}$$

implying an integral over the surface ∂K of the object. κ_1 and κ_2 are the principal curvatures at the various surface points. As for rods and slabs at least one of the curvatures vanishes, the corresponding value of χ is zero as well. On the other hand, entering the phase with hole-like structures, switches the sign of H from positive to negative values. In practice, given a simulation-generated configuration, numerical calculations using the formula (4) first require the actual determination of the surface. In practice this can be done by defining a density threshold which divides clustered regions from the outside. Fig. 3 shows such a calculation of the values of H and χ averaged over the simulation box for different baryon densities as a function of this threshold ρ_{th}. The figure shows that there is a range of densities, where the values of the functionals are reasonably constant allowing for a reliable determination of their value. Using this procedures the averaged values of H and χ are shown in Fig. 4. The different curves show results for changing the isospin-related interaction parameters leading to distinct slope parameters L as indicated in the figure. Overall the isospin dependence turns out to be rather weak. As argued before, one can clearly observe the sign-change of H as function of density. In the right panel it can be observed that the Euler characteristic attains negative values in the intermediate range of densities. This comes about as one can also express the value of χ the as number of connected regions plus the number of cavities minus the number of connecting tunnels [11]. Thus the figure shows the development of tunnels between nuclear slabs [7]. This result stresses the complexity of the pasta region as there are many competing nearly degenerate states of matter, therefore even for low temperatures the system might consist of a superposition of various structures..

One important aspect of studying the pasta phase of the neutron star crust is to investigate the transition from the pasta to the homogeneous matter of the stellar core. This can be done

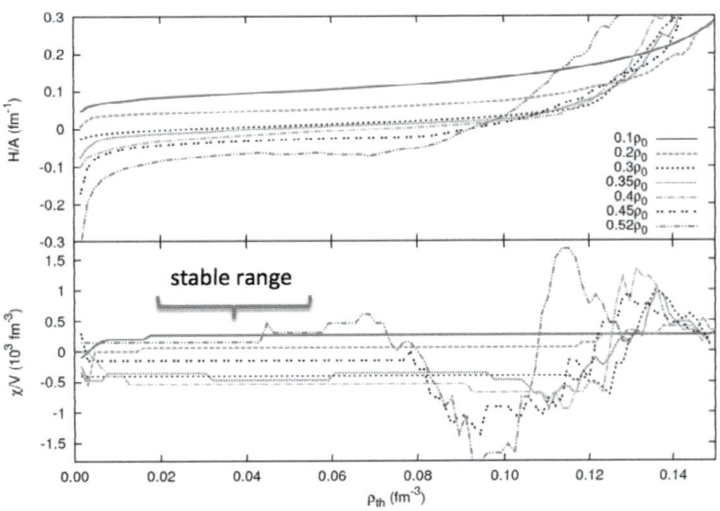

Figure 3. Averaged Minkowski functionals H/A and χ/V as function of the threshold density ρ_{th} as defined in the text. The plot shows results for different baryon densities.

by calculating the two-nucleon correlation function defined as [7]

$$\xi_{NN} = \langle \triangle_N(\mathbf{x})\, \triangle_N\,(\mathbf{x}+\mathbf{r})\rangle \quad . \tag{5}$$

The average is taken over the position $\mathbf{x}$ and the direction of $\mathbf{r}$. $\triangle_N(\mathbf{x})$ denotes the deviation of the nucleon density field $\rho_N(\mathbf{x})$ from the average:

$$\triangle_N = \frac{\rho_N(\mathbf{x}) - \rho_{\mathrm{av}}}{\rho_{\mathrm{av}}} \quad . \tag{6}$$

where $\rho_{\mathrm{av}} = \mathcal{N}/V$ is the average density of nucleons. In Fig. 5 the simulation results of the correlation is shown for densities around the transition point of around $\rho \approx 0.65\,\rho_0$ to the core, comparing different isospin parameters characterized by the indicated values of the slope parameter L given as the change of nuclear symmetry energy with density

$$L = 3\rho_0 \left.\frac{\partial e_{\mathrm{sym}}(\rho)}{\partial \rho}\right|_{\rho=\rho_0} \quad . \tag{7}$$

There is a rather clear and rapid transition for which the correlation changes significantly within a small density window, indicating that in the case of infinite systems the crust-core transition is of first-order type. The two sets of isospin parameters result in slightly different results, but do not affect the overall conclusion.

Correlation functions are important input for determining a variety of transport properties of the matter. In particular correlations in momentum space enter the expressions for electrical and heat conductance as well as shear viscosity [12]. The corresponding structure factor S_i(q) for protons and neutrons (i = p, n) is given by

$$S_i(\mathbf{q}) = <\rho_i^*(\mathbf{q})\rho_i(\mathbf{q})> \tag{8}$$

with the Fourier-transformed densities $\rho_i(\mathbf{q})$. Determining this value for different temperatures gives the results presented in Fig. 6. In this case the numerical runs were performed for a baryon

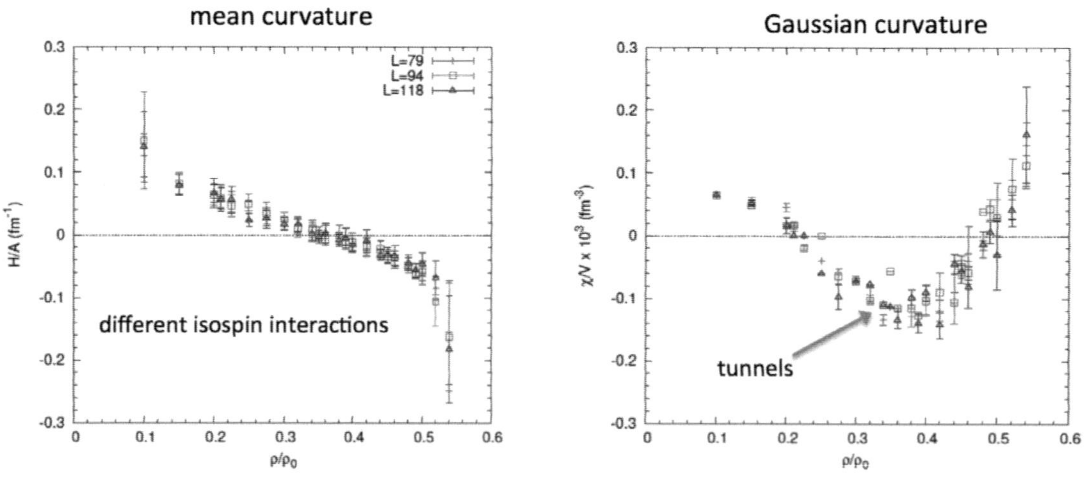

Figure 4. Snapshots of structures obtained in the QMD simulation for various densities for a proton-to-neutron ratio of 0.3. The green and red symbols correspond to neutrons and protons, respectively.

density of $0.1\,\rho_0$, below the onset of the pasta phase, again for a proton fraction $Y_p = 0.3$. The two panels show results for different temperatures $T = 1$ and 4 MeV. Thus, the parameters correspond to values that are relevant during the evolution of the matter in supernova explosions. The structure factor exhibits clear peaks that correspond to the cluster sizes in the system. Note that these structures survive somewhat reduced also for higher temperatures. This can have significant consequences for the heat transport and the supernova evolution, which needs to be investigated in more detail. Work along this line is in progress.

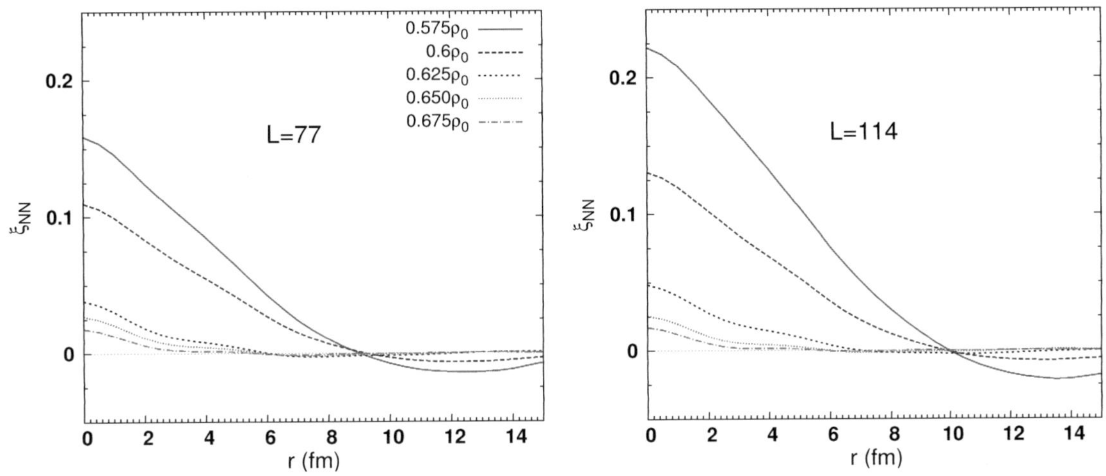

Figure 5. Nucleon-nucleon correlation function for different densities and slope parameters L. The results were obtained for a proton to nucleon ratio of 0.3 .

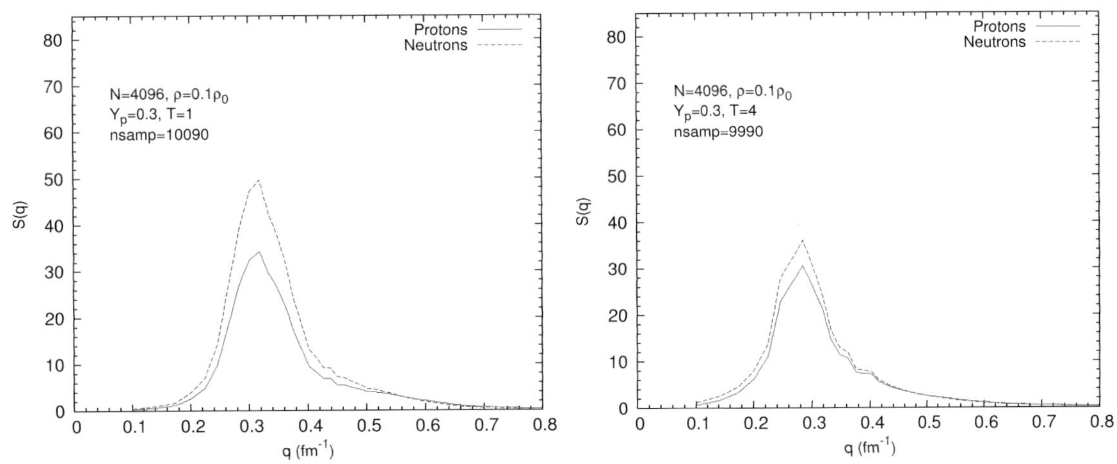

Figure 6. Snapshots of structures obtained in the QMD simulation for various densities for a proton-to-neutron ratio of 0.3. The green and red symbols correspond to neutrons and protons, respectively.

Acknowledgments
The authors acknowledge support by the LOEWE program HIC for FAIR. The numerical calculations were performed on the LOEWE-CSC cluster at the Goethe University in Frankfurt.

References
[1] Rüster S B, Hempel M and Schaffner-Bielich J 2006 *Physical Review C* **73** 035804 ISSN 0556-2813
[2] Schramm S, Gridnev D, Tarasov D V, Tarasov V N and Greiner W 2012 *International Journal of Modern Physics E* **21** 1250047 ISSN 0218-3013
[3] Aichelin J and Stöcker H 1986 *Physics Letters B* **176** 14–19 ISSN 03702693
[4] Aichelin J 1991 *Physics Reports* **202** 233–360 ISSN 03701573
[5] Dorso C, Duarte S and Randrup J 1987 *Physics Letters B* **188** 287–294 ISSN 03702693
[6] Maruyama T, Niita K, Oyamatsu K, Maruyama T, Chiba S and Iwamoto A 1998 *Physical Review C* **57** 655–665 ISSN 0556-2813
[7] Nandi R and Schramm S 2016 *Physical Review C* **94** 025806 ISSN 2469-9985
[8] Chen L W, Cai B J, Ko C M, Li B A, Shen C and Xu J 2009 *Physical Review C* **80** 014322 ISSN 0556-2813
[9] Ravenhall D, Pethick C and Wilson J 1983 *Physical Review Letters* **50** 2066–2069 ISSN 0031-9007
[10] Hashimoto M a, Seki H and Yamada M 1984 *Progress of Theoretical Physics* **71** 320–326 ISSN 0033-068X
[11] Michielsen K and De Raedt H 2001 *Physics Reports* **347** 461–538 ISSN 03701573
[12] Horowitz C J and Berry D K 2008 *Physical Review C* **78** 035806 ISSN 0556-2813

IOP Conf. Series: Journal of Physics: Conf. Series **861** (2017) 012022 doi:10.1088/1742-6596/861/1/012022

A unified description for strange quark matter objects

Cheng-Jun Xia

School of Information Science and Engineering, Ningbo Institute of Technology, Zhejiang University, Ningbo 315100, China and CAS Key Laboratory of Theoretical Physics, Institute of Theoretical Physics, Chinese Academy of Sciences, Beijing 100190, China

E-mail: cjxia@itp.ac.cn

Abstract. A unified description for strange quark matter (SQM) objects ranging from strangelets to strange stars is presented, i.e., the UDS model. The important differences on the properties of SQM objects resulted from introducing the UDS model and conventional treatments are discussed. The previously neglected effects such as charge screening, quark depletion, and electrons in conventional treatments are found to be important for the charge properties and stability of strangelets as well as the surface structures of strange stars, which are now well addressed in the UDS model.

1. Introduction

It was proposed decades ago that strange quark matter (SQM) consisting of roughly equal numbers of u, d, s quarks is the true ground state of QCD [1, 2]. In that case, there may exist stable SQM objects with baryon numbers ranging from a few to $\sim 10^{57}$, e.g., strangelets [3, 4, 5, 6, 7], nuclearites [8, 9], meteorlike compact ultradense objects [10], and strange stars [11, 12, 13].

Extensive efforts were made trying to understand the properties of such objects. For example, adopting the MIT bag model, a strangelet in β-equilibrium is found to be slightly positively charged [4]. Further investigations indicate that their properties depend strongly on model parameters [14]. Greiner et al. considered the possible formation of strangelets via heavy-ion collisions [15]. They found strangelets that are stable against strong decay are highly negatively charged [16]. The stability of strangelets in various shapes was investigated by Mustafa and Ansari, where a spherically shaped strangelet is found to be more stable than others [17]. Moreover, other effects such as color superconductivity [7, 18, 19, 20], perturbative interactions [21, 22], and external magnetic field [23, 24, 25, 26] on SQM objects were discussed extensively as well.

However, significant simplifications were adopted in those studies. For strangelets, electrons were ignored and quarks were assumed to be uniformly distributed. For strange stars, the local charge neutrality condition was required so their structures can be obtained by solving the Tolman-Oppenheimer-Volkov (TOV) equations. Due to the simplicities, these conventional theoretical treatments for SQM objects do not describe the effects such as charge screening, electron-positron pair creation, and nonzero charge densities in strange stars, which could have important implications on the properties of SQM objects [27]. To incorporate all those effects,

IOP Conf. Series: Journal of Physics: Conf. Series **861** (2017) 012022 doi:10.1088/1742-6596/861/1/012022

recently we have proposed a unified description [27, 28, 29] for SQM objects ranging from strangelets to strange stars, called the UDS model [30]. In this paper, we present the UDS model and compare it with the conventional treatments for strangelets and strange stars.

The paper is organized as follows. In Sec. 2 the theoretical framework of the UDS model is presented. The differences of introducing the UDS model and more conventional treatments for SQM objects are examined in Sec. 3. Our conclusion is given in Sec. 4.

2. Theoretical framework

In the UDS model, the particle distributions inside SQM objects are obtained by minimizing the mass at given total numbers of particles. For a spherically symmetric SQM object, the local chemical potential $\mu_i(r)$ follows

$$\bar{\mu}_i = \mu_i(r)\mathrm{e}^{\nu(r)/2} + q_i\varphi(r) = \text{constant}. \tag{1}$$

Here the metric element $\mathrm{e}^{\nu(r)}$ and electric potential $\varphi(r)$ are obtained by integrating

$$\frac{\mathrm{d}\nu}{\mathrm{d}r} = \frac{2G\mathrm{e}^{\lambda}}{r^2}\left[4\pi r^3\left(P - \frac{\alpha Q^2}{8\pi r^4}\right) + M_\mathrm{t}\right] \text{ and } \frac{\mathrm{d}\varphi}{\mathrm{d}r} = -\frac{\alpha Q}{r^2}\mathrm{e}^{(\lambda+\nu)/2}, \tag{2}$$

where $\mathrm{e}^{-\lambda} = 1 - \frac{2G}{r}M_\mathrm{t}$ with the total mass M_t and particle number N_i determined by

$$M_\mathrm{t}(r) = \int_0^r 4\pi r'^2\left(E + \frac{\alpha Q^2}{8\pi r'^4}\right)\mathrm{d}r' \text{ and } N_i(r) = \int_0^r 4\pi r'^2 n_i \mathrm{e}^{\lambda/2}\mathrm{d}r'. \tag{3}$$

Note that the natural system of units is adopted here, while the gravitational and fine-structure constants are given by $G = 6.707 \times 10^{-45}$ MeV^{-2} and $\alpha = 1/137$. In Eq. (3), $n_i(r)$ represents the local number density of particle type i. The total charge can then be obtained with $Q(r) = \sum_i q_i N_i(r)$ ($q_u = 2/3$, $q_d = q_s = -1/3$, and $q_e = -1$). Meanwhile, the pressure $P(r)$ and energy density $E(r)$ indicate the local properties of SQM, which are obtained based on the bag model at a given chemical potential $\mu_i(r)$. The thermodynamic potential density of the bag model reads

$$\Omega = -\sum_i \frac{g_i}{24\pi^2}\left[\mu_i\nu_i(\nu_i^2 - \frac{3}{2}m_i^2) + \frac{3}{2}m_i^4 \ln\frac{\mu_i+\nu_i}{m_i}\right] + B, \tag{4}$$

where $\nu_i = \sqrt{\mu_i^2 - m_i^2}$ and g_i ($g_u = g_d = g_s = 6$, $g_e = 2$) are the Fermi momentum and degeneracy factor. The quark and electron masses are taken as $m_u = 2.3$ MeV, $m_d = 4.8$ MeV, $m_s = 95$ MeV, and $m_e = 0.511$ MeV [31]. Note that for stable SQM objects, the β-equilibrium condition $\mu_u + \mu_e = \mu_d = \mu_s$ needs to be fulfilled. The bag constant B is fixed according to the Witten-Bodmer hypothesis, which lies within the stability window $B^{1/4} = 152 \pm 7$ MeV and we take its central value. Based on the thermodynamic potential density, the basic thermodynamic relations give the particle number density $n_i = -\frac{\partial\Omega}{\partial\mu_i}$, energy density $E = \Omega + \sum_i \mu_i n_i$, and pressure $P = -\Omega$.

On the surface of an SQM object, the surface effects resulted from quark depletion need to be considered, where we have adopted the multiple reflection expansion (MRE) method [5, 32] with a modification to the density of states

$$\frac{\mathrm{d}N_i^\mathrm{surf}}{\mathrm{d}p_i} = n_i^\mathrm{surf}(p_i) = -\frac{g_i R^2 p_i}{\pi}\arctan\left(\frac{m_i}{p_i}\right) + \frac{2g_i R}{3\pi}\left[1 - \frac{3p_i}{2m_i}\arctan\left(\frac{m_i}{p_i}\right)\right]. \tag{5}$$

Table 1. Effects considered in various treatments. For Star II, the electron mass is neglected.

Model	Coulomb interaction	Charge screening	Quark depletion	Electron	Gravity
UDS [27, 28, 29]	✓	✓	✓	✓	✓
Slet I [33, 21]	✓		✓		
Slet II [34, 35]			✓		
Star I [11, 13, 22]				✓	✓
UDS* [29]	✓	✓		✓	✓
Star II [12, 36]	✓			✓	

Here p_i is the momentum of quarks with an upper limit $\nu_i(R)$. The corrections to the total energy and pressure can then be obtained with $\bar{E}_i^{\mathrm{surf}} = \int_0^{\nu_i(R)} \sqrt{p_i^2 + m_i^2} n_i^{\mathrm{surf}}(p_i) dp_i$ and $P^{\mathrm{surf}} = -\sum_i \frac{\mathrm{d}\bar{E}_i^{\mathrm{surf}}}{\mathrm{d}V}\Big|_{N_i^{\mathrm{surf}}}$. Due to the effects of gravity, the mass of an SQM object is modified by $M^{\mathrm{surf}} = \sum_i \bar{E}_i^{\mathrm{surf}} \mathrm{e}^{\nu(R)/2}$. Meanwhile, the quark-vacuum interface should meet the requirement of dynamic stability, i.e.,

$$P(R) - P_e(R) + P^{\mathrm{surf}} = 0. \tag{6}$$

The structure of an SQM object is then determined by solving Eq. (1) while fulfilling the boundary conditions $M_{\mathrm{t}}(0) = 0$, $Q(0) = Q(\infty) = 0$, and Eq. (6). For more details of the calculation, please refer to Refs. [29, 30].

3. Properties of SQM objects predicted by various theoretical treatments

To show the differences resulted from introducing various types of treatments, we investigate the properties of SQM objects obtained with the UDS model and other conventional treatments.

3.1. Small SQM objects

For small SQM objects such as strangelets, the conventional treatments can be attained by ignoring the nonuniform internal distribution of quarks, i.e., adopting the particle density profile $n_i(r) = n_i \theta(R - r)$ instead of solving Eq. (1). The structure of a strangelet can then be determined by fulfilling the β-stability and dynamic stability conditions, i.e., minimizing the mass with respect to the particle numbers and radius at a given total baryon number. Depending on whether or not the Coulomb interaction is considered, we categorize the traditional treatments for strangelets into two types, i.e., Slet I and Slet II. For Slet I, the Coulomb energy $M_{\mathrm{C}} = \frac{\alpha Q_{\mathrm{V}}^2}{10R} + \frac{\alpha Q^2}{2R}$ is included [33], where $Q_{\mathrm{V}} \equiv \frac{4}{3}\pi R^3 \sum_i q_i n_i$ is the volume term of the total electric charge $Q \equiv Q_{\mathrm{V}} + Q^{\mathrm{surf}}$. A detailed demonstration of the effects considered in Slet I(II) is indicated in Table 1. For the cases obtained with constant surface tensions, extensive discussions can be found in Ref. [29].

The important differences resulted from introducing different treatments for strangelets are their charge properties. In Fig. 1 we present the charge-to-mass ratio $f_Z \equiv Q/A$ and surface charge density $\sigma \equiv Q/R^2$ obtained with three types of treatments. When the UDS model is adopted, we assume global charge neutrality with $Q(\infty) = 0$. However, it is meaningful to consider the charge carried by the quark part $Q \equiv Q(R)$, i.e., neglecting the surrounding electrons. The obtained properties of a strangelet with various treatments coincide with each other at $A \lesssim 10$. For larger strangelets, the charge-to-mass ratio f_Z predicted by Slet II

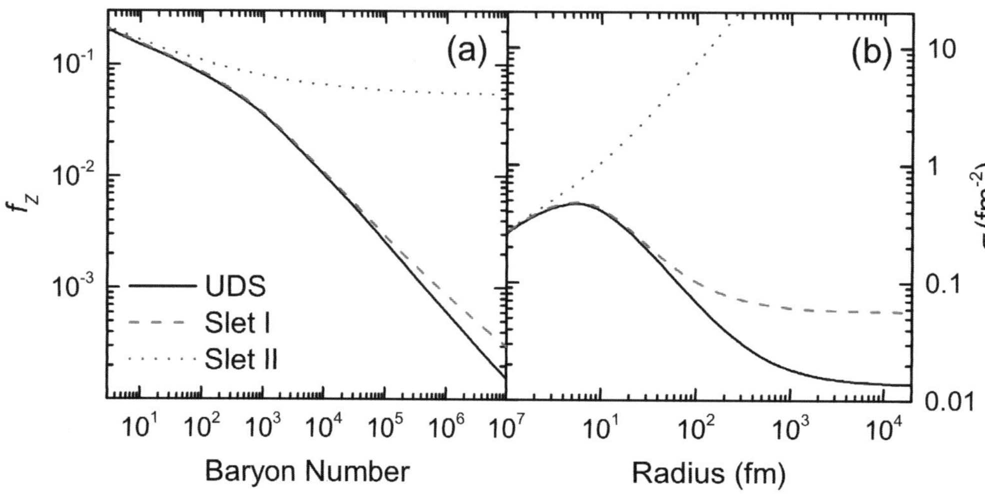

Figure 1. The charge-to-mass ratio f_Z (a) and surface charge density σ (b) of strangelets.

approaches to a constant value at $A \gtrsim 10^5$, which deviates from the results obtained with other treatments including Coulomb interaction ($f_Z \to 0$). The UDS model and Slet I give the same charge number for strangelets up to $A \approx 10^5$ ($R \approx 40$ fm). For larger ones, the surface charge density σ start to approach to constant values, where more positively charged strangelets are obtained with Slet I comparing with the UDS model. The difference is mainly caused by the charge screening effects of SQM, suggesting the important consequences of nonuniform distribution of quarks on the charge properties of SQM objects with $R \gtrsim 40$ fm.

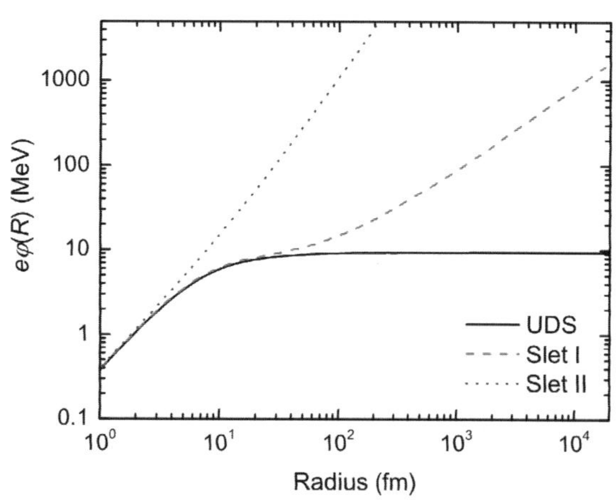

Figure 2. The electric potential on the quark-vacuum interface at $r = R$.

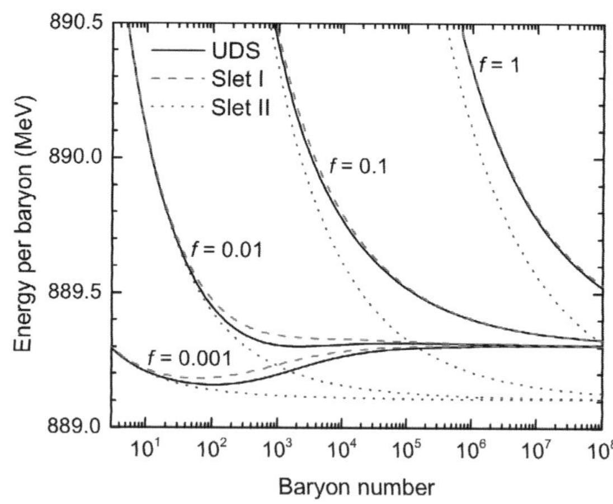

Figure 3. The energy per baryon of strangelets obtained with various treatments.

On the other hand, the effects of electrons should not be neglected for large SQM objects. In fact, as indicated in Ref. [37], the creation of electron-positron pairs in supercritical electric fields plays a crucial role on the maximum net charge an object can carry, which gives $Q_{\max} = 0.71R_{\rm fm}$ ($400 \lesssim R \lesssim 10^4$ fm) and $7 \times 10^{-5} R_{\rm fm}^2$ ($R \gtrsim 10^4$ fm). According to Fig. 1, the charge numbers Q of SQM objects predicted by Slet I and Slet II are much larger than $Q_{\max}$, in which case the

IOP Conf. Series: Journal of Physics: Conf. Series **861** (2017) 012022 doi:10.1088/1742-6596/861/1/012022

electron-positron pair production is inevitable. To show the consequences of neglecting electrons in the conventional treatments, in Fig. 2 we present the electric potential on the quark-vacuum interface at $r = R$. For small strangelets with $R \lesssim 40$ fm, the effects of electrons are not significant, where the electric potentials $\varphi(R)$ predicted by the UDS model and Slet I coincide with each other. For the cases obtained with Slet II, the values of $\varphi(R)$ are much larger because the corresponding strangelets are more positively charged. At $R \gtrsim 40$ fm, $\varphi(R)$ obtained with Slet I starts to increase with radius R, while the one with the UDS model approaches to a constant value. The reason is twofold: (1) the two treatments predict different charge numbers for SQM objects; (2) electrons were not considered in the conventional treatments. This indicates the necessity of including electrons for the study of SQM objects.

Aside from the differences on the charge properties for small SQM objects, other quantities such as the mass, radius, and energy per baryon of a strangelet at a given baryon number are insensitive to different theoretical treatments. This is a natural consequence of the dominant effects of quark depletion, which is on the order of 100 MeV for the energy per baryon of strangelets at $A \approx 10$. Correspondingly, the energy reduction by allowing a strangelet to carry positive charge is around 0.2 MeV, which is negligible in comparison. However, this situation can be easily altered if their energy contributions become comparable under certain circumstances. For example, the energy reduction due to the relocation of charged particles on the quark-baryon interface is greatly enhanced for the quark-baryon mixed phase, which supports the existence of various geometrical structures [38, 39, 40, 41, 42, 43, 44]. Alternatively, if the effects of quark depletion are hindered, the charge properties of a strangelet become important. To show this explicitly, we multiply Eq. (5) by a dampening factor f, i.e., $n_i^{\mathrm{surf}} \to f \times n_i^{\mathrm{surf}}$. The properties of strangelets at $f = 1$, 0.1, 0.01, and 0.001 are then investigated with various theoretical treatments, where the corresponding energy per baryon are presented in Fig. 3. Since Coulomb interaction is neglected in Slet II, the obtained masses of strangelets are always smaller than those with other treatments. Meanwhile, the masses predicted by the UDS model are slightly smaller than Slet I, which is attributed to the nonuniform distribution of quarks. For larger SQM objects, the energy per baryon approach to constant values until gravity starts to reduce the energy, which is approximately $0.85R_{\mathrm{km}}^2$ MeV per baryon for SQM objects with $R \lesssim 3$ km. By decreasing the dampening factor f, the energy contribution from quark depletion is reduced. When f is small enough, there exist local minima for the energy per baryon obtained with the UDS model and Slet I, which resemble the cases with small surface tensions [29, 45, 46]. The existence of the local minima has significant implications on the properties of SQM objects, where large strangelets may undergo fission and a strange star's surface fragments into a crystalline crust [47]. Nevertheless, these cases are quite subtle and sensitive to the theoretical treatments. For example, when Slet II is adopted, there does not exist a strangelet that is more stable than others since the Coulomb energy is neglected. If we take $f = 0.01$, the local minimum in Fig. 3 vanishes for strangelets predicted by Slet I, while it persists when the UDS model is adopted. Moreover, the size of the most stable strangelet is also sensitive to the theoretical treatments. It is thus important to consider the effects of charge screening with nonuniform distributed quarks, which affect the charge properties as well as the stability of small SQM objects.

3.2. Large SQM objects

The most important difference between the UDS model and conventional treatments for large SQM objects such as strange stars is the inclusion of nonzero electric charge inside them. According to our previous investigations [27, 28, 29, 30], there exist mainly two types of charge for a large SQM object, i.e., the charge located on the quark-vacuum interface due to the separation of electrons from SQM and the gravity induced charge throughout the object.

To obtain the structure of a strange star, the conventional treatments adopt the TOV equations with the equation of state (EoS) determined by assuming local charge neutrality for

SQM. We refer to such kind of treatments as Star I, where the effects considered are indicated in Table 1. The predicted values for $e^{\nu(r)/2}$ and $\varphi(r)$ are compared with those obtained with the UDS model in Fig. 4, where the central chemical potential of the strange star is taken as $\mu(0) = \mu_u(0) + \mu_d(0) + \mu_s(0) = 1389$ MeV. Since Star I assumes local charge neutrality, the electric potential $\varphi(r)$ remains zero throughout the strange star. However, nonzero values of $\varphi(r)$ are obtained with the UDS model, suggesting the existence of gravity induced charge. In fact, same situation is expected inside neutron stars [48]. The effects of the gravity induced charge on the structures of strange stars are infinitesimal, which accounts for $\sim 10^{-41}$ of the total mass. Note that the density of gravity induced charge will increase if $\left.\frac{\partial n_{\rm ch}}{\partial \nu}\right|_{\varphi} \Big/ \left.\frac{\partial n_{\rm ch}}{\partial \varphi}\right|_{\nu}$ with $n_{\rm ch} \equiv \sum_i q_i n_i$ is larger, which may become large enough to increase the mass and radius of a strange star [49]. Future investigations on this subject are necessary, where the effects of strong magnetic field and Color Superconducting of SQM may be important.

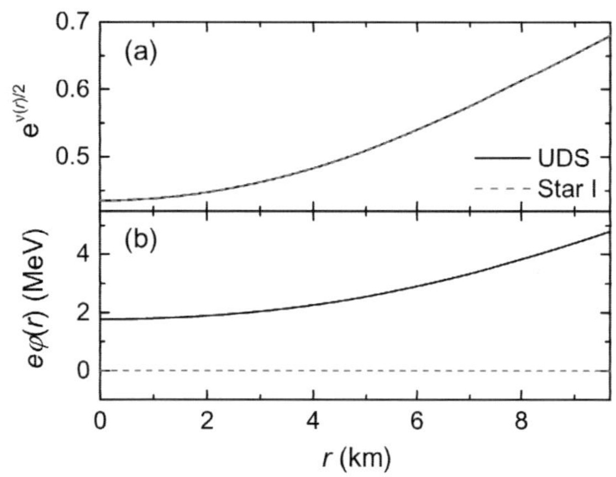

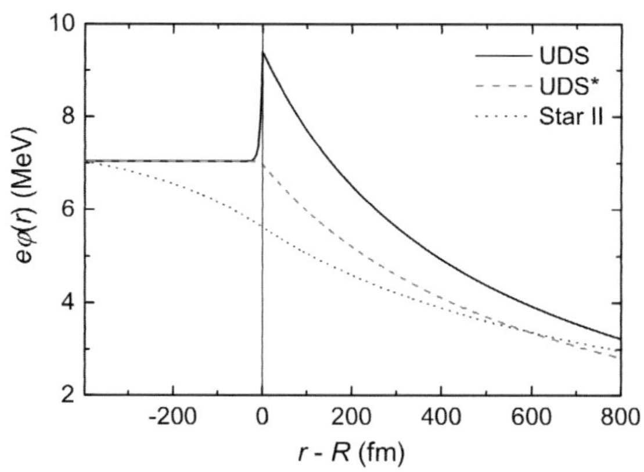

Figure 4. The values of the metric element $e^{\nu(r)/2}$ (a) and electric potential $\varphi(r)$ (b) inside a strange star.

Figure 5. The electric potential $\varphi(r)$ on the surface of a large SQM object.

The structure and charge properties of a strange star's surface were previously investigated assuming constant charge density for quarks and zero mass for electrons [12, 36]. We name such kind of treatments as Star II. For the cases considered here, the charge density of quarks n_q on the surface is obtained at $P = 0$, which gives $n_q \equiv \nu_q^3/3\pi^2$ with $\nu_q = 7.5$ MeV. The electric potential is then determined by solving

$$\varphi'' = \begin{cases} \frac{4\alpha}{3\pi}\left(\varphi^3 - \nu_q^3\right) & (r \leq R), \\ \frac{4\alpha}{3\pi}\varphi^3 & (r > R). \end{cases} \quad (7)$$

To show the differences on the surface charge profiles induced by introducing different treatments, we consider a large SQM object with negligible effects of gravity and present the obtained electric potentials in Fig. 5. The effects considered by Star II, the UDS* and UDS models are indicated in Table 1. Due to the charge screening effects, the electric field predicted by the UDS/UDS* model can penetrate into the SQM by merely ~ 20 fm, while much larger value (~ 400 fm) is obtained with Star II. In such cases, the strength of the electric field obtained with the UDS/UDS* model is much larger. Since quark depletion is considered in the UDS model, an additional potential barrier is formed with the height being ~ 2 MeV, which is caused by the positive surface charge on the quark-vacuum interface. In such cases, negatively charged particles may be trapped and give a distinct photon spectrum when excited, which is relevant to the experimental searches of

SQM. The surface charge densities of the quark part obtained by Star II, the UDS* and UDS models are $\sigma = 0.00438$, 0.00807, and 0.01352 fm^{-2}, respectively. For strange stars, the effects of gravity can not be ignored. According to our calculation [27, 28, 29, 30], the electric potential $\varphi(r)$ and the radial coordinate $(r - R)$ should be multiplied by $e^{\nu(R)/2}$ and $e^{-\lambda(R)/2}$, respectively. The aforementioned differences on the surface structures obtained with various treatments may result in different surface radiation profiles and structures of the thin crusts for strange stars [12]. Note that the masses and radii of strange stars are barely affected by the choice of theoretical treatments, where the differences (within 0.01%) are mainly caused by introducing different conditions for dynamic stability of the surface, i.e., $P = 0$ or Eq. (6). However, a strange star may gain extra mass and radius if it carries net charge on the surface [50]. For a rotating strange star, a magnetic field comparable to the typical strength of pulsars is generated [27, 51]. With the overcritical electric field located on the surface of a strange star, electron-positron pairs can be created, which may be related to some astrophysical events such as supernovae and gamma-ray bursts [52, 53]. Thus, a proper treatment for the surface of a strange star is necessary, where the effects listed in Table 1 should be considered.

4. Conclusion

The UDS model was presented and compared with more conventional treatments on the properties of SQM objects. The bag model was adopted for the local properties of SQM. For strangelets with $R \lesssim 2$ fm, those approaches are equivalent to each other and predict the same results. However, different charge properties were obtained for larger SQM objects. Due to the effects of charge screening, an SQM object obtained with the UDS model is less charged comparing with the conventional treatments. Meanwhile, electrons can not be neglected for the proper determination of the charge properties of SQM objects. When the effects of quark depletion are hindered, nonuniform distribution of quarks become important and will further reduce the mass of an SQM object. For strange stars, on the contrary with previous assumptions of local charge neutrality, we obtained nonzero gravity induced charge. The surface structure of a strange star was found to be sensitive to the theoretical treatments, where the previously neglected effects of charge screening, quark depletion, and gravity are important.

Acknowledgments

This work was supported by National Natural Science Foundation of China (Grant Nos. 11475115, 11575190, 11525524, 11621131001 and 11647601), and National Key Basic Research Program of China (Grant No. 2013CB834400). The computation of this work was supported by the HPC Cluster of SKLTP/ITP-CAS and the Supercomputing Center, CNIC of CAS.

References

[1] Bodmer A R 1971 *Phys. Rev. D* **4**(6) 1601–1606
[2] Witten E 1984 *Phys. Rev. D* **30**(2) 272–285
[3] Chin S A and Kerman A K 1979 *Phys. Rev. Lett.* **43**(18) 1292–1295
[4] Farhi E and Jaffe R L 1984 *Phys. Rev. D* **30**(11) 2379–2390
[5] Berger M S and Jaffe R L 1987 *Phys. Rev. C* **35**(1) 213–225
[6] Gilson E P and Jaffe R L 1993 *Phys. Rev. Lett.* **71**(3) 332–335
[7] Peng G X, Wen X J and Chen Y D 2006 *Phys. Lett. B* **633** 314–318
[8] Rújula A D and Glashow S L 1984 *Nature* **312** 734
[9] Lowder D M 1991 *Nucl. Phys. B (Proc. Suppl.)* **24** 177–183
[10] Rafelski J, Labun L and Birrell J 2013 *Phys. Rev. Lett.* **110**(11) 111102
[11] Itoh N 1970 *Prog. Theor. Phys.* **44** 291–292
[12] Alcock C, Farhi E and Olinto A 1986 *Astrophys. J.* **310** 261–272
[13] Haensel P, Zdunik J L and Schaeffer R 1986 *Astron. Astrophys.* **160** 121–128
[14] He Y B, Chao W Q, Gao C S and Li X Q 1996 *Phys. Rev. C* **54**(2) 857–865

[15] Greiner C and Stöcker H 1991 *Phys. Rev. D* **44**(11) 3517–3529
[16] Schaffner-Bielich J, Greiner C, Diener A and Stöcker H 1997 *Phys. Rev. C* **55**(6) 3038–3046
[17] Mustafa M G and Ansari A 1997 *Phys. Rev. C* **56**(1) 420–425
[18] Madsen J 2001 *Phys. Rev. Lett.* **87**(17) 172003
[19] Alford M G, Schmitt A, Rajagopal K and Schäfer T 2008 *Rev. Mod. Phys.* **80**(4) 1455–1515
[20] Wen X J 2013 *Phys. Rev. D* **88**(3) 034031
[21] Xia C J, Chen S W and Peng G X 2014 *Sci. China-Phys. Mech. Astron.* **57** 1304
[22] Xia C J, Peng G X, Chen S W, Lu Z Y and Xu J F 2014 *Phys. Rev. D* **89**(10) 105027
[23] Felipe R G, Fune E L, Paret D M and Martinez A P 2012 *J. Phys. G* **39** 045006
[24] Wen X J, Su S Z, Yang D H and Peng G X 2012 *Phys. Rev. D* **86**(3) 034006
[25] Hou J X, Peng G X, Xia C J and Xu J F 2015 *Chin. Phys. C* **39** 015101
[26] Isayev A A 2015 *Phys. Rev. C* **91**(1) 015208
[27] Xia C J, Peng G X, Zhao E G and Zhou S G 2016 *Sci. Bull.* **61** 172–177
[28] Xia C J 2016 *Sci. Sin.-Phys. Mech. Astron.* **46** 012021, in Chinese
[29] Xia C J, Peng G X, Zhao E G and Zhou S G 2016 *Phys. Rev. D* **93**(8) 085025
[30] Xia C J and Zhou S G 2017 *Nucl. Phys. B* **916** 669–687
[31] Particle Data Group 2014 *Chin. Phys. C* **38** 090001
[32] Madsen J 1993 *Phys. Rev. Lett.* **70**(4) 391–393
[33] Madsen J 1993 *Phys. Rev. D* **47**(11) 5156–5160
[34] Madsen J 1994 *Phys. Rev. D* **50**(5) 3328–3331
[35] Madsen J 2000 *Phys. Rev. Lett.* **85**(22) 4687–4690
[36] Xu R X and Qiao G J 1999 *Chin. Phys. Lett.* **16** 778
[37] Madsen J 2008 *Phys. Rev. Lett.* **100**(15) 151102
[38] Heiselberg H, Pethick C J and Staubo E F 1993 *Phys. Rev. Lett.* **70**(10) 1355–1359
[39] Voskresensky D, Yasuhira M and Tatsumi T 2002 *Phys. Lett. B* **541** 93–100
[40] Tatsumi T, Yasuhira M and Voskresensky D 2003 *Nucl. Phys. A* **718** 359–362
[41] Voskresensky D, Yasuhira M and Tatsumi T 2003 *Nucl. Phys. A* **723** 291–339
[42] Endo T, Maruyama T, Chiba S and Tatsumi T 2005 *Nucl. Phys. A* **749** 333–336
[43] Maruyama T, Chiba S, Schulze H J and Tatsumi T 2007 *Phys. Rev. D* **76**(12) 123015
[44] Yasutake N, Łastowiecki R, Benić S, Blaschke D, Maruyama T and Tatsumi T 2014 *Phys. Rev. C* **89**(6) 065803
[45] Heiselberg H 1993 *Phys. Rev. D* **48**(3) 1418–1423
[46] Alford M G, Rajagopal K, Reddy S and Steiner A W 2006 *Phys. Rev. D* **73**(11) 114016
[47] Jaikumar P, Reddy S and Steiner A W 2006 *Phys. Rev. Lett.* **96**(4) 041101
[48] Rotondo M, Rueda J A, Ruffini R and Xue S S 2011 *Phys. Lett. B* **701** 667–671
[49] Ray S, Espíndola A L, Malheiro M, Lemos J P S and Zanchin V T 2003 *Phys. Rev. D* **68**(8) 084004
[50] Negreiros R P, Weber F, Malheiro M and Usov V 2009 *Phys. Rev. D* **80**(8) 083006
[51] Negreiros R P, Mishustin I N, Schramm S and Weber F 2010 *Phys. Rev. D* **82**(10) 103010
[52] Ruffini R, Vereshchagin G and Xue S S 2010 *Phys. Rep.* **487** 1–140
[53] Han W B, Ruffini R and Xue S S 2012 *Phys. Rev. D* **86**(8) 084004

IOP Conf. Series: Journal of Physics: Conf. Series **861** (2017) 012023 doi:10.1088/1742-6596/861/1/012023

New aspects of the QCD phase transition in proto-neutron stars and core-collapse supernovae

Matthias Hempel[1], Oliver Heinimann[1], Andrey Yudin[2], Igor Iosilevskiy[3], Matthias Liebendörfer[1] and Friedrich-Karl Thielemann[1]

[1]Department of Physics, University of Basel, Klingelbergstr. 82, CH-4056 Basel, Switzerland
[2]Institute of Theoretical and Experimental Physics, 117218 Moscow, Russia
[3]Joint Institute for High Temperatures (Russian Academy of Sciences), 125412 Moscow, Russia

E-mail: `matthias.hempel@unibas.ch`

Abstract. The QCD phase transition from hadronic to deconfined quark matter is found to be a so-called "entropic" phase transition, characterized, e.g., by a negative slope of the phase transition line in the pressure-temperature phase diagram. In a first part of the present proceedings it is discussed that entropic phase transitions lead to unusual thermal properties of the equation of state (EoS). For example one finds a loss of pressure (a "softening") of the proto-neutron star EoS with increasing entropy. This can lead to a novel, hot third family of compact stars, which exists only in the early proto-neutron star phase. Such a hot third family can trigger explosions of core-collapse supernovae. However, so far this special explosion mechanism was found to be working only for EoSs which are not compatible with the 2 $M_\odot$ constraint for the neutron star maximum mass. In a second part of the proceeding it is discussed which quark matter parameters could be favorable for this explosion mechanism, and have sufficiently high maximum masses at the same time.

1. Introduction

Even after decades of intensive research, it is still not completely understood how massive stars explode. On the one hand, the so-called neutrino-driven mechanism has been shown to be working in multi-dimensional simulations [1, 2], in which hydrodynamic instabilities and non-radial fluid motions increase the neutrino heating efficiency to a sufficient level to lead to shock revival. On the other hand, even the most sophisticated multi-dimensional core-collapse supernova simulations result in rather low explosion energies, which could not explain all of the observations. Furthermore, there is not yet a consensus among different groups about the outcome of simulations and the strength of multi-dimensional effects.

As a consequence, it is not yet known which progenitor stars will end as black holes and which as neutron stars, and also their contributions to nucleosynthesis and galactic chemical evolution are uncertain. Regarding these remaining uncertainties, numerical aspects of both the neutrino transport and the hydrodynamics are of crucial importance. All of the present-day simulations are limited by insufficient computing power, even by using the largest available supercomputers.

However, the supernova dynamics also depend on the employed neutrino-matter interaction rates, and the equation of state (EoS) of hot and dense nuclear matter. At present, these two aspects also carry large uncertainties. For example, we do not know yet the state of matter at the highest densities in the interior of neutron stars. Even if accurate radius measurements of

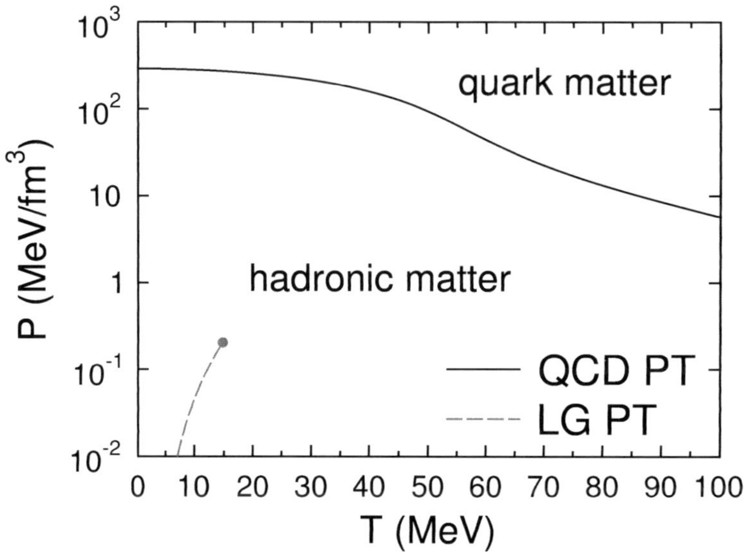

Figure 1. Illustrative phase diagram of isospin-symmetric matter in the temperature-pressure plane. The black solid line shows the QCD phase transition calculated with a simple bag model for up- and down quark matter. The red dashed line shows the nuclear liquid-gas phase transition calculated with the relativistic mean-field model FSUgold [4].

neutron stars will become available in the future, this will not necessarily constrain the interior particle composition, nor the question if a phase transition from hadronic to quark matter takes place in neutron stars. This is known as the so-called masquerade problem [3].

In the present proceeding we investigate the possible role of quark degrees of freedom at high densities and/or temperatures in the context of the explosion mechanism of core-collapse supernovae. In particular, we are interested in the questions how quark matter can influence the dynamics of supernovae, which properties of quark matter are most interesting in this environment, and if there might even be observable signatures of the presence of quark matter in supernovae and proto-neutron stars.

2. The entropic QCD phase transition

In this and the following section we summarize the most important arguments and conclusions of Ref. [5]. When one discusses the state of matter, this is typically done by means of phase diagrams. For the QCD phase diagram usually temperature T and baryon chemical potential μ_B are chosen as state variables. Sometimes instead of μ_B also the baryon number density n_B is used. If one then compares the nuclear liquid-gas phase transition with the QCD phase transition, these two first-order phase transitions have the same qualitative appearance, just the scales are different. Naively one could thus think that they are of the same type.

However, if one compares them in the plane of pressure P and temperature T, as is done in Fig. 1, one sees an obvious qualitative difference: the liquid-gas phase transition line has a positive slope dP/dT, whereas the slope of the QCD phase transition is negative. To distinguish these two types of first-order phase transitions, the terms "enthalpic" ($dP/dT > 0$) and "entropic" ($dP/dT < 0$) were introduced in Refs. [6, 7] and, e.g., used in Ref. [5]. Qualitative differences of the two phase transitions were also noted in Refs. [8, 9, 10, 11]. For a discussion of other interesting aspects of the pressure-temperature phase diagram, see Ref. [12].

As is known from standard textbooks of thermodynamics, the slope of the phase transition line in T-P can be related to the properties of the two coexisting phases by the Clausius-Clapeyron equation:

$$\frac{dP}{dT} = \frac{S^H - S^Q}{1/n_B^H - 1/n_B^Q} \,, \tag{1}$$

where S^H (S^Q) denotes the entropy per baryon of the hadronic (quark) phase, and n_B^H and n_B^Q the corresponding baryon number densities, where $n_B^H < n_B^Q$. The entropic property therefore results from the unusual situation that the more dense phase (in this case the quark phase) has a higher entropy per baryon than the more dilute phase (the hadronic phase). A larger entropy per baryon can in turn be related to a larger specific heat capacity.

It was stressed in Refs. [6, 7] that most of the entropic phase transitions of fluid-fluid type are driven by the same physical mechanism: forced decomposition (delocalization) of some kind of bound complexes under compression (pressure ionization, pressure dissociation, pressure quark deconfinement, etc.). For a phase transition in which "composite" particles are decomposed into their elementary building blocks (such as the QCD phase transition, where hadrons are decomposed into quarks), resulting in a larger number of degrees of freedom, it seems to be a natural consequence that the specific entropy is increased. Indeed, the entropic property has been identified for several different models for the QCD phase transition: in Ref. [10] for the Chiral SU(3) model of Ref. [13], in Fig. 1 for a simple quark bag model, and in Ref. [14] also for the very different "vBag" model. However, it is also known that the heat capacity of quark matter is reduced in color-superconducting or crystalline phases [15]. Therefore it would be quite interesting to see the P-T phase diagrams of such models, to identify whether they also result in entropic phase transitions.

Entropic phase transitions lead to a number of unusual thermodynamical properties, which is connected to negative signs of second-order mixed partial derivatives of the thermodynamic potential that are usually positive [6, 7]. Some of these unusual thermal properties of the EoS are discussed in the following. In case of a simple Maxwellian phase transition, the pressure is independent of density and therefore one has

$$\left.\frac{dP}{dT}\right|_{\rm PT} = \left.\frac{\partial P}{\partial T}\right|_{n_B} \,, \tag{2}$$

where the expression on the left hand side is the derivative along the phase transition line. The EoS of an entropic phase transition therefore has the unusual thermal behavior that $\left.\frac{\partial P}{\partial T}\right|_{n_B} < 0$. This is, for example, opposite to the behavior of all ideal gases. Using standard thermodynamic relationships, one furthermore can show that

$$\left.\frac{\partial P}{\partial T}\right|_{n_B} < 0 \Leftrightarrow \left.\frac{\partial T}{\partial n_B}\right|_{S} < 0 \,. \tag{3}$$

This means that an entropic phase transition gives a decreasing temperature for an adiabatic compression. What is behind Eq. (3) is somewhat more general: in Refs. [6, 7] it was pointed out that an unusual sign of a second-order mixed partial derivative of the thermodynamic potential never occurs isolatedly, but is accompanied by a change of the sign of many other second-order mixed partial derivatives.

These unusual thermal properties are relevant for proto-neutron stars. To describe the hydrostatic structure of relativistic stars, one needs the pressure–energy density relation, $P = P(\epsilon)$. If one considers stars at finite entropy, a useful quantity to characterize thermal effects is therefore the derivative $\partial P/\partial S|_\epsilon$. For $\partial P/\partial S|_\epsilon > 0$ one has an increase of pressure

Table 1. Selected properties of the currently existing general-purpose quark-hadron hybrid EoSs. Listed are the included particle degrees of freedom, the maximum mass $M_{\rm max}$ of cold, spherical (non-rotating) neutron stars and their radii at a fiducial gravitational mass M_G of 1.4 $M_\odot$. Note that for STOSQ184n $M_{\rm max}$ is below 1.4 $M_\odot$ and therefore the value of $R_{1.4M_\odot}$ does not exist. For STOSQ162n the value of $R_{1.4M_\odot}$ could not be found in the literature. This table represents an excerpt from Table III of Ref. [17].

Model Name	Degrees of Freedom	$M_{\rm max}$ ($M_\odot$)	$R_{1.4M_\odot}$ (km)	References
STOSQ209nπ	$n, p, \alpha, (A, Z), \pi, q$	1.85	13.6	[18]
STOSQ162n	$n, p, \alpha, (A, Z), q$	1.54		[19]
STOSQ184n	$n, p, \alpha, (A, Z), q$	1.36	—	[19]
STOSQ209n	$n, p, \alpha, (A, Z), q$	1.81	14.4	[18, 19]
STOSQ139s	$n, p, \alpha, (A, Z), q$	2.08	12.6	[20, 21, 22]
STOSQ145s	$n, p, \alpha, (A, Z), q$	2.01	13.0	[20]
STOSQ155s	$n, p, \alpha, (A, Z), q$	1.70	9.93	[23]
STOSQ162s	$n, p, \alpha, (A, Z), q$	1.57	8.94	[24]
STOSQ165s	$n, p, \alpha, (A, Z), q$	1.51	8.86	[24]

with heating, which can be called "thermal stiffening", and for the opposite sign one has a loss of pressure, which can be called "thermal softening". This derivative can be transformed to

$$\left.\frac{\partial P}{\partial S}\right|_{\epsilon} = -T n_B c_s^2 + \frac{T}{C_V}\left.\frac{\partial P}{\partial T}\right|_{n_B}, \tag{4}$$

where c_s is the speed of sound and C_V is the heat capacity per baryon [16, 5]. The first term in this equation is a relativistic correction and generally smaller than the second term. This implies that an entropic phase transition always has $\left.\frac{\partial P}{\partial S}\right|_{\epsilon} < 0$, corresponding to a softening of the EoS for increasing entropy. In the following section, we discuss the consequences of this softening on the proto-neutron star stability and on the core-collapse supernova dynamics.

3. Consequences for proto-compact stars and core-collapse supernovae

If one wants to explore EoS effects for a realistic setup of proto-neutron stars, core-collapse supernovae, or neutron star mergers, one needs an EoS with full density-, temperature-, and isospin-asymmetry- (or, equivalently, electron fraction Y_e) dependence, covering a wide range of conditions, see Ref. [17]. Usually such "general purpose" EoSs are provided in tabular form. At present just very few general purpose EoSs exist, and even less which consider deconfined quark matter at high densities and/or temperatures. In Table 1 we give an overview about the currently existing general-purpose hybrid EoSs.

All of these EoSs use STOS [25, 26] for the hadronic phase, and a simple bag model of u,d,s-quarks for the quark phase, sometimes including α_S-corrections for strong interactions. All of them assume global charge neutrality during phase coexistence. This assumption, and the choice of T, n_B, and Y_e as state variables, requires to make a Gibbs construction for the phase transition. It results in a so-called non-congruent phase transition, in contrast to a congruent phase transition which one gets for a simple Maxwell construction [27, 10].

From the existing EoSs, only STOSQ139s and STOSQ145s have a sufficiently high maximum mass above 2 $M_\odot$ to be compatible with the pulsar mass measurements of Refs. [28, 29]. They employ bag constants B of $B^{1/4} = 139$ MeV, respectively 145 MeV. However, also these two EoSs

IOP Conf. Series: Journal of Physics: Conf. Series **861** (2017) 012023 doi:10.1088/1742-6596/861/1/012023

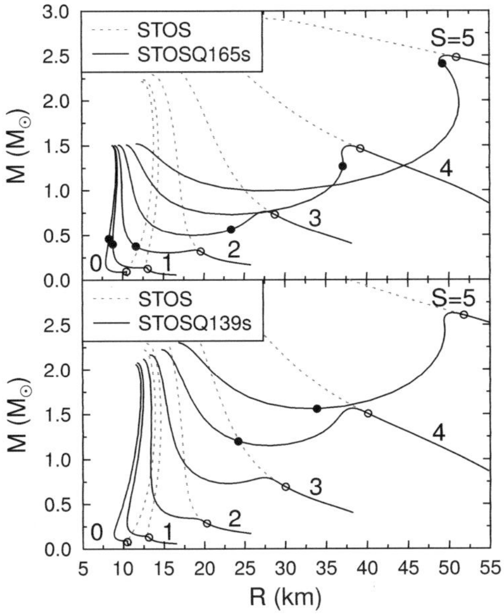

Figure 2. M-R-relations for different entropies per baryon S (indicated by the numbers in the figure) and in beta-equilibrium without neutrinos. Red dotted lines: STOS hadronic EoS. Black solid lines: STOSQ165s hybrid EoS (top panel) and STOSQ139s (bottom panel). Open (full) circles indicate that phase coexistence (pure quark matter) has been reached. Figure adapted from Ref. [5], where further details can be found.

are problematic, as the values of the nuclear symmetry energy and the so-called slope parameter of STOS are in clear contradiction with various experimental constraints [30, 17]. This means there is not a single general-purpose hybrid EoS with good nuclear matter properties and a sufficiently high maximum mass. Nevertheless we use some of the models of Table 1 in the following for further investigation, simply because nothing else is available.

The effects of entropic phase transitions pointed out in Sec. 2 assumed a simple Maxwellian (congruent) phase transition. For the existing general-purpose EoSs one always has a Gibbs (non-congruent) phase transition instead. To investigate whether or not one still has the same effects in this more complicated situation, in Ref. [5] the $P(\epsilon)$-relations of STOS and STOSQ165s were compared with each other. It was found that hadronic matter stiffens when it is heated, i.e., $\left.\frac{\partial P}{\partial S}\right|_{\epsilon} > 0$. In the quark phase, the pressure increases only slightly with entropy, which can be easily understood because the $P(\epsilon)$-relation of ultra-relativistic Fermi gases is independent of the entropy per baryon. For parts of the phase-coexistence regions it was found that $\left.\frac{\partial P}{\partial S}\right|_{\epsilon} < 0$, i.e., matter softens when it is heated. This is the behavior which one expects for an entropic phase transition. However, the unusual thermal properties show up only in parts of the phase-coexistence region.

To investigate the stability of proto-neutron stars, one has to solve the Tolman-Oppenheimer-Volkoff (TOV) equations. Figure 3 shows the resulting mass-radius curves without trapped neutrinos (i.e., in weak equilibrium) for several values of the entropy per baryon, for the hadronic STOS and the hybrid STOSQ165s and STOSQ139s EoSs. One sees that at high enough entropy per baryon a strong third family arises[1]. After the onset of the phase transition, there is a loss of stability which is only regained for much higher central densities, for neutron stars with a sufficiently large quark matter core.

A third family of cold compact stars is already extensively discussed in the literature. The novel aspect which is visible in Fig. 3, and which was first pointed out in Ref. [5], is that this third family exists only in the proto-neutron star stage for high enough entropies. Therefore it is also called the *hot* third family. At zero temperature the third family is only marginal

[1] Different compact star configurations at the same mass but different radii are also called twins.

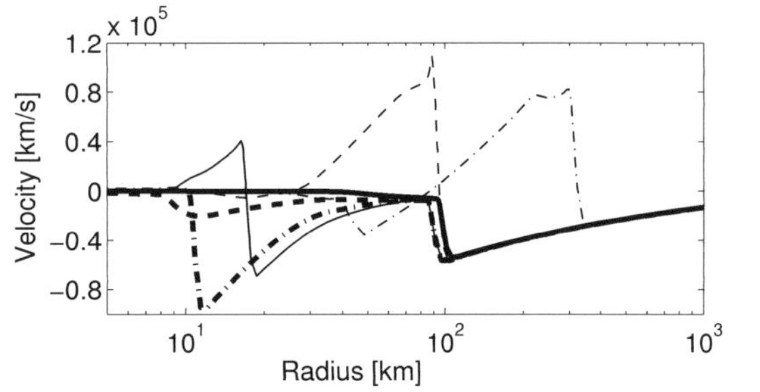

Figure 3. Velocity profiles at different times during the post-bounce evolution of a 10 $M_\odot$ progenitor model based on the STOSQ165s EoS, illustrating the development of the explosion through different stages; thick solid line at 240.5 ms, dashed at 255.2 ms, dash-dotted at 255.5 ms and thin solid line at 255.6 ms, dashed at 256.4 ms, dash-dotted at 258.3 ms. Figure taken from Ref. [24]

or completely absent. The formation of the hot third family is a consequence of the unusual thermal properties of the EoS discussed above. By comparing panel b) with a) of Fig. 3, one sees that for STOSQ139s, which has a high maximum mass, the third family is less pronounced than for STOSQ165s, and occurs only for higher entropies per baryon.

This third family feature is highly relevant for the explosion mechanism of core-collapse supernovae, as explained in the following. During the post-bounce evolution, matter is accreted onto the newly formed proto-neutron star, increasing its mass continuously. In case a local maximum of the mass-radius curve is reached during this accretion, the star will collapse to the more compact configuration and thereby release a significant amount of gravitational binding energy. Exactly this behavior of a collapsing proto-neutron star was observed in the detailed, spherical core-collapse supernova simulations of Refs. [24, 23]. After the collapse of the proto-neutron star, an outgoing shock wave was formed, which merged with the standing accretion shock and eventually resulted in an energetic and robust explosion. Fig. 3 shows the velocity profiles of a simulation using the STOSQ165s EoS.

Interestingly, this explosion mechanism has a clear observational signature. Once the shock runs over the neutrino spheres, a burst of anti-electron neutrinos is released [24, 23, 31]. The time-delay between the first neutrino burst corresponding to bounce and the second neutrino burst induced by the phase transition represents a measure of the phase transition density. Regarding the nucleosynthesis contribution of this special kind of core-collapse supernovae, a weak r-process was found to take place in the ejecta after the onset of the explosion [32].

However, for the very limited set of general purpose hybrid EoSs presently available, explosions could only be obtained for models which have a too low maximum mass. For example the STOSQ139s EoS did not lead to explosions [22]. This is fully consistent with the results presented above that the third family feature is much weaker for STOSQ139s than for STOSQ165s. If the maximum mass is high, the effects of the phase transition are generally weaker, and quark matter behaves similarly as hadronic matter. This can be seen as a dynamic variant of the masquerade problem [3].

As a different consequence of entropic phase transitions in proto-neutron stars, in Ref. [16] it was pointed out that they can lead to a special form of "inverted" convection. In this case the condition for convective stability is opposite as for normal convection, and if convection occurs cold bubbles of matter are transported outwards.

The question whether or not the explosion mechanism of core-collapse supernovae triggered by the QCD phase transition is still possible, can only be answered by developing new quark-hadron general purpose hybrid EoS and testing them in detailed core-collapse supernova simulations. On the one hand, there is the obvious tension between a high maximum mass and a sufficiently strong phase transition. On the other hand, only very few models have been tested so far, regarding both the EoS and the progenitor model. The new insight from Ref. [5], summarized

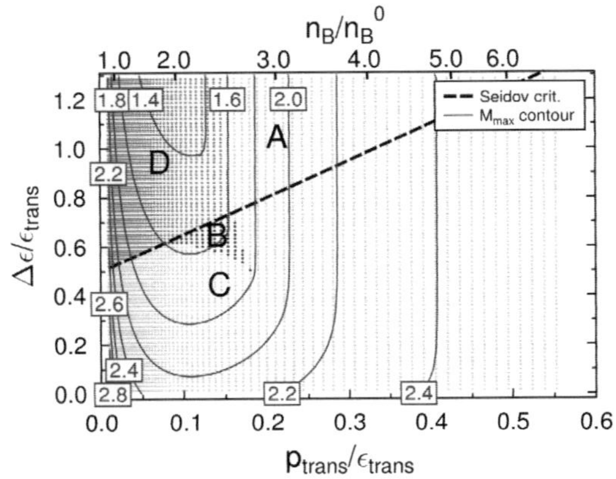

Figure 4. Contour lines for the maximum mass (blue solid lines), for a systematic variation of the phase transition properties $\Delta\epsilon$ and P_{trans}, using HS(DD2) for the hadronic phase, and an EoS with a constant speed-of-sound ($c_s^2 = 1/3$) for the quark phase. The color coding distinguishes the four different types of hybrid stars: "A (absent)", "B (both)", "C (connected)" and "D (disconnected)". Figure taken from Ref. [33].

in the present proceeding, is that the unusual thermal properties of the phase transition and the novel hot third family stand behind this explosion mechanism. This new understanding will help to finally answer the aforementioned question. One direct conclusion is that a third family which is already present in cold neutron stars is helpful, but not strictly required to get a pronounced third family at the proto-neutron star stage. Due to the strong entropy dependence of the third family, one can conclude that the entropy profile of the progenitor star plays an important role.

4. A systematic analysis of cold hybrid stars

With these new conclusions at hand, in Ref. [33] we performed a systematic analysis of the quark matter parameter space, to identify interesting regions favorable for this explosion mechanism. The aim was to fix the quark matter parameters by looking for third-family solutions in cold neutron stars, taking into account the 2 $M_\odot$ constraint. To do so, we employed the method of Ref. [34], where the constant speed-of sound EoS is used for quark matter, so that the phase transition depends only on two parameters, the energy-density discontinuity $\Delta\epsilon$ and the transition pressure P_{trans}. These two parameters are then systematically varied, and for each parameter combination the TOV equations are solved to obtain the mass-radius curve and the maximum mass.

In addition, Ref. [34] introduced four types of hybrid stars for further classification of the solutions. In case a) hybrid stars are absent; all neutron star configurations with quark matter in their interior are unstable. In case c) one has a hybrid star branch which is connected to the branch of hadronic stars. In case d) there is no connected hybrid star branch, but only a disconnected third family. In case b) one has both, i.e., the connected hybrid star branch as in case c), but in addition also the disconnected third family branch as in case d).

For our parameter scan in Ref. [33], we chose the general purpose EoS HS(DD2) [35, 21] for the hadronic phase. This EoS is available in tabular form and can be directly used in simulations of core-collapse supernovae and neutron star mergers, see, e.g., Refs. [36, 37]. HS(DD2) has good nuclear matter properties, in agreement with many experimental constraints, see, e.g., Refs. [21, 38, 17]. Furthermore, its maximum mass of cold neutron stars of 2.42 $M_\odot$ is sufficiently high to allow for additional degrees of freedom at high densities. For example in Ref. [39] the lambda hyperon was added, leading to a still sufficiently high maximum mass of 2.10 $M_\odot$.

Figure 4 shows the results of the parameter scan using a squared speed-of-sound of quark matter of $c_s^2 = 1/3$. As a first result it can be seen in the lower left corner that the phase transition can increase the maximum mass of the hybrid EoS above the value of the purely hadronic EoS. Even values above 2.8 $M_\odot$ are possible. As explained in detail above, it is favorable for the

IOP Conf. Series: Journal of Physics: Conf. Series **861** (2017) 012023 doi:10.1088/1742-6596/861/1/012023

explosion mechanism if one has a third family already for cold neutron stars, which is then enhanced when considering the proto-neutron star stage with finite entropies. Cases b) and d), which both have a third family, are therefore particularly interesting for this scenario. In Fig. 4, case b) occurs only in a very limited parameter region and leads to maximum masses even below 1.8 $M_\odot$. However, there are cases d) with a maximum mass above 2.0 $M_\odot$. These promising configurations require a low phase transition density around saturation density n_B^0 and values of $\Delta\epsilon/\epsilon^{\rm trans} > 0.5$. We found that if a larger value of the speed of sound is considered, the favorable region of cases d) can be significantly extended, but first concentrated on the canonical value of 1/3.

Relatively low phase transition densities are also required for this explosion mechanism for a different reason, namely, because the densities reached in a core-collapse supernova stay below approximately $2n_B^0$ in the early post-bounce phase of an intermediate mass progenitor. If the phase transition density was significantly higher, effects of quark matter could only be expected for very massive progenitors, or at a later stage of the evolution. This would be a different scenario than the one considered in Refs. [24, 23].

Previous parameter scans which used the framework of Ref. [34] did not consider chemical equilibrium in the phase coexistence region explicitly. Pressure equilibrium is automatically achieved, and thermal equilibrium, too, as only zero temperature is used. It is easy to derive the $P(\mu_B)$ relation (with μ_B denoting the baryon chemical potential) of the constant-speed-of-sound EoS formulated as $P(\epsilon)$, where a new free parameter is appearing in form of an integration constant. If one imposes chemical equilibrium at the transition point, this parameter is fixed by the baryon chemical potential of the hadronic phase.

Using this procedure, we found for some parameter combinations that not only one, but several phase transitions occur. For example one possibility is that after a first deconfinement transition a reconfinement from quark to hadronic matter takes place [33]. Such spurious multiple phase transitions were not identified in previous versions of the parameter scan, instead only one deconfinement transition was enforced. This means thermodynamic stability was not taken into account. However, this can also be physically motivated. There is no reason why to expect reconfinement and multiple phase transitions, and they can occur in regimes where either the hadronic or the quark EoS is not reliable any more. On the other hand, they could also point to unphysical parameter combinations of the quark EoS. For some further discussion see Ref. [33] and references therein.

If one takes into account thermodynamic stability and allows for multiple phase transitions in the parameter scan, the maximum masses in the lower left corner of Fig. 4 stay below the value of HS(DD2) [33], which is quite interesting. However, maximum masses above the one of HS(DD2) are still possible, namely for special cases which correspond to absolutely stable quark matter which occur in a different region of the quark matter parameter space [33].

Parts of the small favorable region for the supernova explosion mechanism (case d) with $M_{\rm max} > 2\ M_\odot$ would also be affected from the reconfinement problem. Nevertheless, a slightly reduced region with a third family and a sufficiently high maximum mass is remaining, which does not have this problem.

The constant-speed-of sound EoS of quark matter is not suitable for core-collapse supernova simulations, as it does not provide information about the particle composition or the temperature dependency. However, in Ref. [33] it was shown that the parameters $\Delta\epsilon$ and $P_{\rm trans}$ can be mapped onto a bag model description with strong interaction corrections. In Ref. [33], a particular parameter set from the favorable region was chosen (not affected from the reconfinement problem), and also its bag model parameters were presented. At present, we are working on the generation of a new quark-hadron general-purpose hybrid EoS, employing this parameter set.

IOP Conf. Series: Journal of Physics: Conf. Series **861** (2017) 012023 doi:10.1088/1742-6596/861/1/012023

5. Summary and Conclusions

An entropic phase transition is defined by having a negative slope of the phase transition line in the temperature-pressure plane. Using the Clausius-Clapeyron equation, the negative slope can be related to a higher entropy per baryon of the more dense phase during phase coexistence. In the QCD phase transition hadrons are decomposed into their elementary particles, the quarks, and therefore it seems to be a natural consequence that the QCD phase transition is entropic. So far the authors are not aware of models for the QCD phase transition which predict the opposite (i.e., enthalpic) behavior which one has, e.g., for the nuclear liquid-gas phase transition.

Entropic phase transitions result in unusual thermal properties of the EoS. For example, one has a decrease of temperature for an adiabatic compression, or a decrease of pressure with increasing temperature. They also result in a softening of the $P(\epsilon)$ EoS (i.e., a decrease of pressure) for increasing entropy. This softening can affect the stability of proto-neutron stars. It can lead to a novel, hot third family of compact stars, which exists only at the proto-neutron star stage, if one has high enough entropies. Such a third family is highly relevant for the dynamics of core-collapse supernova, as the collapse from the second to the third family has been shown to be able to trigger explosions. However, so far this mechanism was only achieved for EoSs with too low maximum masses.

To answer the question whether this explosion mechanism is still possible despite the 2 $M_\odot$ constraint, one has to develop new general-purpose hybrid EoSs and test them in numerical simulations. As a first step towards the creation of a new general-purpose hybrid EoS, we performed a systematic parameter scan for the quark matter EoS parameters. For $c_s^2 = 1/3$, there is a small parameter region which leads to sufficiently high maximum masses above 2 $M_\odot$, and has a weak third family feature already for cold neutron stars. This is the parameter region which we consider most promising for the phase-transition induced explosion mechanism of core-collapse supernovae. Besides testing new EoSs, one should also investigate different progenitor stars, because stars with high entropy cores could be particularly favorable for explosions.

References

[1] Janka H T 2012 *Annu. Rev. Nucl. Part. S.* **62** 407–451
[2] Burrows A 2013 *Rev. Mod. Phys.* **85** 245–261
[3] Alford M, Braby M, Paris M and Reddy S 2005 *Astrophys. J.* **629** 969–978
[4] Todd-Rutel B G and Piekarewicz J 2005 *Phys. Rev. Lett.* **95** 122501
[5] Hempel M, Heinimann O, Yudin A, Iosilevskiy I, Liebendörfer M and Thielemann F K 2016 *Phys. Rev. D* **94** 103001
[6] Iosilevskiy I 2013 Enthalpic and entropic phase transitions in high energy density nuclear matter *Physics of Extreme States of Matter* ed V Fortov *et al* (Chernogolovka, Russia: IPCP RAS Publishing) p 136 (*Preprint* `arXiv:1403.8053`)
[7] Iosilevskiy I 2015 Entropic and enthalpic phase transitions in high energy density nuclear matter *Proceedings of the IV Int. Conference Compact Stars in the QCD Phase Diagram, Prerow, Germany, 2014, eConf C140926* ed D Blaschke and T Fischer (*Preprint* `arXiv:1504.05850`)
[8] Satarov L M, Dmitriev M N and Mishustin I N 2009 *Phys. Atom. Nucl.* **72** 1390–1415
[9] Bombaci I, Logoteta D, Panda P K, Providência C and Vidaña I 2009 *Phys. Lett. B* **680** 448–452
[10] Hempel M, Dexheimer V, Schramm S and Iosilevskiy I 2013 *Phys. Rev. C* **88** 014906
[11] Steinheimer J, Randrup J and Koch V 2014 *Phys. Rev. C* **89** 034901
[12] Wambach J 2012 Transport properties of strong-interaction matter *American Institute of Physics Conference Series* (*American Institute of Physics Conference Series* vol 1441) ed Steadman S G, Stephans G S F and Taylor F E pp 794–796 (*Preprint* `arXiv:1111.5475`)
[13] Dexheimer V A and Schramm S 2010 *Phys. Rev. C* **81** 045201
[14] Klahn T, Fischer T and Hempel M 2016 *arXiv:1603.03679*
[15] Buballa M and Carignano S 2016 *Eur. Phys. J. A* **52** 57
[16] Yudin A V, Hempel M, Nadyozhin D K and Razinkova T L 2016 *Mon. Not. R. Astron. Soc.* **455** 4325–4328
[17] Oertel M, Hempel M, Klähn T and Typel S 2016 *Rev. Mod. Phys., accepted* (*Preprint* `arXiv:1610.03361`)
[18] Nakazato K, Sumiyoshi K and Yamada S 2008 *Phys. Rev. D* **77** 103006
[19] Nakazato K, Sumiyoshi K and Yamada S 2013 *Astron. Astrophys.* **558** A50

[20] Sagert I, Fischer T, Hempel M, Pagliara G, Schaffner-Bielich J, *et al* 2012 *Acta Phys. Polon. B* **43** 741
[21] Fischer T, Hempel M, Sagert I, Suwa Y and Schaffner-Bielich J 2014 *Eur. Phys. J. A* **50** 46
[22] Fischer T, Klähn T, Sagert I, Hempel M and Blaschke D 2014 *Acta Phys. Polon. Supp.* **7** 153
[23] Fischer T, Sagert I, Pagliara G, Hempel M, Schaffner-Bielich J, *et al* 2011 *Astrophys. J. Suppl.* **194** 39
[24] Sagert I, Hempel M, Pagliara G, Schaffner-Bielich J, Fischer T, *et al* 2009 *Phys. Rev. Lett.* **102** 081101
[25] Shen H, Toki H, Oyamatsu K and Sumiyoshi K 1998 *Nucl. Phys. A* **637** 435
[26] Shen G, Horowitz C J and Teige S 2010 *Phys. Rev. C* **82** 015806
[27] Hempel M, Pagliara G and Schaffner-Bielich J 2009 *Phys. Rev. D* **80** 125014
[28] Antoniadis J, Freire P C C, Wex N, Tauris T M, Lynch R S, *et al* 2013 *Science* **340** 1233232
[29] Fonseca E, Pennucci T T, Ellis J A, Stairs I H, Nice D J, *et al* 2016 *arXiv:1603.00545*
[30] Hempel M 2015 *Phys. Rev. C* **91** 055807
[31] Dasgupta B, Fischer T, Horiuchi S, Liebendörfer M, Mirizzi A, *et al* 2010 *Phys. Rev. D* **81** 103005
[32] Nishimura N, Fischer T, Thielemann F K, Fröhlich C, Hempel M, *et al* C 2012 *Astrophys. J.* **758** 9
[33] Heinimann O, Hempel M and Thielemann F K 2016 *Phys. Rev. D* **94**(10) 103008
[34] Alford M G, Han S and Prakash M 2013 *Phys. Rev. D* **88** 083013
[35] Hempel M and Schaffner-Bielich J 2010 *Nucl. Phys. A* **837** 210
[36] Pan K C, Liebendörfer M, Hempel M and Thielemann F K 2016 *Astrophys. J.* **817** 72
[37] Bauswein A, Stergioulas N and Janka H T 2014 *Phys. Rev. D* **90** 023002
[38] Hempel M, Hagel K, Natowitz J, Röpke G and Typel S 2015 *Phys. Rev. C* **91**(4) 045805
[39] Banik S, Hempel M and Bandyopadhyay D 2014 *Astrophys. J. Suppl.* **214** 22

IOP Conf. Series: Journal of Physics: Conf. Series **861** (2017) 012024 doi:10.1088/1742-6596/861/1/012024

Pulsar rotation with superfluid entrainment

Marco Antonelli[1] **and Pierre M. Pizzochero**

Department of Physics, Università degli Studi di Milano, and Istituto Nazionale di Fisica Nucleare, Via Celoria 16, 20133 Milano, Italy

E-mail: [1]marco.antonelli@unimi.it

Abstract. Large pulsar glitches (like the ones detected in the Vela) are though to be a consequence of the superfluid component present in the interior of mature neutron stars: this component can rotate differentially with respect to the normal part of the star, storing the angular momentum needed to produce the observed sudden decrease of the pulsar rotational period. However strong entrainment (a non-dissipative effect that couples the superfluid component with the non-superfluid component inside the star) challenges this picture. Here we study the impact of entrainment on the angular momentum that can be exchanged between the normal component and the superfluid during a glitch by means of a consistent global model. This allows to estimate the maximum angular momentum reservoir stored into the superfluid component of the star: the essential ingredient are newly calculated mesoscopic pinning forces that block the superfluid vorticity in the crust of the neutron star. This method can also provide a quantitative test for global models of rotating neutron stars, as well as for microphysical inputs present in literature (like entrainment parameters and pinning forces).

1. Introduction

Glitches are sudden jumps of amplitude $\Delta\Omega$ in the angular velocity Ω of some isolated pulsars with fractional increases that span the range $\Delta\Omega/\Omega \sim 10^{-11} \div 10^{-5}$ in otherwise steadily spinning-down neutron stars (NSs). According to the Jodrell Bank Glitch Catalog[1], more than four hundred glitches have been observed, in more than a hundred objects including rotational-powered pulsars, magnetars and millisecond pulsars. Although there is some evidence that small glitches in the young Crab pulsar are caused by star quakes, the widely accepted explanation for large glitches is based on the presence of a superfluid component (made of paired neutrons) that can rotate differentially with respect to the non-superfluid rest of the NS. The superfluid stores extra angular momentum that can be abruptly exchanged with the normal component, resulting in a little decrease of the observed rotational period of the pulsar (for a review on glitch models see [1]). A rotating superfluid is permeated by many vortex lines (of the order $\sim 10^{19}$ for the Vela pulsar) of quantized circulation. The exchange of angular momentum is mediated by these vortices [2] that induce a mutual friction between the superfluid and normal components [3]. The two physical ingredients that currently challenge the description of pulsar glitches, both at the mesoscopic scale of vortex lines as well as at the macroscopic stellar scale, are entrainment and pinning.

Entrainment - A key feature of superfluidity is the possibility for the normal and superfluid

[1] Glitch data are extracted from the database www.jb.man.ac.uk/pulsar/glitches.html.

components to flow independently although they compenetrate. The formalization of this multi-fluid problem in the interior of NSs is based on the seminal works on the hydrodynamics of He-II that extended the early two-fluid models of Tisza and Landau [4, 5]. Later, with the aim to study superfluid solutions, Andreev and Bashkin developed a three-fluid framework where two super-flows and a normal flow exist simultaneously [6]. They found that entrainment (a non-dissipative effect that couples the two species) can be described in the frame of reference comoving with the normal flow as a "mass-matrix" which relates the momentum of one constituent to the kinematic velocities of both constituents (see also [7–9] for modern formalism with applications to the hydrodynamics of NS interiors). Thus entrainment is also discussed, both in the core and in crust, in terms of effective masses (which definition depend on the reference frame used). Dripped neutrons in NS crusts are analogous to electrons in ordinary metals and their properties can be determined by the band theory of solids [10]: only neutrons in the conduction band can move throughout the crust. Despite the absence of viscous drag, the crust resists the neutron current due to Bragg scattering. This is a non-local and non-dissipative effect: neutrons are scattered by individual clusters and can interfere constructively or destructively. In the latter case neutrons cannot propagate, as if they were very massive [11]. The scenario is different in the core, where neutrons and protons can only be locally scattered by surrounding nucleons [12]. The resulting low effective mass of neutrons in the core can be interpreted in terms of a back-flow of nucleons. This local effect also exists in the crust but it is negligible when compared to Bragg scattering. Entrainment challenges our understanding of pulsar glitches. In particular Vela glitches have been thought to arise from the superfluid in neutron-star crust [13, 14]. However superfluid can be strongly entrained by the crust and this results in a decreased mobility of the crustal superfluid that cannot carry enough angular momentum to explain large glitches [15, 16]. In our numerical examples we use the effective neutron masses given in [12] and in [10] for the core and the crust respectively.

Pinning - Pinning is the mechanism that allows to build up the angular momentum reservoir [2]: when vortices are blocked the normal component slows down due to radiation losses while the pinned superfluid maintains its state of motion, lagging behind and storing angular momentum which can then be released in a glitch.
Pinning is a well known phenomenon that occurs in type-II superconductors, where the flux-tubes can pin to defects of the crystalline structure. Something analogous happens in NSs: when vortex lines are immersed into a background current (of superfluid neutrons) an hydrodynamical lift develops and the vortices tend to drift outside the star (vortex-creep). Strong pinning allows to support higher currents before the dissipative effects generated by the motion of vortices enter the game; this in turns imply that a bigger angular momentum reservoir can be stored into the current of superfluid neutrons. The strength of pinning is described by a mesoscopic quantity f_P that defines the depinning threshold: it is the module of the maximum lift force on vortices that the pinning centers can sustain. At a microscopic scale it is possible to calculate the *individual pinning energies* of a single vortex with a single pinning center [17] but to find the maximal non-dissipative current of neutrons that pinning can sustain we need the *average pinning force* of vortices passing trough many pinning sites [18]. Typically in NSs pinning refers to the interaction of vortices with the crustal lattice and pinning centers are the nuclei of the inner-crust. In this work we will not consider the further possibility of pinning between vortex lines and flux-tubes of the proton superconductor, even if this suggests interesting scenarios [19]. However a vortex-fluxtube pinning profile, that is a consequence of entrainment in the core, can be added to that of crustal pinning we use, without any change in our model.

In this presentation we expand the discussion of the model proposed in [20]. This model, despite its simplicity, can be used to test quantitatively the widespread assumption of a stable

IOP Conf. Series: Journal of Physics: Conf. Series **861** (2017) 012024 doi:10.1088/1742-6596/861/1/012024

vortex array in superfluid NSs. In fact, vortex-creep models (since the seminal work of [21]) do not account for entrainment or stratification. Similarly, global models based on the two-fluid formalism have either neglected entrainment [22] or have been studied under the assumption of uniform entrainment and rigid rotation of the neutron superfluid [23, 24].

2. Application to pulsar glitches: axially symmetric model

The multi-fluid formalism used to describe the superfluid hydrodynamics in NS interiors is well established and in principle enables us to model the glitch and the subsequent relaxation [22, 25–27]. However in a three-dimensional hydrodynamical simulation it is not simple to take into account consistently for stratification, pinning forces and entrainment. We show how to implement consistently these ingredients within a simplified geometry, as a first natural approximation to the full problem. In this way we rule out a priori turbulent motion and its consequences on the vortex-mediated mutual friction [25, 28]: this is a limit of the model to be kept in mind. The model is already detailed in [20], in this section we discuss its features and assumptions:

- The star spins around the z-axis [2] (there is no precession); since we study non-millisecond pulsars, we consider that the equilibrium structure is given by solving the TOV equations. As sketched in Fig. 1, the superfluid is contained into a spherical volume of radius R_d (the radius at which the neutron drip starts, corresponding to a density of $\sim 4 \times 10^{11}$ g cm^{-3}).
- In the absence of a layer of normal matter between the core and the crust [29], the vortex cores can pass continuously through this smooth transition. Leaving out the problem of the unknown composition of the inner core, we consider that the superfluid component (the n-component) is threaded by straight vortices parallel to $\hat{\boldsymbol{e}}_z$. However it is equally simple to consider superfluidity only inside a spherical shell of given radii. The scenario of continuous vortex lines was already suggested in [30] as an alternative to the scenario of distinct vorticity in the crust and in the core.
- The rigid crust and the charged component (i.e. electrons, protons and bounded neutrons, that we call p-component) are locked into the strong magnetic field of the neutron star on short timescales [31]. Electromagnetic losses produce a constant braking torque $-I\dot{\Omega}_\infty \boldsymbol{e}_z$ that acts on the p-component, where I is the total moment of inertia of the star and $\dot{\Omega}_\infty$ the absolute value of the spin-down rate over a long period of time. The velocity field of this rigid p-component is $\boldsymbol{v}_p = x\Omega_p \hat{\boldsymbol{e}}_\theta$, which gives the observed pulsar period $2\pi\Omega_p^{-1}$.
- The presence of vortices induces a mutual friction between the n-component and the p-component: scattering of particles (or quasiparticles) off vortex cores gives rise to mesoscopic drag force $\boldsymbol{f}_D$ [3, 13, 32, 33].
- Vortex lines can pin to nuclei in the solid inner crust. The force per unit length $f_p(\rho)$ expresses the strength of the vortex-lattice interaction [18].

We are thus far from describing the realistic hydrodynamical problem of a NS interior; superfluid turbulence (a well known phenomenon in He-II where it gives rise to a complex tangle of vortex lines [34]) is likely to develop in some regime also in NSs [25], but its effect on macroscopic hydrodynamics and its relevance for glitch models are still debated. In particular the tension of a single line is much smaller than the hydrodynamical force acting on it, which suggests bending of vortices, such that the core is susceptible to become turbulent [28]. However this argument does not account for the stabilizing effect of rotation: the increased effective tension of a bundle of vortices as compared to a single vortex line resists bending and the array of vortex lines could remain parallel to the rotation axis. This can be seen as a generalization of the Taylor-Proudman theorem to axisymmetric and spinnig-down frictionless fluids [35].

[2] We use cylindrical coordinates (x, θ, z). The spherical radius r should be read as $\sqrt{x^2 + z^2}$.

IOP Conf. Series: Journal of Physics: Conf. Series **861** (2017) 012024 doi:10.1088/1742-6596/861/1/012024

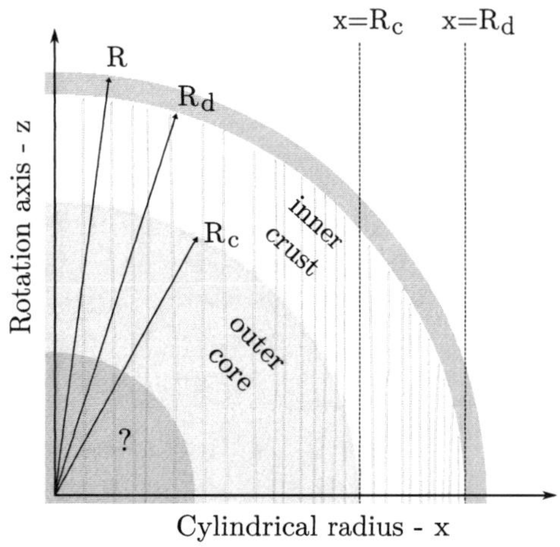

Figure 1. Sketch of the stellar structure (out of scale), with the geometrical definitions used. In the cylindrical shell $R_c < x < R_d$ the vortex lines are completely immersed in the inner crust. The outer crust (the thin blue layer in the figure) is part of the p-component. In principle vortices can stop at any spherical radius and all the input quantities of the model are consistently adjusted by following the prescriptions detailed in ref. [20].

In the zero temperature limit the microphysical quantities of interest, as the the superfluid fraction $x_n(\rho)$ and the dimensionless effective mass $m^*(\rho) = 1 - \epsilon_n(\rho)$ of entrained neutrons, can be easily parametrized as functions of the spherical radius r after solving the TOV equations for a given EOS. We impose a Newtonian dynamics on the general relativistic stellar structure, thus the total moment of inertia is given by

$$I = \frac{8\pi}{3} \int_0^R r^4 \rho(r)\, dr\,.$$

General relativistic corrections to this quantity can be computed using the Hartle slow-rotation approximation, where the background structure is still given by the TOV equations and are expected to be large (see e.g. [9] for a recent description of this approximation). We further discuss this point in the concluding remarks.

Conservation of vorticity - Following the Newtonian formalism outlined in [23], the momentum per particle of the superfluid is a linear superposition of the local kinematic velocities of both components:

$$\boldsymbol{p}_n/m_n = (1-\epsilon_n)\,\boldsymbol{v}_n + \epsilon_n \boldsymbol{v}_p\,,$$

where m_n is the neutron bare mass. At the microscopic scale $\boldsymbol{p}_n$ is irrotational but here we consider it as averaged over a region that contains many vortex lines. In our axially symmetric configuration the azimuthal component of the momentum turns out to be columnar (it has no dependence on z). It is thus convenient to introduce an auxiliary angular velocity $\Omega_v(x)$ defined by

$$\boldsymbol{p}_n(\boldsymbol{x}) \cdot \hat{\boldsymbol{e}}_\theta = m_n\, x\, \Omega_v(x)\,.$$

On the other hand, the angular velocity of the n-component is non-columnar due to the (spherically symmetric) density dependence of the entrainment parameter $\epsilon_n(r)$. Exchange of angular momentum between the superfluid and the p-component is possible only if we allow for radial motion of vortices. Let v_L^x be the radial component of the local mean velocity of vortex lines (the velocity of a vortex is normal to the line), then a simple argument early proposed in [21] gives

$$\dot{\Omega}_v(x,t) = -\left(2\Omega_v(x,t) + x\partial_x \Omega_v(x,t)\right) \frac{v_L^x(x,t)}{x}\,. \tag{1}$$

This equation expresses a conservation law for vortex lines flowing trough a cylindrical shell of radius x. The radial velocity $v_L^x(x,t)$ needs to be expressed in terms of the variables $\Omega_v(x,t)$ and $\Omega_p(t)$ by solving an explicit model of vortex dynamics. In [20] a simple example is proposed: we construct a particular form of the friction functional $\mathcal{B}$ defined as

$$v_L^x(x) \;=\; x\,\mathcal{B}[\Omega_v,\Omega_p,x]\;(\Omega_v(x)-\Omega_p)\,.$$

The explicit form of $\mathcal{B}$ depends on how vortex dynamics is modeled; correlated motion of many lines, pinning and finite tension of macroscopically long vortices make the explicit construction of this functional a non trivial task.

Total angular momentum balance - The isolated pulsar dynamics is slowly driven by angular momentum losses due to radiation emission, namely

$$\int d^3x\,\rho(r)\,\Big[x_n(r)\,\boldsymbol{x}\times\dot{\boldsymbol{v}}_n+(1-x_n(r))\,x^2\,\dot{\Omega}_p\,\hat{\boldsymbol{e}}_z\Big] \;=\; -I\,\dot{\Omega}_\infty\hat{\boldsymbol{e}}_z\,.$$

Expressing the non-columnar velocity of the n-component in terms of the angular velocities, the angular momentum balance takes the form

$$I_p\,\dot{\Omega}_p\,+\,I_v\,\langle\,\dot{\Omega}_v\,\rangle \;=\; -I\,\dot{\Omega}_\infty\,, \tag{2}$$

where Ω_v is averaged over the cylindrical radius x, namely

$$\langle\,\dot{\Omega}_v\,\rangle \;=\; \frac{1}{I_v}\int_0^{R_d} dI_v(x)\;\dot{\Omega}_v(x,t)\,.$$

The partial moments of inertia are $I_p = I - I_v$ and I_v, that is also the normalization factor of the measure dI_v. The physical constraint on entrainment parameter $\epsilon_n < 1 - x_n$ [12] guarantees the positivity of this quantity, namely $I_v < I$. The explicit Newtonian form of dI_v is

$$\frac{dI_v(x)}{dx} \;=\; 4\,\pi\,x^3\int_0^{z(x)} dz\;\frac{\rho_n(r)}{m^*(r)}\,,$$

with $z(x)$ the height of vortices passing through x. The appearance of the effective mass at the denominator is physically reasonable: the effect of entrainment is to redefine the mass of neutrons (i.e. $\boldsymbol{p}_n = m^*\boldsymbol{v}_n$ in the frame corotating with the p-component) that rescales ρ_n.
At the steady-state (i.e when $\dot{\Omega}_p = \dot{\Omega}_v$ for all the x values), Eq. 2 assures that the star responds to the braking torque as a whole with a spin-down rate of $-\dot{\Omega}_\infty$. Thus, provided that the braking torque itself is not evolving, the value of $\dot{\Omega}_\infty$ coincides with the absolute value of the observed mean spin-down rate over a long period of time (even if in this period the pulsar glitched several times). It is also useful to introduce the lag ω, defined as

$$\omega(x,t) \;=\; \Omega_v(x,t)-\Omega_p(t)\,.$$

At the steady state the lag is constant in time and strictly positive for all the x values.

Friction functional - Vortex lines experience a viscous drag force per unit length $\boldsymbol{f}_D$ and an hydrodynamical lift (the Magnus force) $\boldsymbol{f}_M$ expressed by

$$\begin{aligned}\boldsymbol{f}_D &= -\eta(r)(\boldsymbol{v}_L(\boldsymbol{x})-\boldsymbol{v}_p(\boldsymbol{x}))\,,\\ \boldsymbol{f}_M &= \rho_n(r)\boldsymbol{\kappa}(\boldsymbol{x})\times(\boldsymbol{v}_L(\boldsymbol{x})-\boldsymbol{v}_n(\boldsymbol{x}))\,.\end{aligned}$$

As discussed in [3] the vector $\boldsymbol{\kappa}$ is given by

$$\boldsymbol{\kappa}(\boldsymbol{x}) \;=\; \kappa \, \nabla \times \boldsymbol{p_n} \, / \, |\nabla \times \boldsymbol{p_n}| \, .$$

Within our working assumptions, this field is constant and parallel to $\hat{\mathbf{e}}_z$. To implement this infinite rigidity of vortex lines, the equation of motion of vortices is modeled as

$$\boldsymbol{F}_{tot} \;=\; \int_L dl(\boldsymbol{x}) \, [\boldsymbol{f}_M(\boldsymbol{x}) + \boldsymbol{f}_D(\boldsymbol{x})] \;=\; 0 \, ,$$

where the integration is performed over the vortex line. Solving the above equation for the radial component of the vortex velocity gives that the friction functional $\mathcal{B}$ is just a function $\mathcal{B}^x$ of the cylindrical radius

$$v_L^x \;=\; x \, \mathcal{B}^x(x) \, \omega(x,t) \, . \tag{3}$$

At this level we still have to account for pinning: vortex lines that are perfectly pinned are blocked and forced to corotate with the p-component, until the Magnus force induced to vortex lines by the background flow of neutrons overcomes the depinning critical value. This threshold effect introduces explicit and non-smooth dependence of $\mathcal{B}$ on the dynamical variables. The depinning condition can be found by imposing that

$$|\boldsymbol{F}_{tot}|_{\boldsymbol{v}_L=\boldsymbol{v}_p} \;=\; \int_L dl \, \frac{\kappa \, \rho_n}{m*} \, x \, \omega \;=\; \int_L dl \, f_P$$

and a straightforward calculation gives that the second equality is realized when the dynamical lag ω reaches the value

$$\omega_{cr}(x) \;=\; \frac{\int_0^{z(x)} dz \; f_P(r)}{\kappa \, x \, \int_0^{z(x)} dz \; \rho_n(r)/m^*(r)} \, . \tag{4}$$

This function of the cylindrical radius (sketched in Fig. 2 for a typical case) reduces to the critical lag for depinning studied in [36] and [37] in the limit of no entrainment. At the level of equation of motion for rigid lines, pinning could be modeled by introducing a local drag force with a non-constant η parameter that depends on the dynamical lag ω as

$$\begin{array}{lll} \eta(\boldsymbol{x},\omega) \gg \kappa\rho_n(r) & \text{if} & |\omega(x)| \, < \, \omega_{cr}(x) \, , \\ \eta(\boldsymbol{x},\omega) \ll \kappa\rho_n(r) & \text{if} & |\omega(x)| \, \simeq \, \omega_{cr}(x) \, . \end{array}$$

As an order of magnitude estimate, if $\eta \sim 10^{21}\,\mathrm{s\,cm/g}$ in the crust, the $\mathcal{B}^x$ of Eq. 3 ranges from $\mathcal{B}^x(x=0) \sim 10^{-9}$ to $\mathcal{B}^x(x=R_d) \sim 10^{-12}$, thus ensuring that the vortices are expelled on timescales of many years.

However, since we are just interested to estimate the angular momentum reservoir, we can rely to a simpler possibility that is more similar to the early vortex-creep model [21], where the effect of the dynamical lag ω against pinning is to modulate the typical velocity of free vortices given in Eq. 3. This is equivalent to approximate the complete (and unknown) friction functional as

$$\mathcal{B}[\Omega_v, \Omega_p, x] \;\approx\; Y[\omega, x] \; \mathcal{B}^x(x) \, .$$

From the macroscopic point of view, the fraction $Y \in [0,1]$ can be equally interpreted as a parameter that slows the expulsion of vortex lines or the fraction of (unpinned) vortex lines that are free to move under the combined action of $\boldsymbol{f}_M$ and $\boldsymbol{f}_D$. By considering a large population of vortices, Y can also be seen as a local depinning probability. With macroscopically straight vortex lines, the simplest way to encode pinning at the macroscopic scale is thus to construct Y in such a way that

$$\begin{aligned} Y[\omega, x] &\approx 0 \qquad \text{if} \quad |\omega(x)| < \omega_{cr}(x) \,, \\ Y[\omega, x] &\approx 1 \qquad \text{if} \quad |\omega(x)| \simeq \omega_{cr}(x) \,. \end{aligned}$$

Finite temperature, quantum effects and finite vortex tension take part in smoothing the step-like form of Y. As an explicit example, the original vortex-creep model [21] in the regime $0 < \omega < \tilde{\omega}_{cr}$ is realized by

$$Y[\omega, x] \approx \exp\left(\frac{\omega(x) - \tilde{\omega}_{cr}(x)}{\alpha\, \tilde{\omega}_{cr}(x)} \right) ,$$

where α is a dimensionless temperature that tunes the local rigidity of the pinning and $\tilde{\omega}_{cr}(x)$ is an opportune critical lag. In the limit $\alpha \to 0$ the vortex-creep prescription results into a perfect pinning prescription, namely $Y = \theta(\tilde{\omega}_{cr}(x) - |\omega(x)|)$ is the unit step function.

Improving the friction term - Vortices in an hydrodynamical model only serve to construct a particular form of the mutual friction in terms of the dynamical variables of the problem (in our global model we shifted this problem on $\mathcal{B}$). A glitch can then be triggered by perturbing $\mathcal{B}$ or, in order to mimic the sudden unpinning of many vortices, by suddenly increasing the Y fraction in some interval of x. It is thus important to find good prescriptions for the friction functional, even in global or averaged models like the one discussed here. Other contributions to the vortex dynamics (that could have a significant impact on mutual friction) arise from the finite tension force of vortices and their mutual interaction [1, 38]. Self-interaction forces that complicate further the picture are present for curved lines [34]. Moreover the radiation of phonons during the pinning and depinning of a vortex line and proximity effect when two lines are close are important at a microscopic scale [39]: these precesses could provide the physics needed to explain the claimed self-organized criticality of glitching pulsars [40]. Further work on these lines is needed to understand how these issues modify mutual friction.

3. Constraints from large glitches

For any given lag ω, the corresponding reservoir of angular momentum (i.e. the difference between the total angular momentum and that of the star in perfect corotation) is

$$L - I\,\Omega_p = I_v \,\langle \omega \rangle \,.$$

The spin-up phase of glitches is so fast that the braking torque carries out a negligible amount of momentum; the total angular momentum is thus conserved during a glitch

$$\dot{L} = I_v \,\langle \dot{\omega} \rangle + I\,\dot{\Omega}_p = 0\,.$$

Assume now to trigger a *maximal* glitch (a glitch that by definition empties the whole reservoir) at a certain initial time: there will be a subsequent time at which the star is in a state of instantaneous average corotation, namely $\langle \omega \rangle = 0$. This situation can be realized during the spin up phase or just after an overshoot, a transient period during which $\langle \omega \rangle < 0$. Integration of the previous equation between these two times gives

$$I_v \,\langle \Delta\omega \rangle + I \Delta\Omega_p = 0\,,$$

where Δ indicates the difference between a quantity at the particular time when corotation is instantaneously realized and the same quantity at the triggering time.

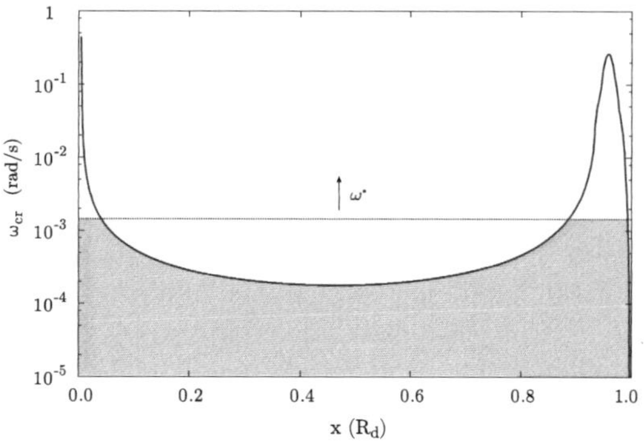

Figure 2. Example of the critical lag for unpinning ω_{cr} for a NS of mass $1.1\,M_\odot$ and the SLy equation of state. The horizontal line indicates the increasing nominal lag ω^*, the shaded area below it is bounded by the corresponding lag developed since corotation. The distance from the rotational axis is expressed in units of the neutron drip radius R_d.

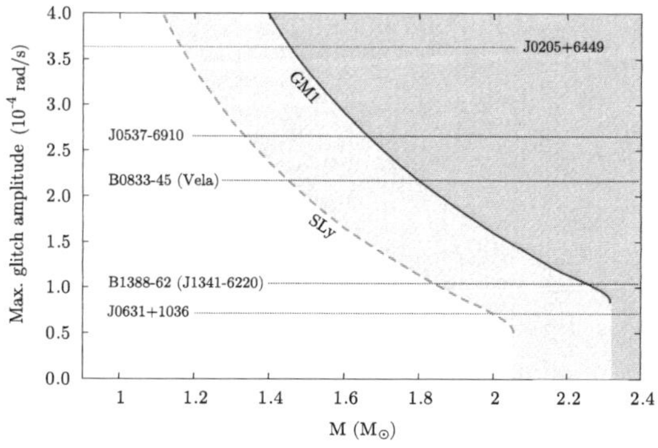

Figure 3. The theoretical maximum glitch $\Delta\Omega_{max}$ is shown as a function of the stellar mass M for the Sly equation of state (green, dashed) and the GM1 (blue, solid). The curves terminate at the maximum mass allowed by each EOS. Horizontal lines indicate the largest glitch $\Delta\Omega$ recorded in the corresponding pulsar. The shaded area defines the forbidden region.

Entrainment independent maximum glitch amplitude - In our model the maximum sustainable lag is the critical depinning lag ω_{cr} (the actual maximum sustainable lag is likely to be less than this quantity, because of thermal effects and dynamical creep) so that we can set an absolute upper limit to the glitch amplitude by imposing $\Delta\omega = -\omega_{cr}$:

$$\Delta\Omega_{max} = \frac{I_v}{I} \langle \omega_{cr} \rangle . \tag{5}$$

Using the explicit forms of dI_v and ω_{cr}, it is trivial to show that $\Delta\Omega_{max}$ depends only on the EOS and on the pinning forces [20]: at the Newtonian level entrainment does not affect this quantity (this ceases to be the case when relativistic corrections are taken into account). Thus, by fixing these two microphysical inputs, it is possible to study how $\Delta\Omega_{max}$ changes with the total mass of the NS. This is shown in Fig. 3, together with the maximum glitch amplitude $\Delta\Omega$ recorded in some pulsars; we thus find an upper limit to the mass of each pulsar by imposing $\Delta\Omega = \Delta\Omega_{max}(M)$. As a general trend, a stiff EOS like GM1 gives less stringent upper mass bounds (e.g. $M_{Vela} < 1.4$ with SLy and $M_{Vela} < 1.8\,M_\odot$ with GM1). Future observations of even larger glitch amplitudes can only lower this upper limit to the mass, leading to stronger constraints. Note also that the possible presence of turbulence in the crust could reduce the maximum glitch amplitude, thus lowering the curves of Fig. 3 and leading to stronger constraints: it seems reasonable to expect that the increase in the vortex length density due to the development of complex vortex tangles weakens the otherwise coherent pinning of the vortex array. Finally we remark that, as long as only crustal pinning is considered, $\Delta\Omega_{max}$ turns out to be the same for vortex lines that stop at the crust-core interface or pass through the core.

Constraints imposed by the reservoir dynamics - It is possible to try to lower the upper bound of the mass by using also observed timing properties other than the largest glitch. Starting from corotation, forward integration in time of Eqs 1 and 2 allows to follow the evolution of the

IOP Conf. Series: Journal of Physics: Conf. Series **861** (2017) 012024 doi:10.1088/1742-6596/861/1/012024

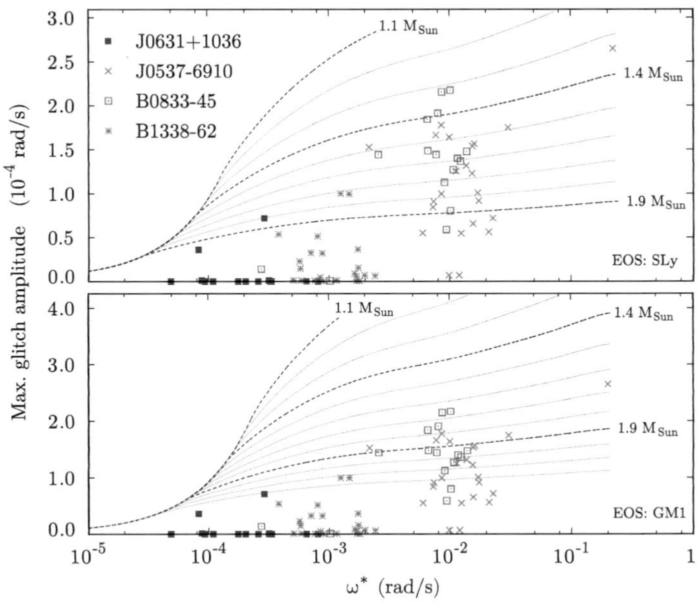

Figure 4. Upper panel: the lines defined in Eq. 7 for different masses relative to the SLy equation of state are shown as a function of the nominal lag ω^*. For comparison we superimpose the single glitches of four different pulsars; the nominal lag of each glitch is given by $-\dot{\Omega}t$, with t the waiting time since the previous event and $\dot{\Omega}$ the spin down rate measured in the corresponding pulsar. Lower panel: the same but using the GM1 equation of state.

quantity

$$\Delta\Omega_m(t) = \frac{I_v}{I}\langle\omega(x,t)\rangle \tag{6}$$

Equation 6 gives the amplitude of a maximal glitch, triggered at time t after the emptying of the reservoir because of an initial large event. This results in a correlation between the amplitude of maximal glitches and the waiting time between them. This of course ceases to be interesting if there are no pulsars that sporadically empty the reservoir: in this case the sequence of amplitudes and waiting times shows no correlations between events, which is certainly true for glitches of small amplitude that comprise the major part of the whole sample among the pulsar population. However J0537-6910 and Vela have been proven to be exceptional objects: systematic study of glitches in individual pulsars seem to indicate that the distribution of waiting times is Poissonian while the distribution of glitch amplitudes is a power law (even though the value of the exponent is not universal across the pulsar population), but Vela and J0537-6910 showed evidence for quasi-periodicity [40]. The main difficulty behind this and similar analysis is the paucity of data in single objects.

In the following we present a more speculative set of ideas: instead of solving dynamical equations with the particular rotational parameters for each pulsar, we use the simple unified model sketched in Fig. 2 and defined as

$$\Delta\Omega_m(\omega^*) = \frac{I_v}{I}\langle\min\left(\omega_{cr}(x),\omega^*\right)\rangle\,. \tag{7}$$

For each pulsar we measure time in terms of a nominal lag $\omega^* = t\,\dot{\Omega}_\infty$. The curves defined by Eq. 7 are shown in Fig. 4 for different masses. On top of this, we plot the single glitches of some pulsars (among which Vela and J0537-6910) versus waiting times to the previous glitch multiplied by the absolute value of the spin-down rate $-\dot{\Omega}$ the object. The largest glitch of J0537-6910 (the only pulsar that display correlation between the amplitude and the time to the next glitch [1]) is also the first one, so we plotted it at the end of the ω^* axis, where $\Delta\Omega_m(\omega^*)$ assumes the value given by Eq. 5. It is interesting to note that is always the largest glitch that touches the curve corresponding to the smaller mass, even if the data are scattered. Note that for each glitch we used the nominal lag corresponding to the time since the previous event, without taking into account the amplitude of this event, even if it is unlikely that small glitches are maximal. This only serves as a clue of the fact that the largest glitch recorded in a pulsar can constrain its mass: at this level no conclusions can be taken on the actual (possibly time-correlated) occurrence of large glitches in pulsars.

A remark on the unified model - We stress that when we simulate the full dynamical equations, starting from perfect corotation at $t = 0$, the lag evolves in a fashion very similar to the lag sketched in Fig. 2: the horizontal plateau (at height ω^*) rises until it locally approaches $\omega_{cr}(x)$, where sub-critical vortex creep shapes the actual lag around ω_{cr}. Thus it is not a bad approximation to recast the dynamical evolution of the lag into the form

$$\omega(x,t) \approx \min\left(\omega_{cr}(x)\,,\,\omega^*(t)\right).$$

However this can only serve to *define* $\omega^*(t)$, a function of time that can be easily extrapolated from $\omega(x,t)$ once the actual evolution is known from the forward integration in time of the dynamical equations (e.g. $\omega^*(t) = \omega(x_m,t)$ where x_m is the cylindrical radius corresponding to the peak of ω_{cr}). On the other hand, in the previous section we assumed $\omega^*(t) = \dot{\Omega}_\infty t$. This choice is convenient since it is possible treat different pulsars within a unified prescription; in this way there is no need to solve the dynamical equations with the specific rotational parameters $\Omega_p(0) = \Omega$ and $\dot{\Omega}_\infty = -\dot{\Omega}$ for every different object. However there are no rigorous arguments for doing this, except for the fact that the slow dynamics of ω toward the steady state is driven by $\dot{\Omega}_\infty$. Moreover this unified prescription based on a linear relation between the actual time and the nominal lag ω^* was already used (in a different context and with a different idealization of the lag dynamics) in some previous works [36, 37, 41] that use the same form of the critical profile given in Eq. 4, but without entrainment.

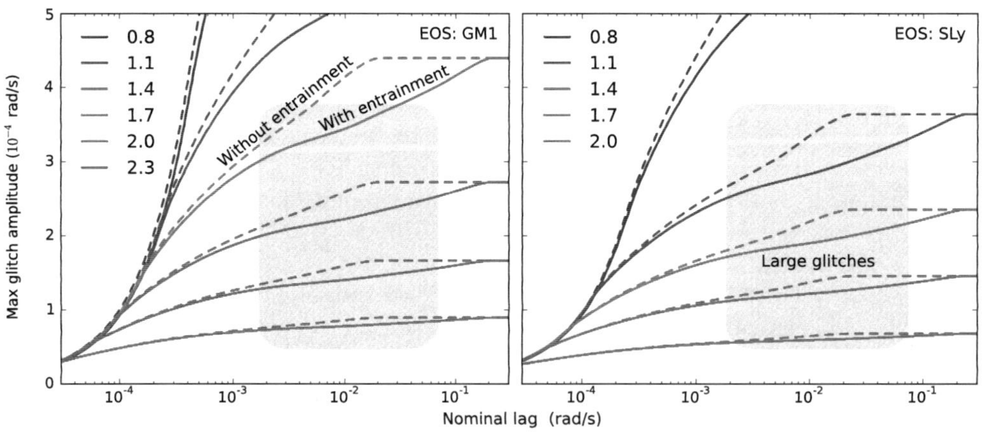

Figure 5. Left: comparison between the maximal glitch amplitudes given by Eq. 7 when entrainment effect is taken into account and when $\epsilon_n = 0$ everywhere. The EOS used is GM1 and the different curves refer to the different mass values (in units of $M_\odot$) listed in the key. The region where large glitches are expected is shaded: it extends circa to the value of the largest glitch recorded in J0205-6449. The ω^* scale on the horizontal axis spans the range $3 \times 10^{-5} \div 0.3\,\mathrm{rad/s}$. Right: the same but using SLy equation of state.

IOP Conf. Series: Journal of Physics: Conf. Series **861** (2017) 012024 doi:10.1088/1742-6596/861/1/012024

4. Effect of entrainment

We switch off the entrainment effect setting $m^*(r) = 1$ everywhere and repeat the integrations needed to calculate the macroscopic input quantities of model. In this limit $\omega \to \Omega_n - \Omega_p$, $I_v \to I_n$, $I_p \to I - I_n$. The curves $\Delta\Omega_m(\omega^*)$ of the unified model (Eq. 7) are found to be considerably steeper than their "entrained" counterparts, as shown in Fig. 5. This is not surprising since the strong crustal entrainment sensibly increase the peak of the critical lag which in turn implies that the plateau at the value $\Delta\Omega_{max}$ is reached for smaller values of the nominal lag. This is in part a drawback of the assumed form of $\omega^*(t)$ that in general is not simply linear in time. Nonetheless the result seems physically reasonable as it is more difficult for strongly entrained superfluid neutrons to lag behind the crust [15]. However a true comparative study of entrained and non-entrained dynamics should be carried by solving the dynamical equations 1 and 2. Despite this caveat this preliminary analysis is encouraging: large glitches of Vela and J0537-6910 seem to have been triggered around $\omega^* \sim 10^{-2}$ rad/s, as can be seen in Fig. 4. For this value of the nominal lag the difference between the case with entrainment and that without entrainment results in at most a $\sim 10\%$ uncertainty on the mass value.

5. Conclusions

We presented a global model for pulsar rotational dynamics that accounts consistently for the layered structure of the star, the differential rotation of the superfluid and the presence of density-dependent entrainment [20]. The Eqs 1 and 2 describe the exchange of angular momentum between two effective components fo the star, characterized by the partial moments of inertia I_v and I_p. All the complex behavior of the rearrangement of the vortex configuration that should appear in Eq. 2 is hidden into the functional $\mathcal{B}$. Even without solving explicitly the dynamical equations, it is possible to draw some general quantitative conclusions; in particular the calculation of the maximum glitch amplitude (that is entrainment independent and is unaffected by the possible presence of a normal layer at the crust-core interface) seems to open two possibilities:

- Test pinning newly calculated forces using observations of large glitches.
- Set an upper limit to the mass of some isolated pulsars (once the EOS and the pinning force profile have been fixed).

Finally we sketched a more speculative set of ideas that could be useful to refine the upper bound to the mass of large glitchers: while the absolute upper limit is fixed by the observation of the largest glitch only, to improve this constraint we have to use some additional information about the temporal occurrence of glitches in a pulsar.
A final remark is due: we used the TOV equations for the structure of the star, but the classical moments of inertia. Fortunately it is quite simple to introduce general relativistic corrections for slow rotation to I and I_v. Preliminary study indicates that (when vortex lines extend into the core, as in this presentation) the quantity I_v/I is nearly unaffected by relativistic corrections. However general relativity does not affect only the moments of inertia but also the measure dI_v and the critical lag; since in this Newtonian model we used the paraxial vortex configuration of a rigidly rotating body, the corrections to the cylindrical profiles could be estimated by resolving the stationary configuration of vortex lines for a slowly and rigidly rotating neutron star. Despite all possible caveats and simplifications that are present in our model, the message we want to vehicle is that also the study of single glitches (and in particular the largest glitch recorded in each pulsar) can robustly constrain the mass of isolated neutron stars: large glitches can provide a test for the consistency of glitch theory that is alternative to the studies based on pulsar activity [14–16].

Acknowledgments

This work consists of a revised version of the presentation given by MA at the *NewCompStar Working Group 2 Meeting* , 23-27 May 2016, GSSI and LNGS (L'Aquila, Italy). Partial support comes from NewCompStar, COST Action MP1304.

MA wishes to express his gratitude to Stella Valentina Paronuzzi Ticco (previously at International School for Advanced Studies in Trieste) for valuable help and thoughtful discussion and also to thank Nicolas Chamel for interesting comments on some aspects of entrainment.

References

[1] Haskell B and Melatos A 2015 *Int. J. Mod. Phys. D* **24** 1530008
[2] Anderson P W and Itoh N 1975 *Nature* **256** 25
[3] Andersson N, Sidery T and Comer G L 2006 *Mon. Not. R. Astron. Soc.* **368** 162
[4] Hall H E and Vinen W F 1956 *Proc. R. Soc. A: Mathematical, Physical and Engineering Sciences* **238** 204
[5] Bekarevich I L and Khalatnikov I M 1961 *Sov. Phys. JETP* **13** 643
[6] Andreev A F and Bashkin E P 1975 *Zh. Eksp. Teor. Fiz* **69** 319
[7] Andersson N and Comer G L 2007, Living Rev. Relativity bf 10 1
[8] Chamel N and Haensel P 2008 *Living Reviews in Relativity* **11** 10
[9] Rezzolla L and Zanotti O 2013 *Relativistic Hydrodynamics* (Oxford: Oxford University Press)
[10] Chamel N 2012 *Phys. Rev. C* **85** 035801
[11] Carter B, Chamel N and Haensel P 2005 *Nuclear Physics A* **748** 675
[12] Chamel N and Haensel P 2006 *Phys. Rev. C* **73**(4) 045802
[13] Alpar M A, Langer S A and Sauls J A 1984 *Astrophys. J.* **282** 533
[14] Link B, Epstein R I and Lattimer J M 1999 *Phys. Rev. Lett.* **83** 5
[15] Andersson N, Glampedakis K, Ho W C G and Espinoza C M 2012 *Phys. Rev. Lett.* **109** 241103
[16] Chamel N 2013 *Phys. Rev. Lett.* **110** 011101
[17] Donati P and Pizzochero P M 2006 *Phys. Lett. B* **640** 74
[18] Seveso S, Pizzochero P M, Grill F and Haskell B 2016 *Mon. Not. R. Astron. Soc.* **455** 3952
[19] Gügercinoglu E and Alpar M A 2014 *Astrophys. J.* **788** L11 2041-8205
[20] Antonelli M and Pizzochero P M 2017 *Mon. Not. R. Astron. Soc.* **464** 721
[21] Alpar M A, Anderson P W, Pines D and Shaham J 1984 *Astrophys. J.* **278** 791
[22] Haskell B, Pizzochero P M and Sidery T 2012 *Mon. Not. R. Astron. Soc.* **420** 658
[23] Prix R, Comer G L and Andersson N 2002 *Astronomy and Astrophysics* **381** 178
[24] Sidery T, Passamonti A and Andersson N 2010 *Mon. Not. R. Astron. Soc.* **405** 1061
[25] Peralta C, Melatos A, Giacobello M and Ooi A 2006 *Astrophys. J.* **651** 1079
[26] Howitt G, Haskell B and Melatos A 2016 *Mon. Not. R. Astron. Soc.* **460** 1201
[27] Sourie A, Chamel N, Novak J and Oertel M 2016 *Mon. Not. R. Astron. Soc.* **464** 4 1
[28] Andersson N and Comer G L 2006 *Classical and Quantum Gravity* **23** 5505
[29] Zhou X R, Schulze H J, Zhao E G, Pan F and Draayer J 2004 *Phys. Rev. C* **70** 2
[30] Ruderman M A 1976 *Astrophys. J.* **203** 213
[31] Easson I 1979 *Astrophys. J.* **228** 257
[32] Jones P B 1990 *Mon. Not. R. Astron. Soc.* **243** 257
[33] Epstein R I and Baym G 1992 *Astrophys. J.* **387** 276
[34] Barenghi C F, Donnelly R J and Vinen W F (eds) 2001 *Quantized Vortex Dynamics and Superfluid Turbulence* (*Lecture Notes in Physics* vol 571) (Berlin, Heidelberg: Springer)
[35] Ruderman M A and Sutherland P G 1974 *Astrophys. J.* **190** 137
[36] Pizzochero P M 2011 *Astrophys. J.* **743** L20
[37] Seveso S, Pizzochero P M and Haskell B 2012 *Mon. Not. R. Astron. Soc.* **427** 1089
[38] Wlazlowski G, Sekizawa K, Magierski P, Bulgac A and Forbes M M N 2016 *Phys. Rev. Lett.* **117** 232701
[39] Warszawski L, Melatos A and Berloff N G 2012 *Phys. Rev. B* **85**(10) 104503
[40] Melatos A, Peralta C and Wyithe J S B 2008 *Astrophys. J.* **672** 1103
[41] Haskell B, Pizzochero P M and Seveso S 2013 *Astrophys. J.* **764** L25

IOP Conf. Series: Journal of Physics: Conf. Series **861** (2017) 012025 doi:10.1088/1742-6596/861/1/012025

From microphysics to dynamics of magnetars

Armen Sedrakian[1], Xu-Guang Huang[2], Monika Sinha[3] and John W. Clark[4]

[1]Frankfurt Institute for Advanced Studies, D-60438 Frankfurt-Main, Germany,
Institute for Theoretical Physics, J. W. Goethe-University, D-60438 Frankfurt-Main, Germany
[2]Physics Department and Center for Particle Physics and Field Theory, Fudan University, Shanghai 200433, China
[3]Indian Institute of Technology Rajasthan, Old Residency Road, Ratanada, Jodhpur 342011, India
[4]Department of Physics and McDonnell Center for the Space Sciences, Washington University, St. Louis, Missouri 63130, USA
Centro de Ciências Matemáticas, University of Madeira, 9000-390 Funchal, Portugal

E-mail: [1]`sedrakian@fias.uni-frankfurt.de`

E-mail: [2]`huangxuguang@fudan.edu.cn`

E-mail: [3]`monikasinha@gmail.com`

E-mail: [4]`jwc@wuphys.wustl.edu`

Abstract. MeV-scale magnetic fields in the interiors of magnetars suppress the pairing of neutrons and protons in the S-wave state. In the case of a neutron condensate the suppression is the consequence of the Pauli-paramagnetism of the neutron gas, i.e., the alignment of the neutron spins along the magnetic field. The proton S-wave pairing is suppressed because of the Landau diamagnetic currents of protons induced by the field. The Ginzburg-Landau and BCS theories of the critical magnetic fields for unpairing are reviewed. The macrophysical implications of the suppression (unpairing) of the condensates are discussed for the rotational crust-core coupling in magnetars and the neutrino-dominated cooling era of their thermal evolution.

1. Introduction

The powerful X-ray and soft γ-ray outburst activity observed in a number of astrophysical point sources has been attributed to the magnetic energy stored in compact objects. The *magnetar* interpretation of these observational phenomena implies magnetic fields larger by a factor 10^3 than the fields deduced for rotationally powered pulsars $B \sim 10^{12}$ G [1]. The interior fields of magnetars cannot be measured, but it has been frequently conjectured that they are stronger than their surface fields by several orders of magnitude; for recent reviews see [2, 3]. Large interior fields affect the equation of state of nucleonic and quark matter, potentially endangering the hydrostatic equilibrium of compact stars. In fact, observationally significant modification of gross parameters of compact stars (mass, radius, moment of inertia, etc.) requires extremely high fields $B \sim 10^{18} - 10^{19}$ G, for which the stars are close to loss of hydrostatic equilibrium as derived from the virial theorem [4, 5] and general relativistic numerical models [6, 7, 8].

Those physical phenomena in dense and strongly interacting matter which are controlled by the form of the quasiparticle spectrum in the vicinity of the Fermi surface are affected by much lower fields of the order $B \sim 10^{16} - 10^{17}$ G. For such fields, electromagnetic interactions become of the order of the nuclear scales $\sim$ MeV characterizing low-energy fermionic excitation spectrum of nucleons (as opposed to the high-energy scale set by the Fermi energy). An example to be addressed below is pairing in nucleonic condensates, which is characterized by the MeV scale. Likewise, neutrino transport and radiation are dominated by this scale throughout the early evolution of compact stars as they cool via neutrino radiation from their interiors. Accordingly, we will also discuss the effects of magnetic fields on the neutrino radiation processes.

It is useful at this point to make the notion of MeV-scale magnetic fields more precise. The interaction energy of the magnetic field with the nucleon spin is $\mu_N B$, where $\mu_N = e\hbar/2m_p$ is the nuclear magneton. Substituting the values of constants we find that $\mu_N B \simeq \pi(B/10^{18}$ Gauss) MeV, i.e., fields of the order of 10^{17} G would substantially affect the MeV-scale pairing via the spin–B-field interaction. In the case of charged particles, the relevant magnetic-field scale follows from the Landau criterion for the critical velocity $v_s \sim \Delta/p_\perp$, where Δ is the pairing gap and $p_\perp$ is the characteristic momentum in the plain orthogonal to the field, restricted to $p_\perp \le p_F$ by the Fermi momentum p_F. The corresponding energy scale is then given by $v_s p_F \sim \pi(v_s/c)(\xi/10 \text{ fm})(B/10^{16}$ Gauss) MeV, assuming that the Larmor radius $p_F c/eB$ is of the order of the coherence length $\xi \simeq 10$ fm and $v_s/c \le 0.3$. Thus, the range of characteristic fields relevant to the MeV-scale physics in the vicinity of the Fermi surface of nucleons is $10^{16} \le B \le 10^{17}$ G. As will be shown, this is indeed confirmed by explicit calculations.

The purpose of this review is to give a concise account of the progress achieved in understanding the effect of MeV-scale magnetic fields on nucleonic pairing [9, 10, 11] and its consequences for neutrino radiation [10] and crust-core coupling in magnetars [12]. The broad and actively pursued subject of the equation of state of various phases of dense matter in much stronger magnetic fields (comparable to the Fermi energy scale of fundamental fermions) is not discussed here.

This review is organized as follows. Section 2.1 summarizes the Ginzburg-Landau theory of charged (proton) and neutral (neutron) superfluid mixtures in the cores of magnetars and addresses the critical magnetic field for destruction of proton superconductivity. Section 2.2 discusses the suppression of pairing in the neutron S-wave condensate (relevant for crusts of magnetars). Section 3 focuses on implications the suppression of the proton and neutron pairing for dynamical coupling between the core and the crust of a magnetar. Finally, Sec. 4 considers the modifications of the neutrino emission induced by the magnetic fields. Section 5 contains some concluding remarks.

2. Unpairing of nucleonic condensates in magnetars

MeV-scale magnetic fields can destroy the superconducting coherence that is required for the formation of condensates in nucleonic matter. The interaction of the magnetic field with the neutron or proton spin induces an imbalance in the number of spin-up and spin-down particles, which implies that the Cooper pairing will be suppressed, because not all spin-up particles will find spin-down “partners” [11]. This so-called *Pauli paramagnetic suppression* acts for both proton and neutron condensates, but is the dominant suppression mechanism only in the case of neutrons. The proton condensate is already suppressed by a smaller magnetic field due to a different mechanism, associated with the Larmor motion of protons in the magnetic field, i.e., originating from the interaction of the charge of the proton with the B-field [9, 10]. The following Sections review these mechanisms in turn.

2.1. Critical unpairing of proton condensate

The BCS superconductors are characterized at least by three distinct length scales: (i) the *London penetration depth* λ, (ii) the *coherence length* ξ, and (iii) the interparticle distance d. It will be assumed henceforth that the last scale is much smaller than the other two in the problem, i.e., the superconductor is in the weakly coupled regime. The ratio of the remaining two scales defines the type of the superconductivity via the Ginzburg-Landau (GL) parameter, $\kappa = \lambda/\xi$ (see, e.g., [13]). In the range $1/\sqrt{2} < \kappa < \infty$, the material is a type-II superconductor; otherwise it is type-I. In type-II superconductors the magnetic field is carried by electromagnetic vortices with quantum flux $\phi_0 = \pi/e$ (here and below $\hbar = c = 1$), while the field forms domain structures in a type-I superconductor. These two scales also define three distinct magnetic-field scales when combined with the flux quantum:

$$H_{c1} \simeq \phi_0 \lambda^{-2}, \quad H_{cm} \simeq \phi_0 (\xi\lambda)^{-1}, \quad H_{c2} \simeq \phi_0 \xi^{-2}.$$

In type-II superconductors the hierarchy of these fields is $H_{c1} \leq H_{cm} \leq H_{c2}$ when $\kappa \geq 1$. At and above H_{c1} the creation of a single flux-tube (Abrikosov quantum vortex) is energetically favorable. The field H_{cm} is the thermodynamical magnetic field whose energy density is equal to the difference in the energy densities of superconducting and normal states. Finally, H_{c2} corresponds to the field at which superconductivity disappears; physically the density of flux-tubes in such a magnetic field is so high that the normal vortex cores overlap. Note that the thermodynamical magnetic field H_{cm} has no physical significance for type-II superconductors, which exist between the lower H_{c1} and upper H_{c2} critical fields.

The GL theory of neutron-proton superfluid mixtures is based on the functional [14, 15, 16]

$$\mathscr{F}[\phi, \psi] = \mathscr{F}_n[\phi, \psi] + \mathscr{F}_p[\phi, \psi] + \frac{1}{4m_p}|\boldsymbol{D}\psi|^2 + \frac{B^2}{8\pi}, \tag{1}$$

where ψ and ϕ are the condensate wave functions for protons and neutrons, $\boldsymbol{D} = -i\boldsymbol{\nabla} - 2e\boldsymbol{A}$ is the gauge-invariant derivative, m_p is the proton mass and indices p and n label the quantities referring to the neutron and proton condensates. If we are interested only in the H_{c2} field, the wave function is small and the functional for the proton condensate can be written at temperature T and critical temperature T_{cp} of the superconducting phase transition as

$$\mathscr{F}_p[\phi, \psi] = \alpha\tau|\psi|^2 + \frac{b}{2}|\psi|^4 + b'|\psi|^2|\phi|^2, \tag{2}$$

where $\tau = (T - T_{cp})/T_{cp}$. The quantities α and b are the coefficients of the GL expansion, while b' describes the density-density coupling between the neutron and proton condensates. In carrying out the Ginzburg-Landau expansion (2) we have taken into account (assuming second order phase transition) that the pre-factor of the $|\psi|^2$ term changes its sign at T_c, therefore it can be expanded in the small parameter τ to leading order.

The current-current coupling, i.e., the *entrainment* of the proton condensate by the neutron condensate, can be absorbed in the effective gauge potential $\boldsymbol{A}$ [10, 14].

The variations $\delta\mathscr{F}[\phi, \psi]/\delta\psi = 0$, $\delta\mathscr{F}[\phi, \psi]/\delta\phi = 0$, and $\delta\mathscr{F}[\phi, \psi]/\delta\boldsymbol{A} = 0$ constitute the coupled equations of motion describing the mixture of condensates. For simplicity, a homogeneous neutron background condensate is assumed. This is a good approximation given the fact that the neutron vortex size for typical neutron star rotation rates is of the order 10^{-4} cm, which by many orders of magnitude exceeds the proton vortex size ~ 100 fm.

Thus, assuming that the neutron condensate is static and homogeneous and that $\boldsymbol{A}$ is locally linear in the coordinates (constant B-field), the linearized GL equations provide the value

$$H_{c2} = \frac{\phi_0}{2\pi\xi_p^2}\left[1 + \beta(b')\right], \tag{3}$$

of the critical field H_{c2} [10], where $m_p|\alpha\tau| = (2\xi_p)^{-2}$. The coupling of the proton condensate to the neutron condensate enhances the critical field by a factor $\beta \simeq 0.2$. The dependence of the

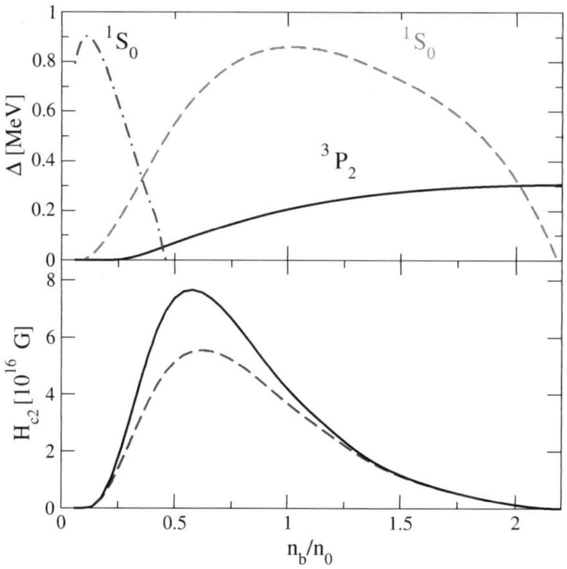

Figure 1. Conjectured layered structure of a magnetar with a constant magnetic field in the core (indicated by the dashed line). To the right of the intersection of this field with H_{c2}, the proton fluid is non-superconducting; to the left, i.e., at $B \leq H_{c2}$, it is superconducting. The crust contains a homogeneous B-field. The density is given in units of the nuclear saturation density n_0.

H_{c2} field on density is illustrated in Fig. 1 according to [10]. The following features are notable. The maximum of H_{c2} is attained close to the crust-core interface at a density $n_b = 0.5n_0$, where n_0 is the nuclear saturation density. Fig. 1 shows the conjectured layered structure of a magnetar: (a) an inner core void of superconductivity, (b) an outer core threaded by flux tubes, and (c) a crust containing a homogeneous field of magnitude B. If the field $B \geq \max[H_{c2}]$ the intermediate flux-carrying region disappears, i.e., superconductivity in a magnetar is completely destroyed. The critical fields in a similar formalism for superfluid-superconducting mixture were studied recently in [17].

2.2. Critical unpairing of neutron S-wave condensate

The nature of the suppression of pairing in the neutron condensate differs from that in the proton condensate, because charge-neutral neutrons interact with the B-field via their spin magnetic moment. This has a destructive effect on S-wave neutron Cooper pairs, which involve spin-up and down partners. Clearly, a large enough magnetic field will quench pairing completely. In the following, this field will be referred to as H_{c2}, as no confusion should arise with the analogous quantity for protons.

S-wave pairing is relevant for the crusts of magnetars, i.e., the low-density regime below the saturation density of symmetrical nuclear matter. At higher densities the dominant pairing state in neutron matter shifts to the $^3P_2$-$^3F_2$ channel, which pairs neutrons in a total spin-1 state (see [18, 19] for a review). In this case, the spin-polarizing effect of the magnetic field on the internal structure of the spin-1 pairs is nondestructive [20, 21, 22].

In Fig. 2 we plot values of H_{c2} for the neutron condensate as a function of density, determined based on a phase-shift-equivalent nucleon-nucleon interaction and numerical solutions of the BCS equations in the case of spin-polarized neutron matter [11]. The shape of the curve reflects the

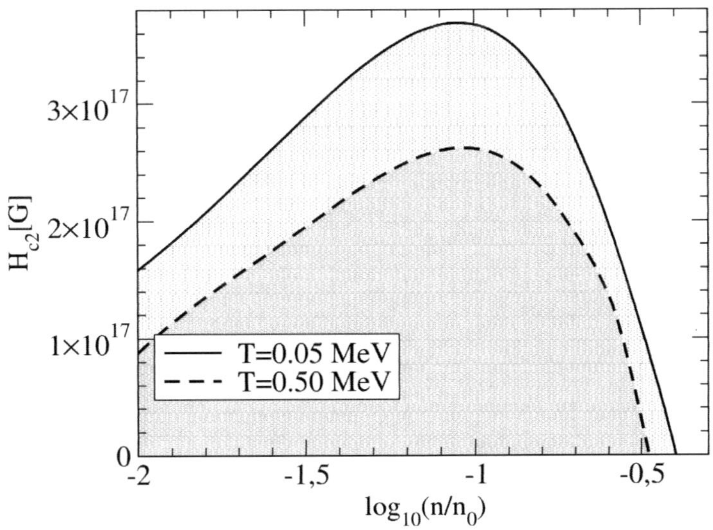

Figure 2. Critical unpairing magnetic field H_{c2} for the neutron condensate due to spin alignment in this field, for two temperatures indicated in the plot [11]. The density is given in units of the nuclear saturation density n_0. The bell-shaped structure of the critical field reflects the analogous shape of the gap function of ${}^1S_0$ paired neutrons, which attains a maximum at a certain density. The maximum in the gap value arises from the interplay between the increase in the density of states at the Fermi surface and decrease in the attraction in the pairing force as the density (and neutron Fermi energy) increases.

corresponding density dependence of the pairing gap, and its temperature dependence follows the BCS prediction: it is largest at $T = 0$ and decreases as the pairing gap decreases with increasing temperature. Thus, if the local field in a magnetar crust exceeds the value H_{c2}, the magnetic field will destroy the condensate. In the specific model of [11], the neutron fluid in magnetars will be non-superfluid (i.e., in a normal phase) for $B > 2.6 \times 10^{17}$ G. Note that at those densities where attractive P-wave interaction exists the suppression of the S-wave pairing may give rise to P-wave superfluid rather than normal spin-polarized fluid [11].

The non-superfluidity or partial superfluidity of magnetars will clearly have important implications for an array of microphysical quantities of neutron star crusts. These include their neutrino emissivities and transport properties, and consequently the thermal relaxation and dynamical coupling time scales that are important, most notably, for the damping of stellar oscillations and the interpretation of rotational anomalies such as glitches and anti-glitches. Finally, it is to be noted that the Pauli paramagnetic destruction mechanism discussed here for S-wave paired neutrons will apply as well to S-wave paired protons; however, the diamagnetic mechanism mentioned in the preceding section is more important for protons.

3. Crust-core coupling time scales in magnetars

We turn now to implications of the unpairing effect for the rotational coupling of a neutron superfluid in magnetar cores. In [12] it was argued that unpairing offers a new channel for coupling of the electron-proton plasma to the neutron vorticity in the core: if protons are unpaired then they are available for scattering off neutron quasiparticles in the vortex cores. This process is much more effective than the scattering of electrons off magnetized neutron vortices by electromagnetic forces, which is the dominant process in the type-I superconducting

IOP Conf. Series: Journal of Physics: Conf. Series **861** (2017) 012025 doi:10.1088/1742-6596/861/1/012025

case. In the more realistic case of type-II superconductivity, the coupling mechanism can be more complicated, because of the non-negligible interactions between the protonic flux-tubes and the neutron vortices.

The time scale of dynamical coupling of the superfluid to the normal plasma is important for the interpretation of rotational irregularities of magnetars, including glitches, anti-glitches, post-glitch relaxation, and non-axisymmetric motions such as precession. The influence of an interior fluid on precession has been discussed extensively in the literature for the case of ordinary neutron stars (see [23] and references therein). In the case of magnetars, the coupling time scale of the P-wave neutron superfluid when the protons form a normal fluid is an important ingredient of such considerations [12].

To illustrate this point, assume a constant field in the core of a magnetar and a fully unpaired proton fluid. The field will then couple to the electron fluid on plasma time scales, which are much shorter than the hydrodynamical time scales. Therefore, the unpaired core of a magnetar can be considered as a two-fluid system consisting of a superfluid neutron condensate component and a normal component formed by the proton and electron fluids.

The neutron superfluid rotates by forming an array of quantized vortices with areal number density

$$N_n = \frac{2\Omega}{\omega_0}, \quad \omega_0 = \frac{\pi}{m_n}, \tag{4}$$

where m_n is the bare neutron mass, Ω the rotation frequency of the star, and ω_0 the quantum of neutron circulation. As seen from Eq. (4), any changes in the rotation frequency of a magnetar must be accompanied by changes in the number of neutron vortices. Because the vortices are created and destroyed at the interfaces, they need to move in the bulk of the superfluid to respond to variations in Ω. The velocity of a vortex $\boldsymbol{v}_L$ is determined by the equation of motion

$$\rho_n\omega_0[(\boldsymbol{v}_S - \boldsymbol{v}_L)\times\boldsymbol{\nu}] - \eta(\boldsymbol{v}_L - \boldsymbol{v}_N) = 0. \tag{5}$$

This equation reflects the balance of forces acting on a vortex segment; the first term of the right is the Magnus force, and the second is the frictional force between the vortices and the normal liquid, with ρ_n denoting the mass density of the superfluid component, $\boldsymbol{v}_N$ is the velocity of the normal component and η is the coordinate-dependent longitudinal friction coefficient and $\boldsymbol{\nu}$ is the unit vector along the vortex circulation. The frictional force due to scattering of normal quasiparticles (electrons, muons, and unpaired protons) is given by

$$\boldsymbol{F} = \frac{2}{\tau N_n}\int f(\boldsymbol{p},\boldsymbol{v}_L)\boldsymbol{p}\frac{d^3p}{(2\pi\hbar)^3} = -\eta\boldsymbol{v}_L, \tag{6}$$

where $f(\boldsymbol{p},\boldsymbol{v}_L)$ is the non-equilibrium distribution function in the frame where $\boldsymbol{v}_N = 0$. Assuming a small perturbation, this function can be expanded about the equilibrium distribution Fermi function f_0 to obtain $f(\boldsymbol{p},\boldsymbol{v}_L) = f_0(\boldsymbol{p}) + (\partial f_0/\partial\epsilon)(\boldsymbol{p}\cdot\boldsymbol{v}_L)$. For strongly degenerate systems, $\partial f_0/\partial\epsilon \simeq -\delta(\epsilon - \epsilon_{Fp})$ and $\eta = m_p^* n_p/\tau N_n$, where n_p and m_p^* are respectively the unpaired quasiparticle number density and effective mass. In our example, the dominant process determining the relaxation time τ is the scattering of unpaired protons off neutron vortex-core quasiparticles; its rate scales with the temperature as $\tau^{-1} \propto T\exp[-\varepsilon^0_{1/2}/T]$ and is proportional to the differential nuclear scattering cross-section $d\sigma/d\Omega$ between neutrons and protons. The exponential factor contains the energy scale $\varepsilon^0_{1/2} = \pi\Delta_n^2/(4\epsilon_{Fn})$, corresponding to the lowest energy state of a neutron quasiparticle confined in the vortex, with Δ_n denoting the neutron pairing gap and ϵ_{Fn} the neutron Fermi energy; (see [24] for further details.) The dynamical coupling time of the superfluid to the plasma is given by

$$\tau_D = \frac{1}{2\Omega}\left(\zeta + \zeta^{-1}\right), \tag{7}$$

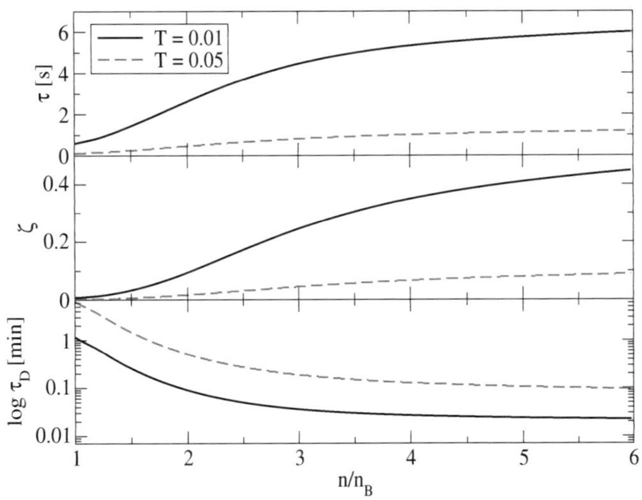

Figure 3. Relaxation time, drag-to-lift ratio, and dynamical coupling time scale (7) as functions of baryon density in the stellar core for temperatures $T = 0.01$ and 0.05 MeV (or equivalently $T = 1.2 \times 10^8$ K and $T = 5.8 \times 10^8$ K). The computations are carried out with an angle-independent neutron-proton cross section $\sigma \simeq 60$ fm^2 and for a rotation frequency $\Omega = 1$ Hz. Note that these results apply only if the proton fluid is unpaired at a given density.

where the dimensionless drag-to-lift ratio ζ is related to the dimensioned friction η via the relation $\zeta = \eta/\rho_n\omega_0$. Figure 3 summarizes the results of computations carried out in [12]. It is seen that for typical magnetar periods of about 10 sec, i.e., for spin rotations of about 1 Hz, the unpaired core couples to the plasma on dynamical time scales from several minutes at the crust-core boundary to a few seconds in the high-density core. Furthermore, the values of ζ obtained imply that the low-density outer core ($\zeta \simeq 0.2$) does not affect free precession, while the high-density inner core ($\zeta \simeq 0.4$) can cause significant damping of precession over a cycle. Thus, magnetar precession cannot be definitely excluded since the values of the drag-to-lift ratio are within the range of the crossover from undamped to damped precession. However, in the inner core, where ζ is large, the condensate vanishes for lower magnetic fields. We conclude that relatively low magnetic fields are sufficient to damp any free precession in magnetars. Of course, this argument applies only to free precession. Magnetic deformations of magnetars can be a continuous source of excitation of precession [25, 26]. The macroscopic arrangement of the magnetic field in the case of type-II superconducting stars and its implications for precession have been discussed elsewhere [27, 28, 29].

4. Neutrino radiation from magnetars

The suppression of pairing by MeV-scale magnetic fields will also have profound consequences on the thermal evolution of magnetars, because the dominant processes of neutrino radiation will not be suppressed by the Boltzmann factor containing the gap in the quasiparticle spectrum of baryons. At the same time, the processes that are intrinsic to condensates, such as the pair-breaking emission of neutrino-anti-neutrino pairs, will not operate by definition.

4.1. Direct Urca process: $n \to p + e + \bar\nu_e$

In strong B-fields the phase-space of nucleons is modified and the Urca process is allowed even below the threshold $x_p \simeq 11\%$, where x_p is the proton fraction in nucleonic matter [30, 31]. The

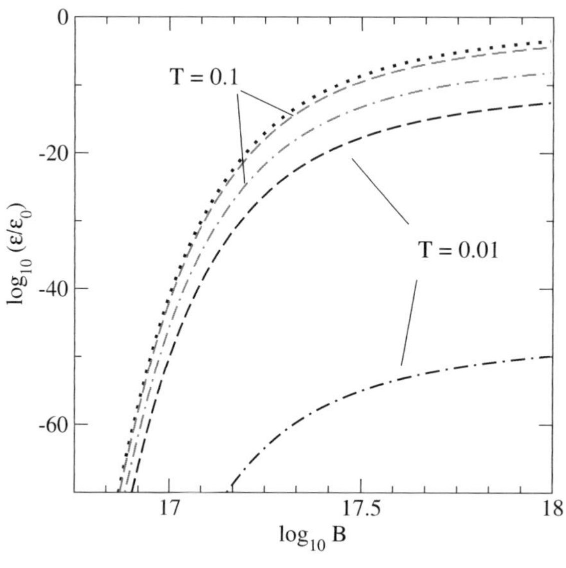

Figure 4. Magnetic-field dependence of the emissivity ϵ of the Urca process (measured in units of the zero-field emissivity ϵ_0), as calculated at temperatures $T = 0.01$ and 0.1 MeV and density $n = n_0$ for the case $x = 0$, i.e., such that emission is forbidden kinematically in the zero-field limit. Three possibilities are considered: (i) normal neutron-star matter (dotted line), (ii) neutrons paired but protons normal (dashed line), and (iii) both neutrons and protons paired (dot-dashed lines). The parameter x of Eq. (4) scales with the field B according to $x \propto N_{Fp}^{2/3} \propto B^{-2/3}$.

kinematics of the Urca process in this case is conveniently characterized by the parameter [31]

$$x = \left[1 - (k_{Fe} + k_{Fp})^2/k_{Fn}^2\right] N_{Fp}^{2/3}, \tag{8}$$

where k_{Fi}, $i = e, p, n$, are the Fermi momenta of electrons (e), protons (p) and neutrons (n) and $N_{Fp} = k_{Fp}^2/2|e|B$ is the number of Landau levels populated by protons. Clearly, for $x > 0$ the Urca process is forbidden in the $B = 0$ case, but in a strong magnetic field some phase space opens up. Then, if the Urca process operates even at a fraction of its strength in the kinematically permitted region, it can become an important factor in the cooling the star's core, because other competing processes are weaker by orders of magnitude. In the case where $x < 0$, i.e., when the Urca process is kinematically permitted, magnetic fields induce *de Haas-van Alfven oscillations* in the emissivity associated with the filling of the Landau levels.

As well known, proton and neutron pairing suppress the Urca process once nucleons make a transition to a superconducting or superfluid state. At asymptotically low temperatures, the emissivity is suppressed by a factor $\exp(-\Delta/T)$ for each participating nucleon, where Δ is the relevant pairing gap and T is the temperature [32, 33, 34]. Clearly, termination of neutron and proton S-wave pairing by the B-field will mitigate this suppression. Because the neutron pairing gap in the P-wave channel is smaller than the proton pairing gap in the S-wave channel, destruction of proton superconductivity by a MeV-scale magnetic field will strongly modify the Urca emissivity. Numerical examples can be found in [10].

4.2. Field-assisted bremsstrahlung processes: $N \to N + \nu_f + \bar{\nu}_f$

The bremsstrahlung $N \to N + \nu_f + \bar{\nu}_f$ of neutrino pairs by a single nucleon (denoted N) is ordinarily prohibited by energy and momentum conservation. Accordingly, the leading charge-neutral mechanism for neutrino-pair production is the so-called modified bremsstrahlung process $N+N \to N+N+\nu_f+\bar{\nu}_f$. However, when the interaction energy of the B-field with the spin of a nucleon becomes of the order of the temperature, the single-nucleon bremsstrahlung process $N \to N + \nu + \bar{\nu}$ becomes kinematically possible because of the paramagnetic splitting of the energies of nucleons with spin-up and spin-down in a strong B field [35]. Pairing will suppress this process exponentially at low temperatures, as discussed above in the case of the Urca process. However, if unpairing by a strong magnetic field takes place, this suppression will be nullified.

4.3. Pair-breaking processes: $[NN] \to N + N + \nu_f + \bar{\nu}_f$

Nucleonic superfluidity results in a new class of neutrino bremsstrahlung processes that owe their existence to the condensate. Symbolically, these can be written as $[NN] \to N+N+\nu_f+\bar{\nu}_f$, where $[NN]$ stands for a Cooper pair. These processes are referred to as pair-breaking-formation (PBF) processes [36, 37, 38, 39, 40, 41]. The rates of neutrino emission via the PBF processes scale as $\epsilon \propto \Delta^2 T^7$, where Δ is the pairing gap. Therefore, the unpairing of the S-wave condensates will have the plain effect of removing the PBF processes from the regions where the field locally exceeds the unpairing fields for protons and neutrons. Consequently, the net neutrino emission rate will be reduced asymptotically to the value which corresponds to the PBF emission by the P-wave condensate.

Assessment of the combined effect of unpairing on the cooling of neutron stars is difficult. While it is clear how the individual processes are affected by the strong magnetic field, their concerted effect needs to be studied in numerical simulations.

4.4. Specific heats

An additional ingredient that modifies the thermal evolution of magnetars is the specific heat of the interior matter, which quantifies the thermal inertia of the star. As well known, in the absence of superconductivity and superfluidity, the heat capacity of nucleonic fluids will scale linearly with temperature. This should be contrasted with the exponential suppression of the heat capacity in the superconducting S-wave state. In fully superconducting/superfluid neutron stars at low B-fields, electrons dominate the heat capacity. One can anticipate that for MeV-scale magnetic fields in which only protons are unpaired, the proton and electron specific heats will decrease linearly with temperature (as in normal Fermi liquids), whereas the heat capacity of the superfluid neutrons will be reduced exponentially in the regions of S-wave pairing and as a power-law in the regions of P-wave pairing. Increase of the specific heat of the interior matter will act to increase the cooling time scale of a magnetar.

In closing, it should be mentioned that other factors such as internal heating due to the Ohmic dissipation of the magnetic fields in the interior of the star will be an important factor in determining the temperature evolution of magnetars. The unpairing of protons will imply that the resistivity due to electron-proton scattering will be larger than in superconducting stars. Consequently, the time scales of the Ohmic dissipation will be shorter.

5. Final remarks

Magnetars pose new challenges at the microphysical level because the electromagnetic interactions (e.g. the magnetic field-nucleon spin coupling) become of the order of the nuclear MeV scale. As a consequence, we find an intimate interplay between the electromagnetic interactions and nuclear and weak processes that take place in the vicinity of the Fermi surfaces of nucleons. This contribution has focused on the recent progress in understanding the mechanisms of suppression of pairing in nucleonic matter by a magnetic field and their implications for macroscopic dynamics of magnetars, such as their rotational and thermal evolutions. These findings call for more detailed studies of the *macroscopic dynamics* of magnetars, which will entail the modification of the pairing properties of nucleonic fluids.

Acknowledgments

AS acknowledges the support by the DFG (Grant No. SE 1836/3-2), by the Helmholtz International Center for FAIR, and by the NewCompStar COST Action MP1304. J.W.C. is grateful for the hospitality of the Center for Mathematical Sciences of the University of Madeira.

References

[1] Thompson C and Duncan R C 1995 *MNRAS* **275** 255–300
[2] Turolla R, Zane S and Watts A L 2015 *Reports on Progress in Physics* **78** 116901
[3] Mereghetti S, Pons J A and Melatos A 2015 *Space Sci. Rev.* **191** 315–338
[4] Chandrasekhar S and Fermi E 1953 *ApJ* **118** 116
[5] Lai D and Shapiro S L 1991 *ApJ* **383** 745–751
[6] Bocquet M, Bonazzola S, Gourgoulhon E and Novak J 1995 *A&A* **301** 757
[7] Frieben J and Rezzolla L 2012 *MNRAS* **427** 3406–3426
[8] Chatterjee D, Elghozi T, Novak J and Oertel M 2015 *MNRAS* **447** 3785–3796
[9] Sinha M and Sedrakian A 2015 *Physics of Particles and Nuclei* **46** 1510
[10] Sinha M and Sedrakian A 2015 *Phys. Rev. C* **91** 035805
[11] Stein M, Sedrakian A, Huang X G and Clark J W 2016 *Phys. Rev. C* **93** 015802
[12] Sedrakian A 2016 *A&A* **587** L2
[13] Tinkham M 1996 *Introduction to superconductivity, McGraw-Hill, New York*
[14] Sedrakyan D M and Shakhabasyan K M 1980 *Astrophysics* **16** 417
[15] Alpar M A, Langer S A and Sauls J A 1984 *ApJ* **282** 533–541
[16] Alford M, Good G and Reddy S 2005 *Phys. Rev. C* **72** 055801
[17] Haber A and Schmitt A 2016 *ArXiv e-prints* (*Preprint* 1612.01865)
[18] Zverev M V, Clark J W and Khodel V A 2003 *Nuclear Physics A* **720** 20–42
[19] Clark J W, Sedrakian A, Stein M, Huang X G, Khodel V A, Shaginyan V R and Zverev M V 2016 *Journal of Physics Conference Series* **702** 012012
[20] Muzikar P, Sauls J A and Serene J W 1980 *Phys. Rev. D* **21** 1494–1502
[21] Sauls J A, Stein D L and Serene J W 1982 *Phys. Rev. D* **25** 967–975
[22] Tarasov A N 2014 *EPL (Europhysics Letters)* **105** 52001
[23] Link B 2007 *Ap&SS* **308** 435–441
[24] Sedrakian A 1998 *Phys. Rev. D* **58** 021301
[25] Wasserman I 2003 *MNRAS* **341** 1020–1040
[26] Lander S K, Andersson N, Antonopoulou D and Watts A L 2015 *MNRAS* **449** 2047–2058
[27] Akgün T and Wasserman I 2008 *MNRAS* **383** 1551–1580
[28] Henriksson K T and Wasserman I 2013 *MNRAS* **431** 2986–3002
[29] Lander S K 2014 *MNRAS* **437** 424–436
[30] Leinson L B and Pérez A 1998 *Journal of High Energy Physics* **9** 020
[31] Baiko D A and Yakovlev D G 1999 *A&A* **342** 192–200
[32] Yakovlev D G, Levenfish K P and Shibanov Y A 1999 *Physics Uspekhi* **42** 737–778
[33] Sedrakian A 2005 *Physics Letters B* **607** 27–34
[34] Sedrakian A 2007 *Progress in Particle and Nuclear Physics* **58** 168–246
[35] van Dalen E N E, Dieperink A E L, Sedrakian A and Timmermans R G E 2000 *A&A* **360** 549–558
[36] Flowers E, Ruderman M and Sutherland P 1976 *ApJ* **205** 541–544
[37] Leinson L B and Pérez A 2006 *Physics Letters B* **638** 114–118
[38] Sedrakian A, Müther H and Schuck P 2007 *Phys. Rev. C* **76** 055805
[39] Kolomeitsev E E and Voskresensky D N 2008 *Phys. Rev. C* **77** 065808
[40] Steiner A W and Reddy S 2009 *Phys. Rev. C* **79** 015802
[41] Sedrakian A 2012 *Phys. Rev. C* **86** 025803

IOP Conf. Series: Journal of Physics: Conf. Series **861** (2017) 012026 doi:10.1088/1742-6596/861/1/012026

Phase Diagram of (Proto)Neutron Star Matter in an Extended Bag Model

Thomas Klähn[1], **Tobias Fischer**[1], **Mateusz Cierniak**[1], **and Matthias Hempel**[2]

[1] Institute of Theoretical Physics, University of Wrocław, Poland
[2] Department of Physics, University of Basel, Switzerland

E-mail: thomas.klaehn@ift.uni.wroc.pl

Abstract. We illustrate the consequence of the simultaneous restoration of chiral symmetry and deconfinement at finite temperature under neutron star conditions in an extended bag model, vBag, which we developed in earlier work.

1. Introduction

Currently, no consistent approach exists to describe medium properties of hadrons and mesons at the level of quarks and gluons at high density. Hence, the deconfinement phase transition is usually constructed from a given hadronic equation of state (EoS) with hadrons and mesons as the fundamental degrees of freedom and an independently computed quark matter EoS, although few attempts exist to improve this situation (c.f. Refs. [1, 2]). A general review of recent developments concerning the EoS in astrophysical applications can be found in Ref. [3].

The novel quark matter EoS vBag was introduced recently [4] with particular focus on applications in astrophysics; vBag explicitly accounts for dynamical chiral symmetry breaking (DχSB) and repulsive vector interactions. The latter property is essential for the stiffening of the EoS towards high densities in order to yield maximum neutron star masses in agreement with the current constraint of about 2 $M_{\odot}$ [5, 6] (the previously quoted value in Ref. [6] was recently revised to 1.928 ± 0.017 $M_{\odot}$ [7]). In order to ensure chiral symmetry restoration and (de)confinement to coincide, which is usually not provided for the 'standard' Maxwell or Gibbs approach for the phase transition, vBag mimics (de)confinement via a phenomenological correction to the EoS. It is determined by the hadronic EoS chosen for the construction of the phase transition.

More recently, vbag was extended to finite temperatures and arbitrary isospin asymmetry to study consequences for the phase diagram [8]. In the following we review the basic foundation of vBag and present the phase diagram of matter under (proto)neutron star conditions.

2. vBag - a Bag Model Extension

We briefly scketch the derivation of vBag [4] as a particular approximation of QCD's in-medium Dyson-Schwinger equations (DSE), which initially contains the standard Nambu–Jona-Lasinio (NJL) model (c.f. [9, 10, 11]) as a particular limiting case. When chiral symmetry is restored it behaves in analogy to the thermodynamic bag model (tdBag). However, vBag accounts for vector repulsion in which aspect it extends beyond the tdBag model.

In general, the in-medium, dressed-quark propagator maintains the structure of a free, relativistic Fermion propagator,

$$S(p^2,\tilde{p}_4)^{-1} = i\vec{\gamma}\vec{p}A(p^2,\tilde{p}_4) + i\gamma_4\tilde{p}_4 C(p^2,\tilde{p}_4) + B(p^2,\tilde{p}_4) \ , \tag{1}$$

with $\tilde{p}_4 = p_4 + i\mu$. Evidently, the gap functions A, B, and C account for non-ideal behaviour due to interactions. Unlike for vacuum studies, the gaps are complex valued and A- and C-gap are degenerate (for the model we introduce in the following A=C holds strictly only in vacuum). In order to obtain the propagator one solves the following gap equation,

$$\begin{aligned} S(p^2,\tilde{p}_4)^{-1} &= i\vec{\gamma}\cdot\vec{p} + i\gamma_4\tilde{p}_4 + m + \Sigma(p^2,\tilde{p}_4) \ , \\ \Sigma(p^2,\tilde{p}_4) &= \int \frac{d^4q}{(2\pi)^4}\, g^2(\mu)D_{\rho\sigma}(p-q,\mu)\frac{\lambda^a}{2}\gamma_\rho S(q^2,\tilde{q}_4)\Gamma^a_\sigma(q,p,\mu) \ , \end{aligned} \tag{2}$$

where m is the bare mass, $D_{\rho\sigma}(k,\mu)$ is the dressed-gluon propagator and $\Gamma^a_\sigma(q,p,\mu)$ is the dressed-quark-gluon vertex. Naturally, at the level of the self energy $\Sigma(p^2,\tilde{p}_4)$ approximations can be made in order to simplify the gap equations. In all following discussions we impose:

$$\Gamma^a_\sigma(q,p) = \frac{1}{2}\lambda^a\gamma_\sigma \ , \tag{3}$$

for the vertex and thus define a rainbow gap equation, which is the leading-order in a systematic, symmetry-preserving DSE truncation scheme [12, 13].

The NJL model is understood as the result of a contact interaction in configuration space provided by the gluon propagator. Transformed into momentum space this reads as a constant, momentum-independent coupling. The model is ultraviolet-divergent if no regularization is performed. Here, we perform a hard cut-off in the ultra-violet (UV) and express the effective gluon propagator as follows,

$$g^2D^{\rho\sigma}(p-q) = \frac{1}{m_G^2}\Theta(\Lambda^2-\vec{q}^2)\delta^{\rho\sigma} \ . \tag{4}$$

The Heaviside function Θ provides a 3-momentum cutoff for space-like momenta $\vec{p}^2 > \Lambda^2$. This is sufficient to regularize all UV divergences inherent to $\Sigma(p^2,\tilde{p}_4)$. Different regularization procedures are available and in fact the regularization scheme does not have to affect UV divergences only. E.g., infra-red (IR) cutoff schemes can remove unphysical implications [14]. However, the chosen hard cut-off scheme reproduces standard NJL model results and allows to match them to tdBAG, i.e. to describe quarks as quasi ideal gas of Fermions. m_G is a gluon mass scale which in this model simply defines the coupling strength. These approximations are sufficient to write the gap equations. For the A-gap follows the trivial, medium independent solution, $A = 1$. The remaining gap equations take the following form,

$$B_p = m + \frac{16N_c}{9m_G^2}\int_\Lambda \frac{d^4q}{(2\pi)^4}\frac{B_q}{\vec{q}^2A_q^2 + \tilde{q}_4^2C_q^2 + B_q^2} \ , \tag{5}$$

$$\tilde{p}_4^2 C_p = \tilde{p}_4^2 + \frac{8N_c}{9m_G^2}\int_\Lambda \frac{d^4q}{(2\pi)^4}\frac{\tilde{p}_4\tilde{q}_4C_q}{\vec{q}^2A_q^2 + \tilde{q}_4^2C_q^2 + B_q^2} \ . \tag{6}$$

The integrals do not explicitly depend on the external momentum p and consequently, both gap solutions are constant at any given μ. Both equations can be recast as follows,

$$B = m + \frac{4N_c}{9m_G^2}n_s(\mu^*,B) \ , \tag{7}$$

$$\mu = \mu^* + \frac{2N_c}{9m_G^2}n_v(\mu^*,B) \ , \tag{8}$$

in terms of the single-flavor scalar and vector densities, n_s and n_v, of an ideal spin-degenerate Fermi gas,

$$n_s = 2\sum_{\pm}\int_\Lambda \frac{\mathrm{d}^3\vec{p}}{(2\pi)^3}\frac{B}{E}\left(\frac{1}{2}-\frac{1}{1+\exp\left(E^{\pm}/T\right)}\right) , \tag{9}$$

$$n_v = 2\sum_{\pm}\int_\Lambda \frac{\mathrm{d}^3\vec{p}}{(2\pi)^3}\frac{\mp 1}{1+\exp\left(E^{\pm}/T\right)} , \tag{10}$$

with $E^2 = \vec{p}^2 + B^2$ and $E^{\pm} = E \pm \mu^*$. The merit of the NJL model is the ability to describe chiral symmetry breaking as the formation of a scalar condensate, and chiral symmetry restoration as the melting of the same. It should be kept in mind though, that it is the scalar density which requires UV regularization and in that sense chiral symmetry breaking can be considered as the most sensitive part of the model.

The next information that Eqs. (9) and (10) provide is well known; the NJL model describes quarks as quasi-ideal particles with corresponding quasi particle poles. However, confinement is not accounted for. In Ref. [4] it has been pointed out how this can be understood as a reason for the fact that NJL models typically provide a larger bag constant than, e.g., the MIT-bag model. The NJL model does not 'bind'. Adding binding energy per volume can lead to interesting consequences for the phase diagram, as illustrated in Ref. [8]. It will be further discussed in the following.

3. Phase Diagram

In order to reproduce the behavior of the NJL model in the chirally restored phase in good approximation we subtract a chiral bag constant $B_{\chi,f}$ from the pressure for each flavor independently,

$$P_f(\mu_f) = P^{kin}_{\mathrm{FG},f}(\mu_f^*) + \frac{K_v}{2} n^2_{\mathrm{FG},f}(\mu_f^*) - B_{\chi,f} . \tag{11}$$

The second term corresponds to the vector condensate, with K_v being related to the vector current-current interaction coupling constant, which in combination with the renormalisation of the chemical potential causes an effective stiffening of the EoS with increasing density. This term is not accounted for in the standard tdBag model. By construction, chiral symmetry is restored for $P_f(\mu_f) > 0$. The critical chiral chemical potential μ_f^{χ} for a single flavor is then defined by $P_f(\mu_f^{\chi}) = 0$. For the two-flavor case we redefine the chiral threshold condition to hold for the total light quark pressure instead for each single flavor independently,

$$\sum_{f=u,d} P_f(T, \mu_{\chi,f}) = 0 . \tag{12}$$

This procedure aims to avoid sequential deconfinement of different quark flavors. We want our model to account for chiral symmetry breaking and the confinement transition simultaneously at $\mu_{B,\chi}$. To do this the critical chemical potential for deconfinement, $\mu_{B,dc}$, has to match $\mu_{B,\chi}$. This is achieved by exploiting that the total pressure of the system is fixed by the previous equations only up to a constant B_{dc},

$$P^Q = \sum_f P_f(T, \mu_f) + B_{\mathrm{dc}} . \tag{13}$$

Introducing B_{dc} thus enables us to ensure that the pressure of hadronic and quark matter equals at the critical chemical potential $\mu_{B,\chi}$ for the chiral transition. In order to match both pressures

IOP Conf. Series: Journal of Physics: Conf. Series **861** (2017) 012026 doi:10.1088/1742-6596/861/1/012026

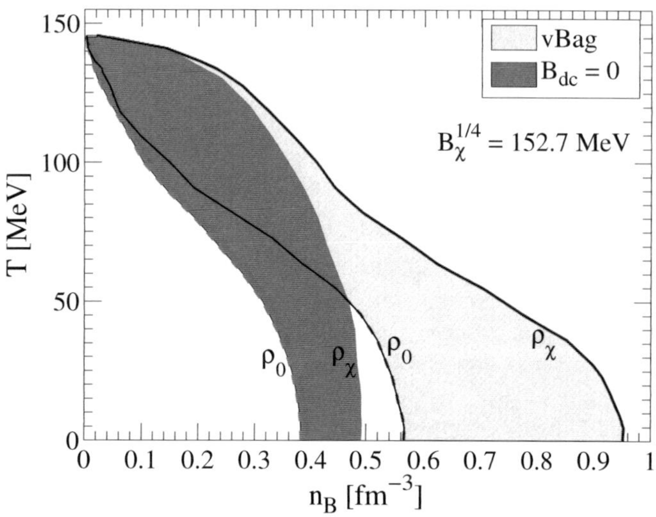

Figure 1. Phase diagram of matter under (proto)neutron star conditions, comparing vBag (blue) with deconfinement bag constant obtained from the underlying nuclear EoS and same parameterization but no deconfinement bag constant added (yellow). At $T = 0$ the latter reproduces NJL model results. Evidently, the vBag prescription lowers and narrows the coexistence region of nuclear and quark matter. The nuclear EoS has been selected from the catalogue of Ref. [15] with the relativistic mean field parametrization DD2 [16, 17]. The general setup of this model is found in Ref. [8].

one easily finds the value of $B_{\rm dc} = P^H(\mu_{B,\chi})$ at given T and μ_C. Note, the baryon chemical potential and the charge chemical potential are defined as follows:

$$\mu_B = \mu_u + 2\mu_d \,, \qquad \mu_C = \mu_u - \mu_d \,. \tag{14, 15}$$

At given T and μ_C we denote our critical chemical potentials for DχSB and confinement as $\mu_{B,\chi}$ and $\mu_{B,\rm dc}$ respectively. As shown in detail in Ref. [8], $B_{\rm dc}$ depends on the nuclear EoS and thus becomes medium dependent. According to our prescription for the phase transition it varies with temperature and isospin asymmetry, i.e. $B_{\rm dc} \to B_{\rm dc}(T, \mu_C)$, while we keep it constant with respect to μ_B. This results in additional contributions to the charge number and entropy density which are derivative terms with respect to the pressure. We will not discuss any of these quantities in the following and rather focus on the effect of simultaneous DχSB and deconfinement on the phase diagram of matter under (proto)neutron star conditions, viz. β-equilibrium, electric charge neutrality and finite temperature.

The yellow region in Fig. 1 corresponds to the result obtained for a model without an additional bag constant which accounts for confinement effects. At $T = 0$ the coexistence region between ρ_0 (onset density of the phase transition) and ρ_χ (onset density of pure quark matter) reproduces exactly NJL model results. However, we treat B_χ as a constant and hence the agreement will be less accurate with increasing temperature as we neglect the melting of the scalar condensate.

As typical for NJL type models, the transition occurs at rather high densities. Diquark condensates can lower this density [18]. Under the assumption of simultaneous DχSB and deconfinement the onset density of quark matter is solely defined by the chiral restoration and, for vBag, the chiral bag constant. The corresponding density follows the right side edge of the blue region in Fig 1. Without the aforementioned assumption this line would change for a different nuclear EoS. In our approach, this is not the case. The nuclear EoS alters the onset of the coexistence region only. We further note, that the transition density in our approach is significantly smaller than for the tdBag model with a bag constant that is adjusted to reproduce NJL model results (yellow region). This behavior of vBag favors the existence of quark matter in the cores of neutron stars.

In Ref. [4] it has been further shown that due to the incorporated vector interaction stable hybrid neutron star configuration of at least 2 $M_\odot$ can be obtained. With vBag we provide a novel quark-matter EoS for a broad spectrum of applications in astrophysics, e.g., simulations of neutron star mergers [19], core collapse supernovae [20, 21, 22] and protoneutron star cooling [23], where commonly the simple tdBag-model has been employed. However, the classical and in astrophysics studies commonly employed tdBag model does not account for chiral physics and without perturbative corrections violates the 2 $M_{\odot}$ neutron star mass constraint due to missing repulsive interactions. Here, vBag provides solutions that cure both problems without losing the simplicity and flexibility of tdBag.

Acknowledgments
We acknowledge support from the Polish National Science Center (NCN) under the grant numbers UMO-2014/13/B/ST9/02621 (TK), UMO-2011/02/A/ST2/00306 (TF), and UMO-2013/09/B/ST2/01560 (MC). We are grateful for the supporting networking activities of the COST Actions NewCompStar MP130 and ChETEC MP16117.

References
[1] Dexheimer V A and Schramm S 2010 *Phys. Rev.* **C81** 045201 (*Preprint* 0901.1748)
[2] Steinheimer J, Schramm S and Stöcker H 2011 *JPhG* **38** 035001 (*Preprint* 1009.5239)
[3] Oertel M, Hempel M, Klahn T and Typel S 2016 (*Preprint* 1610.03361)
[4] Klahn T and Fischer T 2015 *Astrophys. J.* **810** 134 (*Preprint* 1503.07442)
[5] Antoniadis J, Freire P C C, Wex N, Tauris T M, Lynch R S, van Kerkwijk M H, Kramer M, Bassa C, Dhillon V S, Driebe T, Hessels J W T, Kaspi V M, Kondratiev V I, Langer N, Marsh T R, McLaughlin M A, Pennucci T T, Ransom S M, Stairs I H, van Leeuwen J, Verbiest J P W and Whelan D G 2013 *Science* **340** 448
[6] Demorest P, Pennucci T, Ransom S, Roberts M and Hessels J 2010 *Nature* **467** 1081–1083 (*Preprint* 1010.5788)
[7] Fonseca E, Pennucci T T, Ellis J A, Stairs I H, Nice D J, Ransom S M, Demorest P B, Arzoumanian Z, Crowter K, Dolch T, Ferdman R D, Gonzalez M E, Jones G, Jones M L, Lam M T, Levin L, McLaughlin M A, Stovall K, Swiggum J K and Zhu W 2016 *ArXiv* e-prints *astro-ph.HE/1603.00545* (*Preprint* 1603.00545)
[8] Klahn T, Fischer T and Hempel M 2017 *Astrophys. J.* **836** 89 (*Preprint* 1603.03679)
[9] Nambu Y and Jona-Lasinio G 1961 *Phys.Rev.* **124** 246–254
[10] Klevansky S 1992 *Rev. Mod. Phys.* **64** 649–708
[11] Buballa M 2005 *Phys.Rept.* **407** 205–376 (*Preprint* hep-ph/0402234)
[12] Munczek H J 1995 *Phys. Rev.* **D52** 4736–4740 (*Preprint* hep-th/9411239)
[13] Bender A, Roberts C D and Von Smekal L 1996 *Phys. Lett.* **B380** 7–12 (*Preprint* nucl-th/9602012)
[14] Ebert D, Feldmann T and Reinhardt H 1996 *Phys. Lett.* **B388** 154–160 (*Preprint* hep-ph/9608223)
[15] Hempel M and Schaffner-Bielich J 2010 *NuPhA* **837** 210–254 (*Preprint* 0911.4073)
[16] Typel S 2005 *Phys.Rev.* **C71** 064301 (*Preprint* arXiv:nucl-th/0501056)
[17] Typel S, Ropke G, Klahn T, Blaschke D and Wolter H 2010 *Phys.Rev.* **C81** 015803 (*Preprint* 0908.2344)
[18] Klahn T, Blaschke D, Sandin F, Fuchs C, Faessler A, Grigorian H, Ropke G and Trumper J 2007 *Phys. Lett.* **B654** 170–176 (*Preprint* nucl-th/0609067)
[19] Bauswein A, Janka H T, Oechslin R, Pagliara G, Sagert I, Schaffner-Bielich J, Hohle M M and Neuhauser R 2009 *Phys. Rev. Lett.* **103** 011101 (*Preprint* 0812.4248)
[20] Sagert I, Fischer T, Hempel M, Pagliara G, Schaffner-Bielich J, Mezzacappa A, Thielemann F K and Liebendorfer M 2009 *Phys. Rev. Lett.* **102** 081101 (*Preprint* 0809.4225)
[21] Nakazato K, Sumiyoshi K and Yamada S 2008 *PhRvD* **77** 103006 (*Preprint* 0804.0661)
[22] Fischer T, Sagert I, Pagliara G, Hempel M, Schaffner-Bielich J, Rauscher T, Thielemann F K, Käppeli R, Martínez-Pinedo G and Liebendörfer M 2011 *ApJS* **194** 39 (*Preprint* 1011.3409)
[23] Pons J A, Steiner A W, Prakash M and Lattimer J M 2001 *PhRvL* **86** 5223–5226 (*Preprint* astro-ph/0102015)

IOP Conf. Series: Journal of Physics: Conf. Series **861** (2017) 012027 doi:10.1088/1742-6596/861/1/012027

Strangeon and Strangeon Star

Xiaoyu Lai[1] **and Renxin Xu**[2]

[1] School of Physics and Engineering, Hubei University of Education, Wuhan 430205, China
[2] School of Physics, Peking University, Beijing 100871, China and Kavli Institute for Astronomy and Astrophysics at Peking University, Beijing 100871, China

E-mail: [1]`laixy@pku.edu.cn`

E-mail: [2]`r.x.xu@pku.edu.cn`

Abstract. The nature of pulsar-like compact stars is essentially a central question of the fundamental strong interaction (explained in quantum chromo-dynamics) at low energy scale, the solution of which still remains a challenge though tremendous efforts have been tried. This kind of compact objects could actually be strange *quark* stars if strange quark matter in bulk may constitute the true ground state of the strong-interaction matter rather than ^{56}Fe (the so-called Witten's conjecture). From astrophysical points of view, however, it is proposed that strange cluster matter could be absolutely stable and thus those compact stars could be strange *cluster* stars in fact. This proposal could be regarded as a general Witten's conjecture: strange matter in bulk could be absolutely stable, in which quarks are either free (for strange quark matter) or localized (for strange cluster matter). Strange cluster with three-light-flavor symmetry is renamed *strangeon*, being coined by combining "strange nucleon" for the sake of simplicity. A strangeon star can then be thought as a 3-flavored gigantic nucleus, and strangeons are its constituent as an analogy of nucleons which are the constituent of a normal (micro) nucleus. The observational consequences of strangeon stars show that different manifestations of pulsar-like compact stars could be understood in the regime of strangeon stars, and we are expecting more evidence for strangeon star by advanced facilities (e.g., FAST, SKA, and eXTP).

1. Introduction: *What is a strangeon*?

The baryonic matter in the Universe is well described by the Standard Model of particle physics. Six flavors of quarks are fundamental Fermions that constitute baryonic matter, among which up (u), down (d) and strange (s) quarks are light flavors (the other heavy flavors: c, t, b). Atomic nucleus, the most familiar baryonic matter to us, is composed of nucleons which is further composed of only two favors of constituent quarks, i.e. u and d quarks. In other words, the nucleon is the constituent of a 2-flavored micro-nucleus. Similarly, the "strangeon" we are explaining in this contribution is the constituent of a 3-flavored macro-nucleus, and is simply an abbreviation of "strange nucleon" (with strangeness degree of freedom there).

The existence of 3-flavored nuclei might be superficially understood by analysis of energy scale. Quarks are often described by mass parameters that are measured indirectly since they are confined inside hadrons rather than free particles, and the mass parameters are the key parameters to make a judgment on the quark-flavor degrees of freedom at a certain energy scale. The current mass of both u and d quarks are only a few MeV while that of s quark is around 100 MeV. For nuclei or nuclear matter with baryon density $n_0 \simeq 0.16$ fm^{-3}, the separation between quarks is $\Delta x \sim 0.5$ fm, and the energy scale is then in the order of $E_{\rm nucl} \sim 400$ MeV according to Heisenberg's relation $\Delta x \cdot pc \sim \hbar c \simeq 200$ MeV $\cdot$ fm. All of the heavy flavors of quarks have masses larger than 1 GeV, and would then not participate in dense matter with density $n \sim n_0$ as valence quarks. One could thus naively expect that 3-flavored nuclei could exist in case of density around n_0 because u and d quarks are the lightest and $E_{\rm nucl}$ is much larger than the mass difference between s and u/d quarks. However, it is a fact that the valence strangeness degree of freedom is absolutely missing in stable nuclei with sizes $\sim$ fm. On one hand, the mass of s quark is larger than that of u/d quarks, so s quarks would convert to u and d quarks via weak interaction. But, on the other hand, the light-flavor (u, d and s) symmetry might be restored if the flavor symmetry leads to lowering the energy (as for an example of nuclear analogue, the symmetry energy), especially in case that the strong-interaction matter is very big. We are trying to explain in this proceeding that 3-flavored matter could be manifested in the form of "gigantic nucleus", and that "strangeon" is the constituent of such a 3-flavored "nucleus".

To demonstrate the difference between two- and three-flavored nuclei, we could first look at the structure of an atom. Before the era of Ernest Rutherford who was the father of nuclear physics, the structure of atoms was described in J. J. Thomson's model, which states that an atom is composed of positively charged medium (occupying the whole atomic volume) as well as electrons embedded. In 1911, E. Rutherford proposed that positively charged nucleus, carrying almost all atomic mass, is actually a tiny part of an atom while electrons are not inside but around the nucleus, according to the experimental results of deflecting α-particles passing through a thin gold foil [1]. In the Standard Model of particle physics, ordinary nuclei, composed of nucleons, have only two flavors of quarks (u and d) in fact. Conceptually, why is our baryonic matter symmetric with two flavors rather than three?

A philosophical explanation could be simple: micro-nuclei are too small to have three-flavor symmetry, but bigger is different [2]. The electric charges of u and d quarks are $+2/3$ and $-1/3$ respectively, so two-flavor symmetric strong-interacting matter should be positively charged, and electrons are needed to maintain electric neutrality. The probability of electrons inside a micro-nucleus is negligible because the nucleus radius (~ 1 fm) is much smaller than the Compton wavelength of electrons, $\lambda_{\rm e} \sim 10^3$ fm. Therefore, electrons contribute negligible energy there, and two-flavored micro-nuclei would be energetically economic. However, for a gigantic nucleus with size approximating to or even larger than $\lambda_{\rm e}$ (corresponding to number of nucleons $A > 10^9$), it might be three-flavor symmetric. If the two-flavor symmetry still stays in macro-nucleus, it would become a huge Thomson atom with electrons inside it. Electron could then contribute a significantly high Fermi energy to the total energy of the system, being $E_{\rm F} \sim 100$ MeV. The situation becomes different if strangeness (i.e., three-flavor symmetry) is included: no electrons exist if the matter is composed of equal numbers of u, d and s quarks! In this case, the three-flavor symmetry, an analogy of two-flavor symmetry in micro-nucleus, may result in a ground state of gigantic nuclei. Strangeon, the constituent of a gigantic nucleus, is also an analogy of nucleon which is the constituent of a micro-nucleus. Certainly micro-physical calculations (e.g., an extended version of relativistic mean field theory applicable to strangeon matter) are really necessary and welcome in order to have a solid foundation of the starngeon conjecture.

It is worth noting that, rational thinking about stable strangeness datas back to 1970s and bulk strange object is speculated to be the absolutely stable ground state of strong-interacting matter, which is known as Witten's conjecture [3]. Although the conjecture was proposed based on the matter composed of almost free quarks, we can make an extension that it still reasonably holds no matter whether quarks are free or localized.

In some of our previous papers, we use a word, "strange quark-cluster" [4], which has actually the same meaning as strangeon. This paper is organized as following. The gravity-compressed dense matter is introduced in Section 2 in order to make sense of realistic "gigantic nucleus" in astrophysics. The observational consequences of strangeon stars, including the surface and global properties, are discussed in Section 3. A brief summary is given in Section 4.

IOP Conf. Series: Journal of Physics: Conf. Series **861** (2017) 012027 doi:10.1088/1742-6596/861/1/012027

2. Compact stars with strangeness

"Gigantic nucleus" is not a new concept. In the year 1932, L. Landau published a paper in which an idea of gigantic nucleus was presented [5]: "*We expect that this must occur when the density of matter becomes so great that atomic nuclei come in close contact, forming one gigantic nuclues.*" Although the main motivation for Landau to write this paper was to explain the origin of stellar energy (and this explanation is certainly wrong), he recognized for the first time a kind of compact stellar object at about nuclear density, which is surely necessary for us to understand various extreme phenomena through astronomical observations. Two years later, in 1934, W. Baade and F. Zwicky conceived the idea that forming neutron stars could be the energy source of supernova rather than of stars [6]. Neutron stars theoretically predicted were finally discovered when Hewish and Bell detected radio pulsars in 1967 [7]. The gigantic nucleus proposed by Landau was then identified by astronomers to be "neutron stars", as neutrons and protons at that time were believed to be elementary particles and gigantic nucleus should be neutron-rich. After physicists realized that the more fundamental baryonic particles are quarks, some models were built about compact stars including quark degree of freedom, either in the core of neutron star (i.e. mixed or hybrid stars) or as the whole star (quark stars).

One may speculate the nature of dense matter inside pulsar-like compact stars as following. After core-collapsing of an evolved massive star, the supernova-produced rump left behind where normal nuclei are intensely compressed by gravity to form "compressed baryonic matter" which could manifest the behaviors of pulsar-like compact stars. The average density of compressed baryonic matter, or compact stars, should be supra-nuclear density (a few nuclear saturation densities) due to gravitational force. Dense matter may change from a hadronic phase to a deconfined phase as baryon density increases, but a very serious problem is: can the density of realistic compact stars be high/low enough for quarks to become deconfined/confined?

We argue that gravitationally compressed baryonic matter might be in a state of strangeon matter by starting from deconfined quark state with the inclusion of strong interaction between them. With the Dyson-Schwinger-Equation approach to the non-perturbative quantum chromodynamics (QCD), one would estimate the strong coupling constant $\alpha_{\rm s}$ [8],

$$\alpha_{\rm s} = \frac{\alpha_{\rm s}(0)}{\ln(e + a_1 x^{a_2} + b_1 x^{b_2})}, \quad (1)$$

where $\alpha_{\rm s}(0) = 2.972$, $a_1 = 5.292\ {\rm GeV}^{-2a_2}$, $a_2 = 2.324$, $b_1 = 0.034\ {\rm GeV}^{-2b_2}$, $b_2 = 3.169$, $x = p^2$ with p the typical momentum in GeV. In compact stars with quark number density $n \sim 3n_0$, $p^2 \simeq 0.16\ {\rm GeV}^2$, so the coupling parameter $\alpha_{\rm s}$ would be even larger than 2. This means that a weakly coupling (i.e., non-perturbative) treatment could be inadequate for realistic dense matter in compact stars. It is also worth noting that the dimensionless electromagnetic coupling constant is about $1/137 < 0.01$, which makes quantum electromagnetic dynamics (QED) tractable in a perturbative way. That is to say, a weakly coupling strength comparable with that of QED is possible only if the baryon number density $n_{\rm B} > 10^{123} n_0$, being unbelievable and unrealistically high. From this point of view, although some efforts have been made to understand the state of pulsar-like compact stars in the framework of conventional quark stars, including the MIT bag model with almost free quarks [9] and the color-superconductivity state model [10], realistic stellar densities cannot be high enough to justify the use of perturbative QCD which most of compact star models rely on. The strong coupling between quarks may naturally render quarks grouped in quark-clusters, and each quark-cluster/strangeon is composed of several quarks condensating in position space rather than in momentum space [4].

What's new about the concept of gigantic nucleus? In a word, the answer is that: Landau's gigantic nucleus is 2-flavored, while ours 3-flavored. The compact star composed of quark-clusters could also be regarded as a gigantic nucleus, then why should the gigantic nucleus be necessarily 3-flavored? A 3-flavored gigantic nucleus consists u, d and s quarks grouped in

IOP Conf. Series: Journal of Physics: Conf. Series **861** (2017) 012027 doi:10.1088/1742-6596/861/1/012027

strangeons, while a 2-flavored micro-nucleus consists u and d quarks grouped in nucleons. The only difference between gravitationally compressed baryonic matter and micro-nucleus could be a simple change from non-strange to strange: "2"→"3". Certainly, the state of compressed baryonic matter in compact stars is essentially a non-perturbative QCD problem and is difficult to answer from first principles. Nevertheless, strangeon matter in bulk may constitutes the true ground state of strong-interacting matter rather than ^{56}Fe, and this could be seen as a generalized Witten's conjecture, while the traditional Witten's conjecture focuses on the matter composed of almost free u, d and s quarks. Consequently, we suggest that a gigantic nucleus is actually condensed matter of strangeons which is 3-flavored.

The weak equilibrium among u, d and s quarks is possible, instead of simply that between u and d quarks. At the late stage of stellar evolution, normal baryonic matter is intensely compressed by gravity in the core of massive star during supernova. The Fermi energy of electron is significant in compressed baryonic matter, and it is very essential to cancel the electrons by weak interaction in order to make a lower energy state. There are two ways to kill electrons as shown in Fig. 1: one is via the conventional *neutronization*, $e^- + p \to n + \nu_e$, where the fundamental degree of freedom could be nucleons; the other is through the unconventional *strangeonization*, where the degrees of freedom are quarks. While neutronization works for removing electrons, strangeonization has both the advantages of minimizing the electron's contribution of kinetic energy and maximizing the flavor number, with the latter perhaps related to the flavor symmetry of strong-interaction matter. These two ways to kill electrons are relevant to the nature of pulsars, corresponding to neutron star and strangeon star, respectively, as summarized in Fig 1.

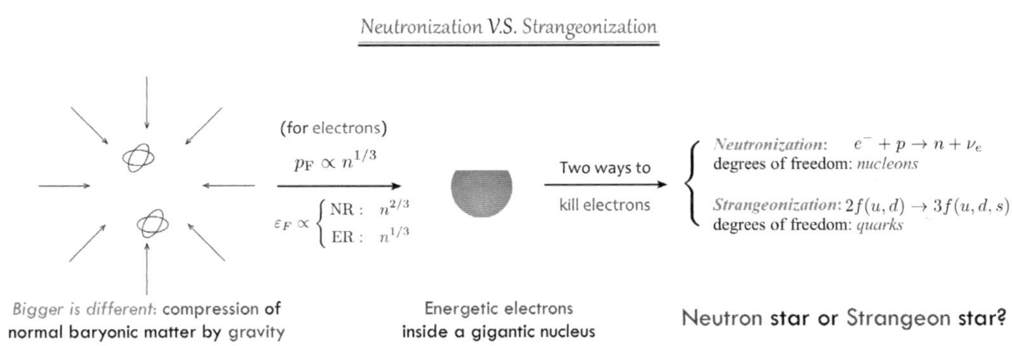

Figure 1. Neutronization and strangeonization are two competing ways to cancel energetic electrons [2].

In a word, strangeon matter is conjectured to be the state of matter inside compact stars, where strangeons form due to both the strong and weak interactions and become the dominant components inside those stars. Then, what could be a realistic strangeon? We know that Λ particle (with structure of $\{uds\}$) possesses light-flavor symmetry, and one may think that a kind of strangeons would be Λ-like. However, the interaction between Λ's could be attractive, which would render more quarks grouped together. Motivated by recent QCD simulations of the H-dibaryons (with structure $uuddss$), a possible kind of strangeons, H-clusters, is proposed [11], and it is found that H-cluster matter could be more stable than nuclear matter when the in-medium stiffening effect is taken into account. Actually we are not certain about the number

IOP Conf. Series: Journal of Physics: Conf. Series **861** (2017) 012027 doi:10.1088/1742-6596/861/1/012027

N_q of valence quarks inside a strangeon, but one could speculate that $N_q = 6$, 9, 12, and even 18. Moreover, because of their classical behavior (as a large mass may result in a small quantum wavelength), strangeons that exist in compact stars could locate in periodic lattices (i.e., in a solid state) when temperature becomes sufficiently low. Pulsars are in the low temperature limit, and they could then be solid strangeon stars.

Different models of pulsar inner structure are summarized in Fig. 2. The conventional neutron stars (hadron star and hybrid/mixed star) are gravity-bound, while strange stars (strange quark stars and strangeon stars) are self-bound on surface by strong force. Let's explain in details, with attention paid to the differences among these models. In the neutron star picture, the inner and outer cores and the crust keep chemical equilibrium at each boundary, so neutron star is bound by gravity. The core should have a boundary and is in equilibrium with the ordinary matter because the star has a surface composed of ordinary matter. There is, however, no clear observational evidence for a neutron star's surface. Being similar to traditional strange quark stars, strangeon stars have almost the same composition from the center to the surface, and the strangeon matter surface could be natural for understanding some different observations, as shown in the next section. Strangon stars are self-bound by the residual interaction between strangeons, while strange quark stars by bag-like strong interaction.

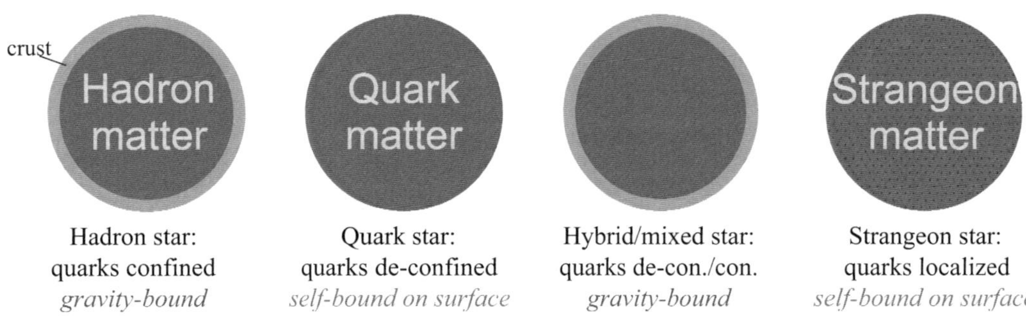

Figure 2. Different models of pulsar's nature. Hadron star and hybrid/mixed star belong to conventional neutron stars. Strangeness plays an important role for strange quark star and strangeon star as a result of three-light-flavor (u, d and s) symmetry.

3. Observational consequences expected in a strangeon star model

Observations of compact stars, depending upon surface and global properties, could provide hints for the state of compressed baryonic matter, as discussed in the following subsections.

3.1. Surface properties

(1) Drifting sub-pulses. Although pulsar-like stars have many different manifestations, they are populated by radio pulsars, and it seems that most of radio sub-pulses are drifting. Among the magnetosphere-dynamics models for pulsar radiative process, the Ruderman-Sutherland [12] model is a virtue not shared by others, and the drifting sub-pulses were first explained. The RS model suggests strong binding of particles on pulsar polar caps, but the calculated binding energy in neutron star models could not be so high. This problem could be naturally solved in bare strangeon star scenario due to the strong self-bound of strangeons on surface.

The magnetospheric activity of bare strangeon star was investigated in quantitative details [13]. Since strangeons on the surface are confined by strong color interaction, the binding energy of strangeons can be even considered as infinity compared to electromagnetic interaction.

IOP Conf. Series: Journal of Physics: Conf. Series **861** (2017) 012027 doi:10.1088/1742-6596/861/1/012027

As for electrons on the surface, calculations have shown that the huge potential barrier built by the electric field in the vacuum gap above the polar cap can usually prevent electrons from streaming into the magnetosphere. Therefore, in the strangeon star model, both positively and negatively charged particles on the surface are usually bound strongly enough to form a vacuum gap above its polar cap, and the drifting (even bi-drifting) sub-pulses can be understood naturally [14]. Certainly, more researches, both theoretical and observational, on this topic are welcome since this subpulse drifting phenomenon could hold the key to understanding the magnetospheric activity as well as the surface material of pulsar.

(2) Clean fireball for SNE/GRBs. It is still an unsolved problem to simulate supernovae successfully in the neutrino-driven explosion models of neutron stars. Nevertheless, in the strangeon star scenario, the bare strangeon matter surface could be essential for successful explosions of both core and accretion-induced collapses [15]. A nascent strange star born in the center of GRB or supernova would create a thermal fireball due to its ultrahigh surface temperature [16], and the photon luminosity is not constrained by the Eddington limit since the surface of strange star could be bare and chromatically confined [17, 18]. Therefore, in this photon-driven scenario the strong radiation pressure caused by thermal emission from strangeon star might play an important role in promoting core-collapse supernovae [19] and even in explaining long-lived plateau of light curves of GRB afterglow [20].

The neutrino burst observed during supernova 1987A could also be understood in the regime of strangeon star [21]. Huge internal energy is stored in a newborn strangeon star after collapse, and then the energy is released by photons and neutrinos, but being dominated initially by neutrino radiation. A liquid-solid phase transition at temperature ~ 1 MeV may occur only a few ten-seconds after core-collapse, and the thermal evolution and the neutrino emission of a strangeon star could then be modeled, as illustrated in Fig. 3.

(3) Non-atomic spectral features. The existence of atomic features in the thermal X-ray emission of neutron star atmosphere is expected in neutron star models. Differently, a bare strangeon star does not have atomic features in the thermal X-ray emission because no atom exists on its surface. Note that slight plasma may cover on a strangeon star surface, which is optically thin in X-ray bands (see "*(5)*" below for discussions). The featureless spectrum predicted by strangeon star model is consistent with observations of *Chandra* and *XMM-Newton*, both of which have not detected certain atomic feature predicted in neutron star atmosphere models. In neutron star scenario, the spectrum determined by radiative transfer in atmosphere should differ substantially from Planck-like one.

(4) Absorption lines in thermal X-ray spectrum. Although the three-light-flavor symmetry breaking is tiny (with the electric charge per baryon $\sim 10^{-5}$), the electron number density is still high and the Fermi energy of electrons is $E_{\rm F} \sim 10$ MeV, one order smaller than that in case of two-light-flavor symmetry. Strangeon matter is bound by the strong interaction, while electrons are bound to strangeon matter by the electromagnetic interaction, so some of the electrons in the surface region of a strangeon star reside outside of strangeon matter boundary, leading to a fairly thin (thousands of femtometers thick) electron sea on the surface of strangeon matter [9]. The global motion of the electron sea on the magnetized surface was investigated [24], and it is found that hydrodynamic surface fluctuations of the electron sea would be greatly affected by the magnetic field. Some observations did show absorption lines of pulsar-like stars, and the best absorption features are detected for the central compact object (CCO) 1E 1207.4-5209, at ~ 0.7 keV and ~ 1.4 keV [25, 26, 27]. The absorption feature of 1E 1207.4-5209 could be understood in this hydrocyclotron oscillations model [24]. Besides the absorption lines in 1E 2107.4-5209, the detected lines around (17.5, 11.2, 7.5, 5.0) keV in the burst spectrum of SGR 1806-20 and those in other dead pulsars (e.g. radio quiet compact objects) would also be of hydrocyclotron origin [24]. It seems that X-ray absorption feature could be natural in strangeon star model.

(5) Strangeness barrier. The constituent quarks in a strangeon are of three flavors (u, d and

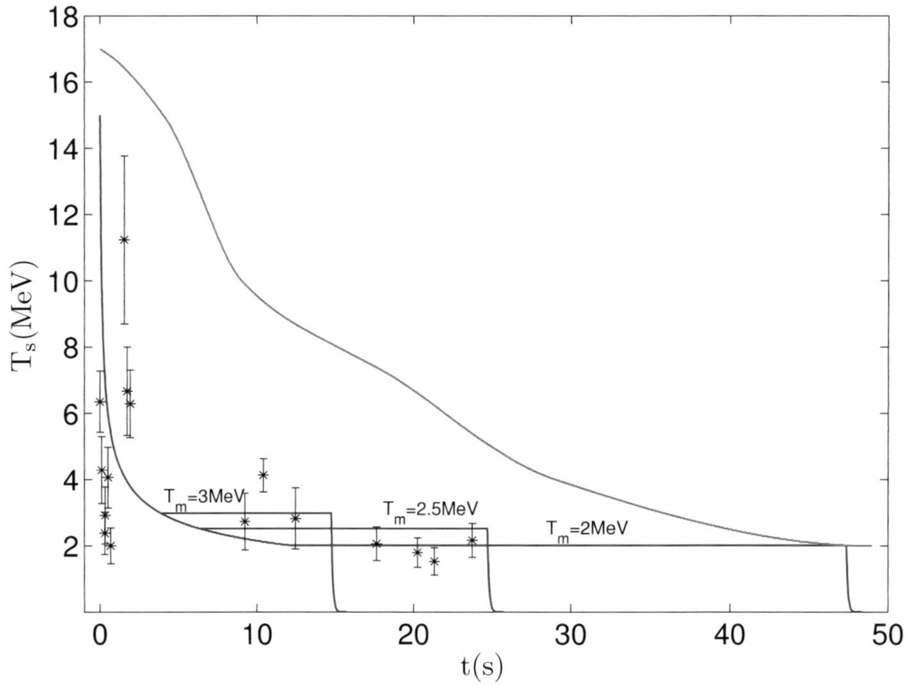

Figure 3. This is the T(temperature)-t(time) relation of a proto-strangeon star with mass $1.4 M_\odot$ and radius 10 km [21]. We take logarithm of age t in order to see the details. The blue lines are $T-t$ curves for strangeon star with different parameters of melting temperature $T_{\rm m}$, and the upper red line is for a normal proto-neutron star taken from [22]. The 15 black dots with error bars are 15 neutrino events, of which the neutrino energy (E_ν) has been represented by proto-strangeon star's temperature (T) by a relation of $E_\nu = 3.15T$. It is evident that temperature T of a proto-strangeon star drops down rapidly in millisecond timescale, and then the star will keep homothermal during a liquid-solid phase transition, as the straight lines indicate. After the phase transition, solid strangeon stars cool down more drastically than ever before. In this case, consequently, the emission intensity decreases quickly, and hence leading to a cut-off of this neutrino burst. On the contrast, normal neutron stars cool down smoothly all the time, and have no interceptive indication during supernova neutrino burst. It indicates that strangeon star model with melting temperature ~ 1 MeV coincides with the burst events, either the duration or the energy of neutrino radiation, detected from SN1987A.

s) rather than of two flavors (u and d) for normal nucleons. Consequently, the fundamental weak interaction does play an essential role to convert normal 2-flavored matter (i.e., nucleons) to 3-flavored one (i.e., strangeons) during an accretion phase. The weak-conversion, however, is not easy, and could be successful only after frequent collisions (order of $\gg 1$), similar to the famous $pp-$reaction with flavor change. We would then introduce a term of *strangeness barrier* to describe this kind of difficulty [28]. Because of this barrier above strangeon star surface, which separates two-flavor matter from three-flavor matter, most of the ordinary nuclei falling onto a strangeon star might bounce back. Therefore, a strangeon star may be surrounded by a hot corona or an atmosphere, or even a crust for different accretion rates. The strangeness barrier could help us to understand the redshifted O VIII Ly-α emission line (only with $z = 0.009$) and the change in the blackbody radiation area of 4U 1700+24 if it is a low mass strangeon star [28], as well as the low mass function measured for the puzzling X-ray binary.

Additionally, the strangeness barrier could also be meaningful to understand Type-I X-ray bursters as well as to constrain their masses and radii [29]. Despite these, the barrier could also be necessary to solve the optical/UV excess puzzle with a strangeon star atmosphere. X-ray dim isolated neutron stars (XDINSs) are characterized by Planckian spectra in X-ray bands, but show optical/ultraviolet (UV) excesses exptrapolated from X-ray spectra. In the regime of strangeon stars, the ISM-accreted matter could form a plasma atmosphere due to strangeness barrier. A radiative model of bremsstrahlung emission from the plasma atmosphere could fit well the spectra of the seven XDINSs, from optical/UV to X-ray bands, as shown in Fig. 4 for the famous one, RX J1856.5-3754 [30]. It is worth noting that this strangeon star

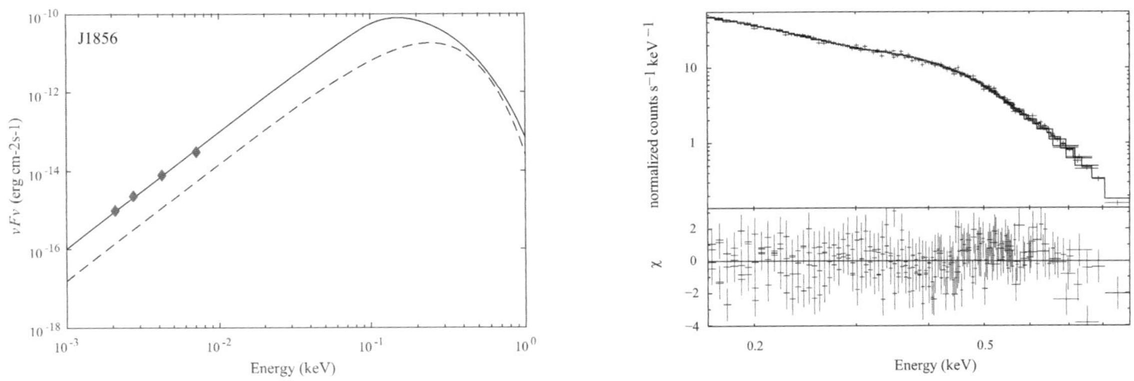

Figure 4. The bremsstrahlung emission (left, solid line) and X-ray data fitting (right, with $\chi^2/\text{dof} = 1.17/298$) from a strangeon star atmosphere are shown for RX J1856.5-3754 [30]. By comparison, a dashed line of pure black-body model extrapolated from its X-ray spectrum is also drawn. The red diamonds are of optical observations from the *HST* photometry. The extracted spectra are binned with 1000 counts per bin in each observation at least. The X-ray data are fitted by the bremsstrahlung plus a Gaussian function which indicates an absorption line around 0.21 keV ($\sigma = 0.02$ keV).

atmosphere could simply be regarded as the upper layer of a normal neutron star atmosphere, but with almost homogeneous electron (or ion) temperature. Therefore, in the strangeon star atmosphere, thermal X-rays from lower layer of normal neutron star atmosphere are prohibited, and relatively more optical/UV photons are then radiated. Hard X-ray cut-off (i.e., without a hard tail) would also be natural in our model. Furthermore, the optical/UV spectral deviations from a Rayleigh-Jeans distribution could come from non-uniformities of the atmosphere [31].

3.2. Global properties

(1) Stiff equation of state and massive strangeon star. The equation of state of strangeon matter would be stiffer than that of nuclear matter for two main reasons: 1) strangeons should be non-relativistic particles for its large mass, and 2) there could be strong short-distance repulsion between strangeons (an analogy for the hardcore of nuclear force). It is also worth noting that, both the problems of hyperon puzzle and quark-confinement do not exist in strangeon star model, whereas those two are still challenging in conventional neutron star and quark star models.

It has been addressed that strangeon stars could have high maximum masses ($> 2M_\odot$) inferred by the stiff equation of state, as well as very low masses ($< 10^{-2}M_\odot$) as a direct consequence of self-bound surface [32]. Later observations of PSR J1614-2230 imply a mass of $1.97 \pm 0.04 M_\odot$ [33], indicating a stiff equation of state. It is conventionally though that the state of dense matter softens and thus cannot result in high maximum mass if pulsars are quark stars. However, strangeon stars could not be ruled out by massive pulsars, and the observations of pulsars with higher mass (e.g. $> 2.5 M_\odot$) would even be a strong support to strangeon star model [11, 34]. New observation indicates that the mass of PSR J2317+1439 would even be larger than $2.04 M_\odot$ (at 90% confidence) [35], being expected in strangeon star model.

(2) Self bound and low mass strangeon stars. As discussed previously, similar to strange quark stars, strangeon stars are bound by strong force instead of gravity. Self-bound strangeon stars have non-zero surface density, and their radii usually increase as masses increase, while for gravity-bound neutron stars the radii decrease as mass increase. Gravity only play a significant role when the mass of the strangeon star approaches the maximum mass, beyond which the (repulsive) interaction between strangeons could no longer resist the gravity and the star would collapse to become a black hole. Specially, $M \propto R^3$ for low-mass strangeon stars with negligible gravity, so the mass of strangeon stars could be very low with masses as low as $\sim 10^{-2} M_\odot$ [15] (being even manifested in the form of strangeon planet [23]).

(3) Anisotropic pressure and extra-free energy besides powers of spin, accretion and heat. Strangeon stars could be in a solid state at low temperature. The local pressure could be anisotropic in elastic matter, so for solid strangeon stars the radial pressure gradient could be partially balanced by the tangential shear force. Release of the elastic energy as well as gravitational energy would not be negligible, and may have significant astrophysical implications.

A sudden change of tangential force may result in a starquake, with release of both gravitational and elastic energy. There are actually two kinds of starquakes in the solid strangeon star model: bulk-invariable (type I) and bulk-variable (type II) starquakes, with energy release negligible for the former but significant for the latter. Typical energy of 10^{44-47} erg is released during superflares of soft gamma-ry repeaters (SGRs), and a type II giant starquake could reproduce such a flare [36]. Both types of glitches with (type II) and without (type I) X-ray enhancement could be naturally understood in the starquake model of solid strangeon stars [37].

Significant stellar oscillations could usually accompany a starquake, especially for the Type II quakes. The magnetospheric activity in the polar cap region of pulsars could be excited under such oscillations. The toroidal oscillation of the star propagates into the magnetosphere, which provides additional voltage due to unipolar induction, changes Goldreich-Julian charge density from the traditional value due to rotation, and hence influences particle acceleration [38]. The onset of radio emission after glitches/flares in SGRs or AXPs (anomalous X-ray pulsars) could be the result of oscillation-driven magnetic activities of solid strangeon stars. Within that model, transient radio signals of AXPs/SGRs may be caused by activation of the pulsar inner gap from below the radiative death line, due to an oscillation-induced voltage enhancement.

(4) Rigidity. Rigid body precesses naturally when spinning, either freely or by torque, but fluid one can hardly. The observation of possible precession or even free precession of B1821-11 [39] and others could suggest a global solid structure of compact star. It is suggested that the gravitational wave behaviors should be mass-dependent [40], and no gravitational wave originated from r-mode instability could be detected for solid strangeon stars. Moreover, the bumpy distortion, i.e. local mountains on star surface, can lead to asymmetry about a pulsar's rotation axis, which could be tested by gravitational wave detectors in the future. By precision pulsar timing, either in radio or in X/γ-ray bands, we may discover more and more candidates with precession, that might provide a reliable way to test the strangeon star model. We are looking forward to these observations and discoveries.

The discovery of the gravitational waves [41] opens a new window for exploring the Universe. It is worth noting that, strangeon star with rigidity is quite likely to be tested by kilo-Hz gravitational wave observations of two kinds of events at least: 1) merger of pulsar-pulsar/pulsar-black hole binaries, during which the predicted waveform and the tidal effects in inspiral depend on the equation of state of supra-nuclear matter; 2) starquake of pulsar-like compact stars, during which the induced gravitational waves of compact stars may be discovered by sensitive detectors. Certainly, electro-magnetic diagnostics is important for both events of gravitational wave radiation, to remove the degeneracy of physical parameters.

4. Summary

Normal micro-nuclei with two flavors, formed initially during the Big Bang nucleosynthesis, make our Universe rich and colorful. We conjecture, from an astrophysical point of view, that a gigantic nucleus in which three-flavor symmetry is restored could form by compression of gravity during a supernova, and its constituent are strangeons made of almost equal numbers of u, d and s quarks. That is to say, all the pulsar-like compact stars could actually be strangeon stars. Here we improve the original idea of gigantic nuclei presented by L. Landau over 85 years ago, with the main difference being from two-flavor to three-flavor, or from non-strangeness to strangeness. We demonstrate that the strangeon stars could be necessary to understand different manifestations of pulsar-like compact stars, and are expecting to test this model by future observation with advanced facilities, including either ground radio telescopes Five-hundred-meter Aperture Spherical radio Telescope (FAST) and Square Kilometre Array radio telescope (SKA), or space-based high energy observatories Lightweight Asymmetry and Magnetism Probe (LAMP) and enhanced X-ray Timing and Polarimetry (eXTP).

Let's summarize the paper by a famous sentence of P. W. Anderson (1932 -): "The ability to reduce everything to simple fundamental laws does not imply the ability to start from those laws and reconstruct the Universe". As for the state of supra-nuclear matter in pulsar-like compact stars, we are still quite embarrassed because, besides the many-body problem, the fundamental law of strong interaction is still unclear in the low-energy scale.

Acknowledgments

We are grateful to the members at the pulsar group of Peking University. This work is supported by the National Natural Science Foundation of China (11673002 and U1531243) and the Strategic Priority Research Program of CAS (No. XDB23010200).

References

[1] Rutherford E 1911 *Phil. Mag.* **21** 669
[2] Xu R X and Guo Y J 2017, in: *Centennial of General Relativity: A Celebration*, p. 119-146 (arXiv:1601.05607)
[3] Witten E 1984 *Phys. Rev. D* **30**, 272
[4] Xu R X 2003 *ApJ* **596** L59
[5] Landau L 1932 *Sov. Phys.* **1** 285
[6] Baade W and Zwicky F 1934 *Phys. Rev.* **46** 76
[7] Hewish A, Bell S J, Pilkington J D H, et al. 1968 *Nature* **217** 709
[8] Fischer C S 2006 *J. Phys. G: Part. Nucl, Phys.* **32** R253
[9] Alcock C, Farhi R and Olinto A 1986 *ApJ* **310** 261
[10] Alford M J and et al. 2008 *Rev. Mod. Phys.* **80** 1455
[11] Lai X Y, Gao C Y and Xu R X 2013 *MNRAS* **431** 3282
[12] Ruderman M A and Sutherland P G 1975 *ApJ* **196** 51
[13] Yu J W and Xu R X 2011 *MNRAS* **414** 489
[14] Xu R X, Qiao G J and Zhang B 1999 *ApJ* **522** L109
[15] Xu R X 2005 *MNRAS* **356** 359
[16] Haensel P, Paczynski B and Amsterdamski P 1991 *ApJ* **375** 209
[17] Ouyed R, Rapp R and Vogt C 2005 *ApJ* **632** 1001
[18] Paczynski B and Haensel 2005 *MNRAS* **362** L4
[19] Chen A B, Yu T H and Xu R X 2007 *ApJ* **668** L55
[20] Dai S, Li X L and Xu R X 2011 *Science China: Physics, Mechanics & Astronomy* **54** 1541
[21] Yan M, Lu J G, Yang Z L, Lai X Y and Xu R X 2017 *RAA* in press

IOP Conf. Series: Journal of Physics: Conf. Series **861** (2017) 012027 doi:10.1088/1742-6596/861/1/012027

[22] Pons J A, Reddy S, Prakash M, Lattimer J M and Miralles J A 1999 *ApJ* **513** 13
[23] Xu R X and Wu F 2003 *Chin. Phys. Lett.* **20** 80
[24] Xu R X, Bastrukov S I, Weber F, Yu J W and Molodtsova I V 2012 *Phys. Rev. D* **85** 023008
[25] Sanwal D, Pavlov G G, Zavlin V E and Teter M 2002 *ApJ* **574** 61
[26] Mereghetti S, De Luca A, Caraveo P, Becker W, Mignani R and Bignami G F 2002 *ApJ* **581** 1280
[27] Bignami G F, Caraveo P A, Luca A De and Mereghetti S 2003 *Nature* **423** 725
[28] Xu R X 2014 *RAA* **14** 617
[29] Li Z S, Qu Z J, Chen L, Guo Y J, Qu J L and Xu R X 2015 *ApJ* **798** 56
[30] Wang W Y, Lu J G, Tong H, Ge M Y, Li Z S, Men Y P and Xu R X 2017 *ApJ* **837** 81 (arXiv:1603.08288)
[31] Wang W Y, Feng Y, Lai X Y, Lu J G, Chen X L and Xu R X 2017 submitted (arXiv:1705.03763)
[32] Lai X Y and Xu R X 2009 *MNRAS* **398** L31
[33] Demorest P, Pennucci T, Ransom S, Roberts M and Hessels J 2010 *Nature* **467** 1081
[34] Guo Y J, Lai X Y and Xu R X 2013 *Chin. Phys. C* **38** 055101
[35] Dai S., Smith M. C., Wang S., et al. 2017, *ApJ* in press (arXiv:1705.02593)
[36] Xu R X, Tao D J and Yang Y 2006 *MNRAS* **373** L85
[37] Zhou E P, Lu J G, Tong H and Xu R X 2014 *MNRAS* **443** 2705
[38] Lin M X, Xu R X, and Zhang B. 2015, *ApJ* **799** 152
[39] Stairs I H, Lyne A G and Shemar S L 2000 *Nature* **406** 484
[40] Xu R X 2006 *Astropart. Phys.* **25** 212
[41] Abbott and et al. 2016 *PRL* **116** 061102

IOP Conf. Series: Journal of Physics: Conf. Series **861** (2017) 012028 doi:10.1088/1742-6596/861/1/012028

Supernovae and neutron stars: playgrounds of dense matter and neutrinos

Kohsuke Sumiyoshi

Numazu College of Technology, Ooka 3600, Numazu, Shizuoka 410-8501, Japan

E-mail: sumi@numazu-ct.ac.jp

Abstract. Core-collapse supernovae are vital as the birthplace of compact objects, where one expects various phases of the dense matter. The current status of supernova studies with the nuclear data for dense matter and neutrino reactions is overviewed with a focus on recent progress of the neutrino-radiation hydrodynamics in two- an three-dimensions and remaining mysteries. In addition to its importance for the explosion mechanism, the equation of state is also essential to predict the neutrino bursts, which can be used to probe deep inside the compact objects. It is valuable to discuss variations of the extreme conditions for hyperons and quarks in central objects during explosive phenomena by looking into the pattern of neutrino signals in relation with dynamics and dense matter.

1. Introduction

Core-collapse supernovae occur through dynamical stages: collapse, bounce and explosion with the birth of compact objects. Despite extensive studies for decades, the explosion mechanism of core-collapse supernovae is still elusive [1, 2]. Starting with a central Fe-core of massive stars of about $20M_{\odot}$, electron captures and photo dissociation trigger the gravitational collapse with production of neutrinos. Those neutrinos frequently scatter due to high density and are trapped inside the core during the collapse. Further collapse leads to the core bounce and a shock wave is launched. If the shock wave successfully propagates, it leads to the explosion with gigantic energy of 10^{51} erg. In the center, a hot proto-neutron star is born, which contains plenty of trapped neutrinos. The proto-neutron star emits a bunch of neutrinos, carrying away most of the gravitational energy of 10^{53} erg. A cold neutron star is born afterward and evolves through the long cooling phase.

One of the main questions in the conference is where the exotic phase may appear in these processes. To answer this important question, we need to solve a difficult problem from the launch of shock wave, the propagation in the Fe core for the explosion and the following history of compact star. It is rather difficult to get explosions in numerical simulations, though, because the shock wave stalls on the way during the propagation of shock wave. The shock wave must propagate through the outer part of Fe core, where the initial shock energy is used up by Fe dissociation. About 100 ms after the bounce, the shock wave stalls at around 100 km above the nascent proto-neutron star, which emits neutrinos from the reservoir of trapped neutrinos. A portion of the outgoing neutrinos are absorbed by the material just behind the shock wave through neutrino absorption on nucleons. The amount of energy transfer from neutrinos to matter is about 10^{51} erg, which is comparable to the explosion energy. This idea of the neutrino heating mechanism [3, 4] is one of the main scenarios for the supernova explosion. However, this effect is sensitive to the neutrino energy and flux from the central object and depends on the dynamics such as the amount and duration of accretion for the heating [5]. After the longstanding development of first-principle-type simulations of neutrino-radiation hydrodynamics, recent simulations under the spherical symmetry (1D) show that no explosion is obtained for most of massive stars with detailed nuclear physics except for specific cases [6, 7].

The essential factor to tackle the obstacle to the explosion is the effect of hydrodynamical instabilities. The combination of neutrino heating and hydrodynamical instabilities assists to obtain successful explosions. For example, the shock wave is launched further away in some directions due to the deformation of shock wave and this results in a longer falling time of material for the heating. The convection makes the fluid flow stay long around the heating region in addition. These effects seen in two- and three-dimensional (2D and 3D) simulations contribute to turn the stalled shock wave into a successful explosion even when the 1D simulation fails [8, 9, 10, 11, 12].

There are remaining issues in supernova studies, though, having successful 2D and 3D simulations: the main trigger in the complicated dynamics with neutrinos, preference in 2D or 3D, reason for too small explosion energy (e.g. [11]). It is necessary to clarify the remaining mysteries by performing the full simulations by combining nuclear physics (equation of state and neutrino reactions) and astrophysics (hydrodynamics, neutrino transfer and general relativity) into the supercomputing technology. In this contribution, I focus on the two topics: neutrino transfer to evaluate the neutrino heating and equation of state to examine the influence of nuclear physics.

2. Neutrino Transfer

Neutrino transfer is important to determine the precise amount of neutrino heating in the 2D/3D hydrodynamics. It is necessary to evaluate the neutrino flux and heating through trapping, emission and absorption of neutrinos. This is a difficult task since one must describe the detailed transport and reactions of neutrinos in the intermediate region for the neutrino heating covering the two limiting regimes from diffusion to free-streaming. In the central part of supernova core, frequent reactions and scattering occur at high density and temperature so that the diffusion approximation can be applied. In outer layers of the core at low density, neutrinos propagate without any reaction so that the free-streaming limit can be applied. However, the important region for neutrino heating lies between the two regimes [13]. Therefore, it is mandatory to solve the Boltzmann equation (or its equivalent form), which has been a formidable task so far in 2D/3D.

Recently, we have developed the numerical code to solve the Boltzmann equation in six-dimensions (6D) [14]. 6D means 3D space and 3D neutrino momentum (energy and two angles to designate the propagating direction). We solve the Boltzmann equation directly to handle the time evolution of neutrino distribution in 6D with advection and reactions. Most important and difficult part is the collision term having energy- and angle-dependent reaction rates of neutrinos. The equation is stiff having largely different time scales for the reactions, non-linear with pair processes and frame dependent. It is necessary to handle conversion of the reaction rates between co-moving and laboratory frames. Therefore, it needs a huge computational load to handle the whole framework and various approximations have been adopted. For example, the diffusion approximation used inside the core is often extended to its outer region. Another popular approximation is the ray-by-ray method for 2D/3D. In this scheme, one solves the 1D spherical problem, which can be done in the basis of the first principle calculation, along the radial ray independently for various directions. However, the supercomputing power nowadays enables us to handle the 6D Boltzmann equation [14]. Using this new code, we have made

IOP Conf. Series: Journal of Physics: Conf. Series **861** (2017) 012028 doi:10.1088/1742-6596/861/1/012028

comparison of the neutrino transfer in 3D supernova cores with that in the ray-by-ray method [15]. We found that there are underestimate and overestimate, depending on the radial direction, of about 20% in local neutrino heating rate due to the approximation.

Utilizing this Boltzmann solver, we have developed the numerical code for neutrino-radiation hydrodynamics in multi-dimensions [16]. The 6D Boltzmann equation is coupled with hydrodynamics and gravity in 2D with axial symmetry. The extension of the code to 3D supernovae is now under development. The Boltzmann part contains special relativistic effects such as the Dopper effect and angle aberration. The framework is partially general relativistic having the moving mesh method to handle wobbling motion of central compact object [17]. This code enables us to follow the evolution of neutrino transfer and fluid flow in the whole regime from diffusion to free-streaming. Recently we have successfully performed the long term evolution of two supernova models with the massive stars of $11M_{\odot}$ and $15M_{\odot}$ [18, 19]. These simulations are done with the Furusawa EoS table with the mixture of nuclei under the nuclear statistical equilibrium (NSE) [20] (extended version of Shen EoS [21, 22, 23]) and updated electron capture rates [24]. Although the neutrino heating works with hydrodynamical instability, the shock wave does not revive even after long evolution. We have found that no explosion occurs up to 400 ms and 600 ms for $11M_{\odot}$ and $15M_{\odot}$ models, respectively. Behavior of the shock propagation after the bounce is different in the two models depending on the progenitor profiles with different compactness (i.e. density profile). It is necessary to study further the dependence on nuclear physics under our framework. The results of additional simulations with the Lattimer-Swesty EoS [25] and other progenitors will be published elsewhere [19]. The full treatment of the neutrino-radiation hydrodynamics in 2D with the Boltzmann equation is a big step forward to the goal of the general relativistic neutrino-radiation hydrodynamics for 3D supernovae.

3. Equation of state

Data tables of equation of state (EoS) are necessary ingredient for supernova simulations. To cover the wide range of density, temperature and electron fraction, one needs to utilize a consistent framework for matter and nuclei with constraints from experimental data of nuclei and observational data of neutron stars. The number of available EoS tables for supernovae is increasing, but still limited as compared to that for neutron stars. The two major EoS tables: Lattimer-Swesty EoS (LS) [25] and Shen EoS (Shen) [21, 22, 23] have been used for many simulations. Recently, improvement of EoS tables is in progress with updates of nuclear interactions, treatment of multi-composition of nuclei under NSE and extensions to include hyperons and quarks [26].

I discuss the influence of EoS on the explosion with LS and Shen as examples of EoS tables [27]. LS and Shen have different symmetry energy and stiffness, being rather extreme cases in a modern sense, but using these two sets in simulations is a kind of benchmark test for the supernova problem. Maximum neutron star mass for cold neutron stars is $1.8M_{\odot}$ and $2.2M_{\odot}$ with LS and Shen, respectively. Here I choose the case of the incompressibility of 180 MeV for LS to illustrate the effects. When these sets have been applied to 1D supernova simulations, it turned out that both cases do not lead to the explosion with a small difference in the evolution of the shock wave [7]. The difference in the shock motion after the core bounce is not so large because the two cases have similar profiles at the core bounce.

In Fig. 1, the typical profiles of supernova core from the simulation of a $15M_{\odot}$ star with Shen are shown in the plane of temperature and electron fraction versus density. Due to the gravitational collapse, the central density increases up to the nuclear matter density and, consequently, the temperature increases along the adiabatic curve. The core bounce occurs just above the nuclear matter density and the temperature becomes high where the shock wave forms (0 ms of the left panel). At 150 ms after the bounce, the temperature increases high further due to the passage of shock wave through the outer layers at low densities, but the central density does not change very much. Since the density at this stage is not yet extremely high, the difference of the EoSs turns out to be still small. Comparison using more recent EoS tables shows similar tendency [28, 29].

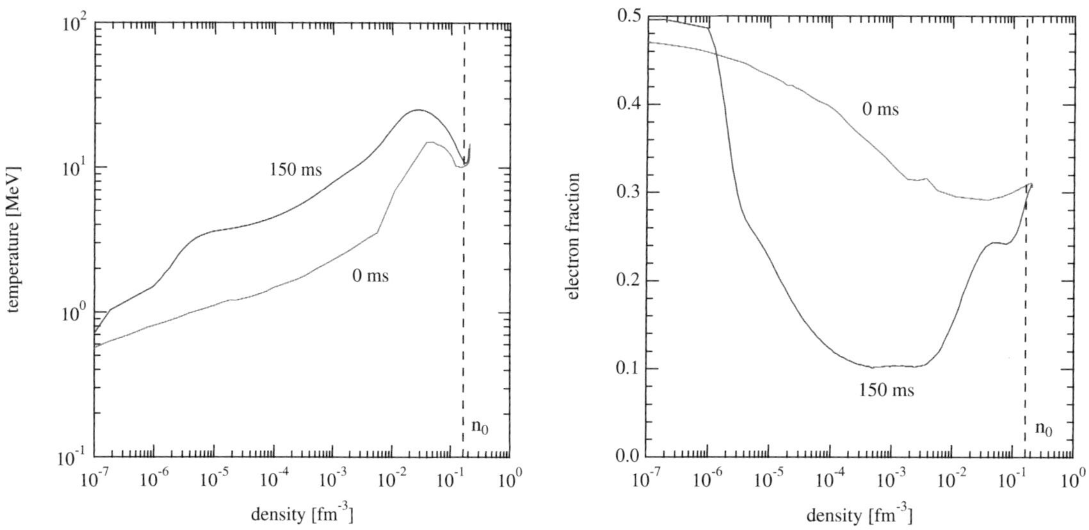

Figure 1. Conditions of dense matter in a supernova core in the plane of temperature (left) and electron fraction (right) versus density. The snapshots are taken from the supernova core of $15M_\odot$ star with Shen EoS [7]. The conditions are shown for the fluid elements along the radial coordinate at 0 and 150 ms after the core bounce.

It is significant to see that neutron fraction is not yet large enough to have exotics around the core bounce as shown in the right panel of Fig. 1. During the collapse, the electron (proton) fraction decreases through the process of electron captures from the initial value (0.46 for ^{56}Fe) up to the neutrino trapping. The proton fraction stays large around 0.3 at the center, because of the blocking due to neutrino degeneracy. The proton fraction gradually decreases in the limited region where neutrinos can escape through electron captures on protons. The proton fraction becomes $\sim$0.1 at low densities as seen in the curve of 150 ms, however, the proton fraction in the central region stays around 0.3. Therefore, the neutron fraction is not large enough, being different from the situation of neutron stars with the proton fraction less than 0.1.

Although it seems that the condition of dense matter in supernovae is not favorable, if exotic phases can appear, its effect may be drastic. For example, the appearance of quark degrees of freedom may bring successful explosions in 1D [30, 31, 32]. During the shock stall, there is a second collapse deep inside due to the phase transition and an additional shock wave is launched for the explosion. Although this scenario of the explosion by the second collapse is restricted to the soft EoS cases, which are not in accord with the $2M_\odot$ neutron stars, it attracts attention to quark physics as a possible way to get explosions.

In contrast to the detailed studies of EoS effects in 1D simulations, there are only a few studies by sophisticated 2D simulations to explore the influence on explosions [33, 34]. It seems that softer EoSs are favorable to have higher luminosity for neutrino heating. However, complicated system with hydrodynamical instabilities and neutrino transfer makes us still away from the clear understanding. One needs more systematic studies in 2D/3D by surveying the range

IOP Conf. Series: Journal of Physics: Conf. Series **861** (2017) 012028 doi:10.1088/1742-6596/861/1/012028

of parameters in nuclear physics such as the symmetry energy and the incompressibility, for example.

4. Neutrino burst

The neutrino burst in proto-neutron stars and black hole formation is another subject to explore EoSs including exotic phases. In a successful explosion, a newly born proto-neutron star emits a plenty of neutrinos over ~10 seconds. About 10^4 events of neutrinos will be detected at the terrestrial detectors such as Super-Kamiokande and LVD at Gran Sasso for a next Galactic supernova. Those neutrinos carry the information from hot and dense matter inside the compact object. Therefore, detailed studies on the signal of neutrino burst are invaluable as the probe of EoS.

The difference of EoS in stiffness and composition is reflected in the properties of the neutrino bursts. The neutrino signal evolves according to the shock passage, the thermal evolution and the accretion of material. The difference due to EoSs right after the bounce (for example, LS and Shen) is rather small due to the similar environment as discussed above [7]. If quarks appear as in [30, 31, 32], there is a clear feature in the signal of neutrino burst. In addition to the first peak from the usual shock wave, there is a second peak due to the second collapse and bounce shock. This can be a clear signal of the quark-hadron phase transition for the EoS, which has a low density threshold of the quark appearance.

Further evolution of the proto-neutron star leads to firm oppotunities to probe EoS. The EoS differences become larger as the trapped neutrinos escape. Because of reduction of pressure, the proto-neutron star becomes more compact: density and temperature become higher. The difference in density profiles becomes large at late stages and central density can be different by a factor of 2–3. Temperature profiles become different accordingly. These differences are reflected to the properties of supernova neutrinos such as the average energies and luminosities [35, 36, 37]. They affect also the cooling speed and the corresponding decrease of luminosities for ~20 seconds. These differences of the neutrino signals are solid EoS probes by the terrestrial detection of neutrinos.

During the cooling evolution of proto-neutron stars, the matter becomes neutron rich due to deleptonization by emitting neutrinos. This opens up possibilities to have exotic phases such as hyperons and quarks. For example, hyperons appear at a late stage, ~50 seconds after the birth [36]. The average energy and luminosity are different from those in the nucleonic cases depending on the choice of hyperon interactions. Note that many of EoSs used so far in the previous simulations do not accord with the $2M_\odot$ neutron star mass constraint. Further careful studies are necessary to explore this delicate effect seen at the late stage where the luminosity becomes low.

Windows to see exotics may be open wider in case of more massive stars, which lead to *failed*-supernovae. When the Fe core mass is very large $\sim 2M_\odot$ for massive stars of $40M_\odot$, for example, there is no chance to have a successful explosion due to the total failure of the shock propagation through intense accretion [38, 39]. Matter from the outer layer of Fe core accretes onto a nascent proto-neutron star and its mass increases rapidly after the recession of shock wave. When the mass exceeds the critical mass of proto-neutron stars, which are hot and neutrino-rich, the dynamical collapse occurs and eventually the black hole is formed [40, 41]. During this evolution in ~1 second, neutrinos are emitted abundantly and will be detected by the terrestrial detectors. The event number of neutrinos from the black hole forming core-collapse is amount to 10^4 and comparable to that of ordinary supernova neutrinos [42]. Although there is no bright optical display, there is a short neutrino burst, which informs us the black hole formation. Since the neutrinos are emitted from deep inside the collapsing central object toward the black hole, the neutrinos are probe of matter at very high density and temperature.

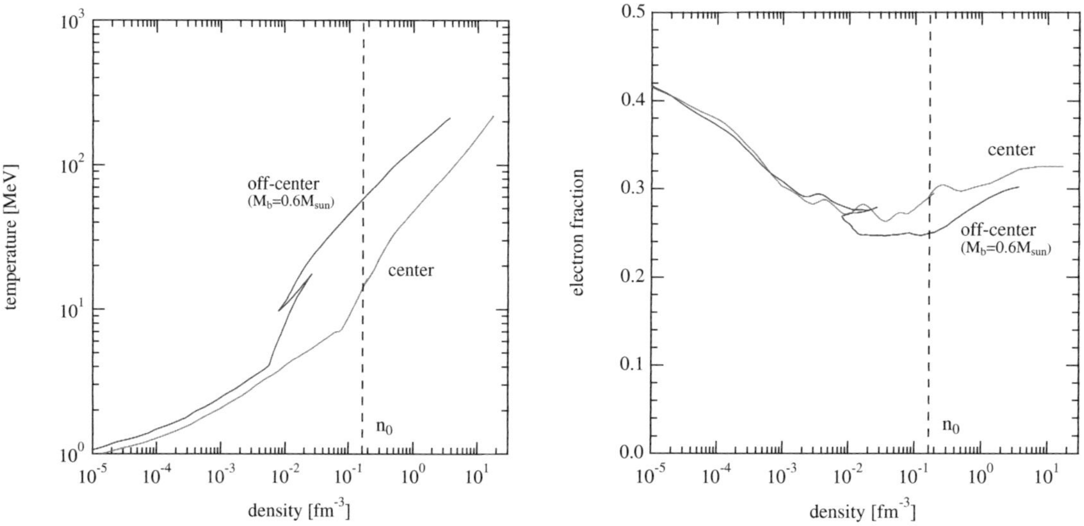

Figure 2. Conditions of dense matter in a *failed*-supernova core in the plane of temperature (left) and electron fraction (right) versus density. Trajectories of the dense matter are taken from the numerical simulation which follows the initial collapse to the black hole formation of $40M_\odot$ star with Shen EoS [40]. The evolutions of conditions are shown for the fluid elements at the fixed baryon mass coordinate at the center and off-center (M_b=$0.6M_\odot$).

Figure 2 shows conditions of dense matter taken from the numerical simulations of gravitational collapse of a massive star of $40M_\odot$ [40]. After the core bounce, the density and temperature increase rapidly beyond the nuclear matter density, $n_0 \sim 0.16$ fm^{-3}, due to the compression to support a massive proto-neutron star. The temperature at off-center location becomes high due to the shock passage with high entropy values. Because of the intense accretion of matter, the mass of proto-neutron star increases very fast and exceeds $2M_\odot$. For example, it reaches the threshold for the black hole at $2.1M_\odot$ for LS and $2.6M_\odot$ for Shen [40]. The dynamical collapse to the black hole formation leads to extreme conditions beyond $10n_0 \sim 1.6$ fm^{-3} and 100 MeV. These are suitable places for exotic phase, where hyperons and quarks can appear. Note that the electron fraction stays ~0.3 due to the neutrino trapping and the matter is not very neutron rich in the black hole formation.

The neutrino emission till the black hole formation has a unique feature. Short duration of ~1 second and rapid increase of energy and luminosity are clearly different from the feature in ordinary supernova neutrinos. Average energy and luminosity increase rapidly due to the temperature increase and the persistent matter accretion. The duration is determined by the EoS stiffness, which controls the threshold mass of proto-neutron stars to the black hole formation. For example, the duration is 0.6 and 1.3 second for LS and Shen, respectively, leading to a factor of two difference between the two EoSs [40].

During the collapse toward the black hole, there is a chance to see hyperons or quarks [42, 43, 44]. When the density and temperature become high above the threshold for exotic particles, the new particles such as hyperons and quarks appear abundantly and the EoS becomes soft immediately due to the new degrees of freedom. The appearance of exotics shortens the duration of neutrino burst in general, therefore, this is a probe of exotic matter [42, 43]. Note again that the exotic EoS must be in accord with the $2M_\odot$ constraint. When one adopts the EoS tables with the $2M_\odot$ constraint satisfied, there may be difficulties to distinguish the exotics from

IOP Conf. Series: Journal of Physics: Conf. Series **861** (2017) 012028 doi:10.1088/1742-6596/861/1/012028

the *soft* nucleonic EoS cases [28]. In order to judge the appearance of exotic phases, it is necessary to examine the neutrino bursts toward the black hole formation using the new EoS tables with hyperons and quarks obtained with constraints from nuclear physics and astrophysics.

5. Extreme conditions

A wide variety of extreme conditions appear in supernovae, neutron stars and black hole formation as we have seen in the previous sections. I discuss further the characteristics of environment (density, temperature and electron fractions) in supernova explosions and black hole formations and extend discussion to the case of neutron star mergers.

The condition of hot and dense matter in supernova cores is extreme with respect to the one in our daily life, but is rather an extension from those of unstable nuclei. In fact, the density at the core bounce is just beyond the nuclear matter density and the proton fraction is around 0.3 as seen in Fig. 1. Therefore, the effort to obtain information extrapolated from experimental data of neutron rich nuclei is important. For example, there is a connection between the supernova conditions and the environment in heavy ion collisions [45, 46]. In the phase diagram of density-temperature plane, there is a region typically probed by heavy ion collisions near the boundary between nuclei and nucleonic gas or liquid. Supernova trajectories as in Fig. 1 go through near this boundary between nuclei and nucleonic gas [47], i.e. the region of cluster appearance.

When the temperature increases at a certain density, light clusters such as helium appear at the border before they are dissolved into nucleon gas from nuclei. This environment is similar to the one in heavy ion collisions. In supernova core, the light nuclei abundantly appear in such environment [48, 49]. By comparing the composition of dense matter from supernova EoS tables with the abundance of light clusters produced in heavy ion collisions, experimental data can be used to examine the composition of matter in supernovae [46]. Those light clusters including deuterons, tritons, ^{3}He and ^{4}He may also have impact on neutrino reactions in supernova cores. New channels of emission and absorption on light clusters may have influences on supernova explosions [50, 51, 52, 53].

In the case of black hole formation from very massive stars, conditions of dense matter attain very high density and temperature as seen in Fig. 2. Dynamical collapse to the black hole leads to extreme density beyond 10 times the nuclear matter density and temperature beyond 100 MeV. Therefore, exotic phase is expected at least just before the black hole formation. The trajectories of matter condition in collapsing dynamics move through interesting region of the QCD phase diagram. The thermal conditions may cross the phase boundary and/or the critical point depending on quark-hadron models [54].

In principle, any type of black hole forming phenomena such as collapsars for gamma ray bursts may have a chance to reach exotic phase. Density and temperature become extreme anyway just before the black hole formation because of dynamical collapse. One of such interesting scenarios is the black hole formation from a hypermassive object in merging binary neutron stars. The neutron star merger gets attention recently as a target of gravitational waves, which provides the patterns to probe the neutron star EoS, and a candidate of r-process nucleosynthesis. From inspiral motions of two neutron stars, the ensuing merging process may lead to the formation of a hypermassive neutron star, which is rapidly rotating, hot neutron star. This object eventually turns into the black hole formation. Therefore, the hypermassive neutron star is another interesting object to explore the exotic phase.

The environment inside the hypermassive object is fascinating for the exotic phase. Since the mass of the hypermassive neutron star is larger than the cold neutron star mass limit, the condition is more extreme than that of neutron stars. In the simulation of neutron star merger [55], for example, the hypermassive neutron star has the mass of $2.6M_\odot$, which is larger than the maximum mass of cold neutron stars, $2.4M_\odot$, using DD2 EoS table [28]. In Fig. 3, the condition of dense matter in the hypermassive neutron star [56] is shown in the plane of temperature and electron fraction versus density. The density at center reaches three times the nuclear matter density to support the very massive object and the temperature becomes high beyond 20 MeV at the nuclear matter density. This situation is more extreme than the one in the supernova core seen in Fig. 1. It is remarkable that the matter is neutron rich for the wide range of densities including the high density at the center because of the neutron-richness originated from cold neutron stars. This is clearly different from the condition inside the proto-neutron star, which is less neutron rich with the trapped neutrinos as in Figs. 1 and 2. Therefore, the extreme condition in the hypermassive object is advantageous to have the mixture of exotic matter with high neutron chemical potentials.

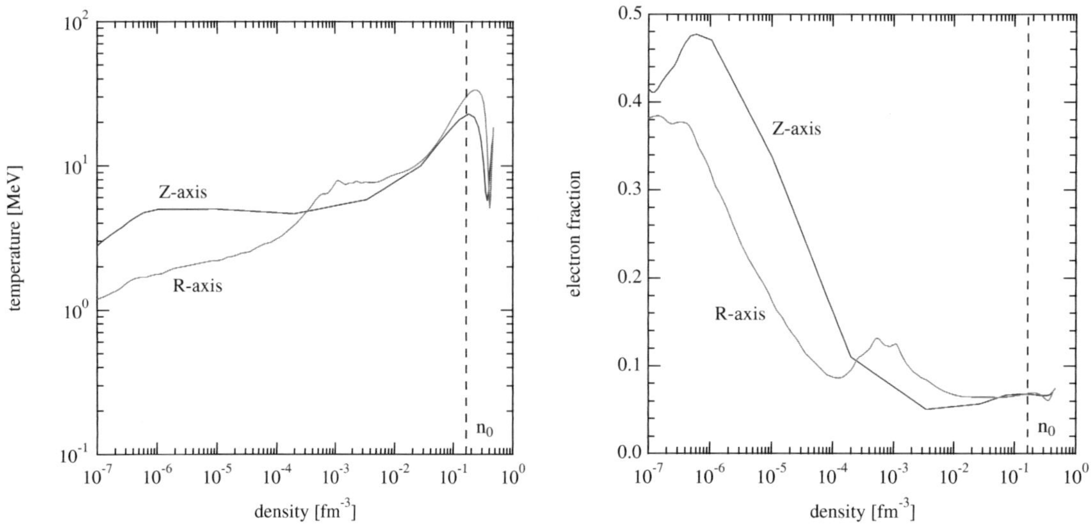

Figure 3. Conditions of dense matter in a hypermassive neutron star in the plane of temperature (left) and electron fraction (right) versus density. The snapshots of hypermassive neutron star are taken from the neutron star merger simulation [55, 56]. The conditions are shown for the fluid elements along the polar direction (Z-axis) and on the equatorial plane (R-axis).

Through comparisons of the three cases of matter conditions in Figs. 1–3, one can see the dynamical situations of supernovae and neutron stars are playgrounds with a variety of the combination of density, temperature and electron fraction, being different from those in static and cold neutron stars. The extreme conditions in the dynamics toward black hole formation, which contain very massive objects, are fascinating for exotic phases among them. The neutron rich situation in neutron star mergers is very favorable having neutron-richness, which enhances the appearance of new degrees of freedom. Appearance of the exotic phases can be a trigger of dynamical collapse in the evolution of massive objects and may provide observational signatures of neutrinos bursts and gravitational waves. In order to pursue these possibilities, it is essential to combine the knowledge of extreme conditions with the astrophysical simulations by bridging various research fields.

6. Summary

The supernova mechanism relies on the nuclear physics such as the equation of state and the neutrino reactions. This fact becomes more apparent since it is now possible to perform neutrino-radiation hydrodynamics with the 6D Boltzmann equation. Detailed descriptions of the neutrino transfer provide us with precise evaluation of the neutrino heating for the explosion mechanism. The equation of state is an important ingredient of physics to determine the outcome of the shock wave and the following evolution of compact objects. It also affects the neutrino bursts from the compact object born in the collapse of massive stars. The properties of neutrino signals are the valuable information to probe the EoS including possible appearance of exotic phases. The characteristics of neutrino burst in the supernova explosion and the black hole forming case are different each other in time duration and spectra, reflecting the dynamical evolution. Moreover, the neutrino signal at late stages may carry the information of appearance of exotics. Thermodynamic conditions of explosive phenomena may contain variety of extreme conditions and have a chance to access exotic phases especially at the moment toward the black hole formation. It is now important to merge efforts both from simulations and microphysics to explore interesting physics of quarks and hadrons in the QCD phase diagram.

Acknowledgments

This work is supported by the HPCI and Post-K projects by MEXT and JICFuS. I acknowledge the usage of K-computer (hp150225, hp160071, hp160211, hp170031, hp170230) and HPC resources at KEK (14/15-17, 15/16-08, 16/17-11), JLDG, YITP, UT and RCNP. This work is also supported by Grant in Aid for Scientific Research (24103006, 24105008, 26104006, 24244036, 15K05093, 16H03986).

References

[1] Kotake K, Sumiyoshi K, Yamada S, Takiwaki T, Kuroda T, Suwa Y and Nagakura H 2012 *Prog. Theor. Exp. Phys.* **2012** 01A301
[2] Janka H T, Melson T and Summa A 2016 *Ann. Rev. Nucl. Part. Sci.* **66** 341–375
[3] Bethe H A and Wilson J R 1985 *Astrophys. J.* **295** 14–23
[4] Bethe H A 1990 *Rev. Mod. Phys.* **62** 801
[5] Janka H and Müller E 1996 *Astron. Astrophys.* **306** 167
[6] Liebendörfer M, Mezzacappa A, Thielemann F, Messer O E, Hix W R and Bruenn S W 2001 *Phys. Rev. D* **63** 103004
[7] Sumiyoshi K, Yamada S, Suzuki H, Shen H, Chiba S and Toki H 2005 *Astrophys. J.* **629** 922
[8] Marek A and Janka H 2009 *Astrophys. J.* **694** 664
[9] Suwa Y, Kotake K, Takiwaki T, Whitehouse S C, Liebendörfer M and Sato K 2010 *Publ. Astron. Soc. Japan* **62** L49
[10] Takiwaki T, Kotake K and Suwa Y 2012 *Astrophys. J.* **749** 98
[11] Takiwaki T, Kotake K and Suwa Y 2014 *Astrophys. J.* **786** 83
[12] Lentz E J, Bruenn S W, Hix W R, Mezzacappa A, Messer O E B, Endeve E, Blondin J M, Harris J A, Marronetti P and Yakunin K N 2015 *Astrophys. J.* **807** L31
[13] Yamada S, Janka H T and Suzuki H 1999 *Astron. Astrophys.* **344** 533
[14] Sumiyoshi K and Yamada S 2012 *Astrophys. J. Suppl.* **199** 17
[15] Sumiyoshi K, Takiwaki T, Matsufuru H and Yamada S 2015 *Astrophys. J. Suppl.* **216** 5
[16] Nagakura H, Sumiyoshi K and Yamada S 2014 *Astrophys. J. Suppl.* **214** 16
[17] Nagakura H, Iwakami W, Furusawa S, Sumiyoshi K, Yamada S, Matsufuru H and Imakura A 2017 *Astrophys. J. Suppl.* **229** 42
[18] Sumiyoshi K, Nagakura H, Iwakami W, Furusawa S, Matsufuru H, Imakura A and Yamada S 2016 *Proceedings of the 14th International Symposium on Nuclei in the Cosmos XIV* ed Kubono S (Tokyo: JPS Conference Proceedings) in press
[19] Nagakura H, Iwakami W, Furusawa S, Okawa H, Harada A, Sumiyoshi K, Yamada S, Matsufuru H and Imakura A 2017 *ArXiv e-prints* (*Preprint* 1702.01752)
[20] Furusawa S, Sumiyoshi K, Yamada S and Suzuki H 2013 *Astrophys. J.* **772** 95
[21] Shen H, Toki H, Oyamatsu K and Sumiyoshi K 1998 *Nucl. Phys. A* **637** 435–450
[22] Shen H, Toki H, Oyamatsu K and Sumiyoshi K 1998 *Prog. Theor. Phys.* **100** 1013–1031
[23] Shen H, Toki H, Oyamatsu K and Sumiyoshi K 2011 *Astrophys. J. Suppl.* **197** 20
[24] Juodagalvis A, Langanke K, Hix W R, Martínez-Pinedo G and Sampaio J M 2010 *Nucl. Phys. A* **848** 454–478
[25] Lattimer J M and Swesty F D 1991 *Nucl. Phys.* **A535** 331
[26] Oertel M, Hempel M, Klähn T and Typel S 2016 *ArXiv e-prints* (*Preprint* 1610.03361)

[27] Sumiyoshi K, Suzuki H, Yamada S and Toki H 2004 *Nucl. Phys.* **A730** 227
[28] Hempel M, Fischer T, Schaffner-Bielich J and Liebendörfer M 2012 *Astrophys. J.* **748** 70
[29] Steiner A W, Hempel M and Fischer T 2013 *Astrophys. J.* **774** 17
[30] Sagert I, Fischer T, Hempel M, Pagliara G, Schaffner-Bielich J, Mezzacappa A, Thielemann F K and Liebendörfer M 2009 *Phys. Rev. Lett.* **102** 081101
[31] Fischer T, Sagert I, Pagliara G, Hempel M, Schaffner-Bielich J, Rauscher T, Thielemann F K, Käppeli R, Martínez-Pinedo G and Liebendörfer M 2011 *Astrophys. J. Suppl.* **194** 39
[32] Nakazato K, Sumiyoshi K and Yamada S 2013 *Astron. Astrophys.* **558** A50
[33] Janka H T 2012 *Annu. Rev. Nucl. Part. Sci.* **62** 407
[34] Suwa Y, Takiwaki T, Kotake K, Fischer T, Liebendörfer M and Sato K 2013 *Astrophys. J.* **764** 99
[35] Sumiyoshi K, Suzuki H and Toki H 1995 *Astron. Astrophys.* **303** 475
[36] Pons J A, Reddy S, Prakash M, Lattimer J M and Miralles J A 1999 *Astrophys. J.* **513** 780
[37] Suzuki H, Kogure H, Tomioka F, Sumiyoshi K, Yamada S and Shen H 2003 *Nucl. Phys. A* **718** 703
[38] Liebendörfer M, Messer O E B, Mezzacappa A, Bruenn S W, Cardall C Y and Thielemann F K 2004 *Astrophys. J. Suppl.* **150** 263
[39] Sumiyoshi K, Yamada S, Suzuki H and Chiba S 2006 *Phys. Rev. Lett.* **97** 091101
[40] Sumiyoshi K, Yamada S and Suzuki H 2007 *Astrophys. J.* **667** 382
[41] Sumiyoshi K, Yamada S and Suzuki H 2008 *Astrophys. J.* **688** 1176
[42] Nakazato K, Sumiyoshi K, Suzuki H and Yamada S 2010 *Phys. Rev. D* **81** 083009
[43] Nakazato K, Sumiyoshi K and Yamada S 2008 *Phys. Rev.* **D77** 103006
[44] Sumiyoshi K, Ishizuka C, Ohnishi A, Yamada S and Suzuki H 2009 *Astrophys. J.* **690** L43
[45] Qin L, Hagel K, Wada R, Natowitz J B, Shlomo S, Bonasera A, Röpke G, Typel S, Chen Z, Huang M, Wang J, Zheng H, Kowalski S, Barbui M, Rodrigues M R D, Schmidt K, Fabris D, Lunardon M, Moretto S, Nebbia G, Pesente S, Rizzi V, Viesti G, Cinausero M, Prete G, Keutgen T, El Masri Y, Majka Z and Ma Y G 2012 *Phys. Rev. Lett.* **108** 172701
[46] Hempel M, Hagel K, Natowitz J, Röpke G and Typel S 2015 *Phys. Rev. C* **91** 045805
[47] Buyukcizmeci N, Botvina A S, Mishustin I N, Ogul R, Hempel M, Schaffner-Bielich J, Thielemann F K, Furusawa S, Sumiyoshi K, Yamada S and Suzuki H 2013 *Nucl. Phys. A* **907** 13–54
[48] Sumiyoshi K and Röpke G 2008 *Phys. Rev. C* **77** 055804
[49] Arcones A, Martínez-Pinedo G, O'Connor E, Schwenk A, Janka H T, Horowitz C J and Langanke K 2008 *Phys. Rev. C* **78** 015806
[50] Nakamura S X, Sumiyoshi K and Sato T 2009 *Phys. Rev. C* **80** 035802
[51] Furusawa S, Nagakura H, Sumiyoshi K and Yamada S 2013 *Astrophys. J.* **774** 78
[52] Nasu S, Nakamura S X, Sumiyoshi K, Sato T, Myhrer F and Kubodera K 2015 *Astrophys. J.* **801** 78
[53] Fischer T, Martínez-Pinedo G, Hempel M, Huther L, Röpke G, Typel S and Lohs A 2016 *Euro. Phys. J. Web Conf.* vol 109 p 06002 (*Preprint* `arXiv:1512.00193`)
[54] Ohnishi A, Ueda H, Nakano T Z, Ruggieri M and Sumiyoshi K 2011 *Phys. Lett. B* **704** 284–290
[55] Sekiguchi Y, Kiuchi K, Kyutoku K and Shibata M 2015 *Phys. Rev. D* **91** 064059
[56] Fujibayashi S, Sekiguchi Y and Sumiyoshi K 2017 in preparation

IOP Conf. Series: Journal of Physics: Conf. Series **861** (2017) 012029 doi:10.1088/1742-6596/861/1/012029

Oscillation modes of strange quark stars with a strangelet crust

Jessica Asbell and Prashanth Jaikumar[1]

Department of Physics & Astronomy, California State University Long Beach, 1250 Bellflower Blvd., Long Beach, CA 90840 USA

E-mail: [1]`prashanth.jaikumar@csulb.edu`

Abstract. We study the non-radial oscillation modes of strange quark stars with a homogeneous core and a crust made of strangelets. Using a 2-component equation-of-state model (core+crust) for strange quark stars that can produce stars as heavy as 2 solar masses, we identify the high-frequency l=2 spheroidal (f, p) in Newtonian gravity, using the Cowling approximation. The results are compared to the case of homogeneous compact stars such as polytropic neutron stars, as well as bare strange stars. We find that the strangelet crust only increases very slightly the frequency of the spheroidal modes, and that Newtonian gravity overestimates the mode frequencies of the strange star, as is the case for neutron stars.

1. Introduction

The recent direct detection of gravitational waves from the inspiral and merger of a pair of massive Black Holes by Advanced LIGO heralds a new era of compact star observations [1, 2]. It is hoped that systematic analysis of Advanced LIGO data will be able to pin down the equation of state of dense matter relevant to compact stars. The merger and ring down phases of colliding neutron star and black-hole binaries [3, 4] are the strongest transient events, but they leave in their wake many different oscillations of the stellar fluid that carry information on the equation of state (EoS) of dense matter [5, 6]. The classification of these modes is a well-developed field, and their study is promising because the spectrum of such modes is connected to underlying stellar structure [7, 8]. The even-parity or spheroidal modes result from density and pressure perturbations to the star, while the odd-parity axial modes are non-trivial only for rotating stars. The spectrum of spheroidal f, p, g modes was recently studied by Lugones and Vasquez-Flores [9] for stars made of hadronic, self-bound and hybrid matter, in order to find discriminating features among them. Other works have also discussed the differences in mode frequencies between neutron stars and quark/hybrid stars [10, 11, 12, 13]. The work reported in these proceedings considers the effect of a crust made of non-superconducting quark matter on the spectrum of spheroidal modes in a strange star. Our 2-component model for strange stars includes a crust made of strangelets [14, 15] and therefore differs in an essential way from previous studies, where only homogeneous quark phases or mixed phases of quarks and hadronic matter [16] were considered.

IOP Conf. Series: Journal of Physics: Conf. Series **861** (2017) 012029 doi:10.1088/1742-6596/861/1/012029

2. The 2-Component Model for Strange Quark Stars

The term quark star (QS), or strange quark star (SQS), commonly refers to a compact star composed completely of self-bound strange quark matter, and is a possible consequence of the strange quark matter hypothesis [17]. One can then describe a compact star with an appropriate equation of state for deconfined quarks at high density, and determine if the oscillation spectra of such a star is different from that computed with a nuclear matter EoS.

2.1. Core EoS

The core is assumed to consist of homogeneous and unpaired charge neutral 3-flavor interacting quark matter, which we describe using the simple thermodynamic Bag model EoS [18] with $\mathcal{O}(m_s^4)$ corrections to account for the moderately heavy strange quark. Perturbative interactions to non-interacting quark matter [19] are absorbed into a parameter $(1 - a_4) \sim \mathcal{O}(\alpha_s^2) \approx 0.3$ as suggested in [18] to extend the applicability of the model to stars as heavy as $\approx 2M_\odot$. The core EoS is

$$P_{\rm core} = \frac{1}{3}(\epsilon - 4B) - \frac{m_s^2}{3\pi}\sqrt{\frac{\epsilon - B}{a_4}} + \frac{m_s^4}{12\pi^2}\left[2 - \frac{1}{a_4} + 3\ln\left(\frac{8\pi}{3m_s^2}\sqrt{\frac{\epsilon - B}{a_4}}\right)\right], \tag{1}$$

where ϵ is the energy density of homogeneous quark matter (also to $\mathcal{O}(m_s^4)$ in the Bag model) and B the Bag constant which we fix by requiring that the first-order transition between neutral quark matter and the vacuum (P=0) occur at a quark chemical potential μ_q=$\mu_{\rm crit} \leq 310$ MeV [15]. This ensures that the hypothesis of absolute stability for 3-flavor quark matter is not violated. Corrections due to the superconducting gap Δ can be included in the EoS as in [20], but are omitted here to avoid proliferation of parameters. In Fig.1, we show the mass-radius relationship for homogeneous quark stars for different $\mu_{\rm crit}$ values at m_s=100 MeV, with B determined from the criterion for absolute stability.

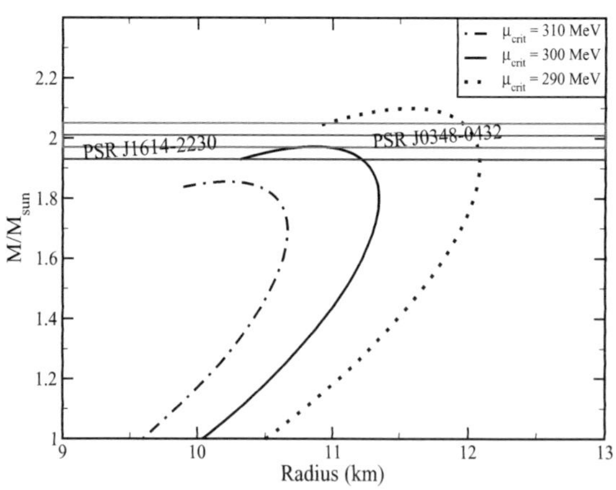

Figure 1. Mass-Radius relation for homogeneous quark stars described by the EoS (Eq. 1). The interaction parameter a_4=0.7 and m_s=100 MeV. Measured limits on the mass of PSR J1614-2230 (1.97±0.04 $M_\odot$) and PSR J0348-0432 (2.01±0.04 $M_\odot$) are shown.

Usually, theoretical models of quark stars consist of entirely homogeneous stars comprised of strange quark matter or a thin nuclear crust on top of a pure strange star. These cases predict steep drops in density as the surface is approached and the pressure goes to zero. They also predict large electric fields at the surface (or just below, in the case of a nuclear crust). The model we use differs from these in that it considers a heterogeneous crust on top of a fluid quark star. When the condition of local charge neutrality is relaxed, quarks and electrons can form a mixed phase. This model has the advantage that it reduces the density gradient near the crust and produces negligible electric field energy at the stars surface. However, there will be short-range (screened) electric fields inside the mixed phase, which can be tolerated if the Gibbs free energy is lowered sufficiently by forming a mixed phase. The thickness of such a crust would be small but on the same order as the crust of a neutron star, approximately 10^4 cm. This amounts to no more than 1% of the stellar radius.

2.2. Crust EoS

We assume that the crust is in a non-superconducting, globally neutral mixed phase of strangelets and electrons, so we can choose the configuration described in [14] which has ungapped 3-flavor quark matter with massive strange quark. The composition of the crust changes with depth as the quark phase fraction x increases from zero at the surface to one in the homogeneous phase. Phase coexistence with stable strangelets requires that (in the absence of surface tension) the pressure inside and outside the strangelet be the same, hence the quark pressure $P_q(\mu_q, \tilde{\mu}_e)$ in the mixed phase is zero, and the pressure is only due to electrons which depends on the electron chemical potential $\tilde{\mu}_e$ as

$$P_{\text{crust}} = \frac{\tilde{\mu}_e^4}{12\pi^2}, \quad \tilde{\mu}_e = \frac{n_Q}{\chi_Q}\left(1-\sqrt{1-\xi}\right), \quad \xi = \frac{2P_0\chi_Q}{n_Q^2} \tag{2}$$

where $n_Q(\mu_q, m_s)$ and $\chi_Q(\mu_q, m_s)$ represent the quark charge and quark susceptibility (both slowly varying in the mixed phase) and P_0 is the pressure of quarks without electrons, which varies considerably in the mixed phase. These generic relations can be applied to a specific model of quark matter, so long as we work to second order in the (small) electron chemical potential. This is a well-justified and useful approximation since $\tilde{\mu}_e/\mu_q \approx 0.05$. Just as for the core, we adopt the Bag model EoS to describe the crust strangelets with $\mathcal{O}(m_s^4)$ corrections, from which it follows that to the same order: $n_Q(\mu_q) = m_s^2\mu_q/(2\pi^2)$ and $\chi_Q(\mu_q) = (2\mu_q^2)/(\pi^2)$. Homogeneous quark matter gives way to the mixed phase crust at a radius $r{=}r_c$ where $\xi|_{r=r_c}{=}1$. The models with crust have their surface at $\mu_q = \mu_{\text{crit}}$, so they have very nearly the same mass and radius as the bare strange star models. A sketch of the constituents in the mixed phase is given in Fig.2.

3. Method for Computing Oscillation Spectra

In order to perform a perturbative analysis of compact star oscillations, we first must model their equilibrium structure. The Tolman-Oppenheimer-Volkoff (TOV) equations are used to provide the equilibrium structure of a compact star. We approach the non-radial oscillation problem in Newtonian gravity and employ the Cowling approximation, which allows one to neglect perturbations in the gravitational potential [21]. Under the Cowling approximation, the metric is stationary and matter is decoupled from space-time. While it is impossible to generate gravitational waves in Newtonian gravity, this approximation is justified as a first approximation since it simplifies the system considerably. However, the f and p mode frequencies computed from general relativistic calculations may differ significantly. To describe the fluid pulsations, we choose a reference frame in which the unperturbed fluid velocity is zero and search for oscillating solutions by writing the displacement as $\vec{u} = \vec{\xi}(\mathbf{x})e^{i\sigma t}$ which describes a wave with frequency σ

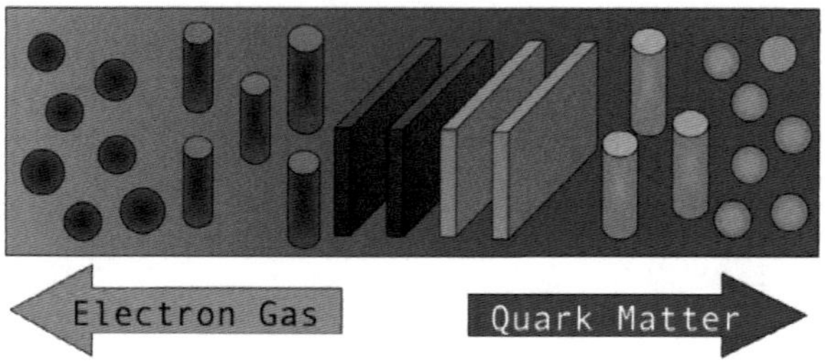

Figure 2. Diagram of a mixed-phase crust with net-positive nuggets of quark matter in a neutralizing electron background (voids). Pressure and density increase toward the center of the star, causing the nuggets to deform and fuse, and nuggets of negatively charged electrons embed in a positive quark background (voids). Deeper into the star, at very high pressure, quark matter becomes homogeneous.

and amplitude $\vec{\xi}(\mathbf{x})$. In spherical coordinates we make the radial and angular dependencies of the displacement explicit $\vec{\xi}(\mathbf{x}) = (\xi_r, \xi_\theta, \xi_\phi)$ and identify the fluid pulsation variables

$$\xi_r = U(r)Y_{lm}\,, \quad \xi_\theta = V(r)\frac{\partial Y_{lm}}{\partial \theta}\,, \quad \xi_\phi = \frac{V(r)}{\sin\theta}\frac{\partial Y_{lm}}{\partial \phi} \tag{3}$$

where $U(r)$ and $V(r)$ are the radial and transverse displacement eigenfunctions. In general, the perturbation equations we solve for a 1-component (homogeneous) star are [22]:

$$(1+\widetilde{V})\frac{dy_1}{dx} = \left(\frac{\widetilde{V}}{\Gamma} - 3\right) y_1 + \left(\frac{l(l+1)}{c_1\Omega^2} - \frac{\widetilde{V}}{\Gamma}\right) y_2 \tag{4}$$

$$(1+\widetilde{V})\frac{dy_2}{dx} = (c_1\Omega^2 + \mathcal{A}r)y_1 + (1 - \widetilde{U} - \mathcal{A}r)y_2 \tag{5}$$

with the following definitions: $y_1 = \frac{U(r)}{r}$, $y_2 = \frac{\delta P}{\rho g r}$, $\widetilde{V} = \frac{\rho g r}{p}$, $\widetilde{U} = \frac{d\ln(M_r)}{d\ln(r)}$, $c_1 = \left(\frac{r}{R_\star}\right)^3 \frac{M_\star}{M_r}$, $\Omega^2 = \frac{\sigma^2 R_\star^3}{M_\star}$ where δP is the Eulerian pressure perturbation, M_r is the mass contained in radius r, $M_\star$ and $R_\star$ are the mass and radius of the star and $\mathcal{A}$ is the Schwarzschild gradient for local convective stability. In our analysis, we set $\mathcal{A}$=0 for the bare strange star, although finite temperature effects or composition gradients can change the situation. This system of pulsation equations is solved numerically to find the oscillation eigenfrequencies (σ) that obey the boundary conditions that solutions y_1 and y_2 are regular at the origin (r=0) and the Lagrangian variation of the pressure vanishes at the surface (r=$R_\star$). In addition, the solutions are normalized such that the relative radial displacement $y_1(R_\star)$=1. For a 2-component star, the crust eigenfunctions are connected to those in the core through suitable continuity conditions. The complete and lengthier set of 4 equations (2 for the fluid displacement and 2 for the tractions) and the relevant boundary conditions are written down in [22, 23], so we do not reproduce them here. We use these to obtain results for the 2-component strange star with a crust.

4. Results: Oscillation Spectra for Strange Quark Stars

In Table 1, we show the mode period values for the l=2 f, p and core g modes for a neutron star modeled as a single component, fluid, non-rotating core described by a polytropic EoS. We consider three different values of the polytropic index (n=1.0, 1.5, 2.0), keeping stellar mass and radius fixed at M=1.4$M_\odot$ and R=10 km. Mode frequencies and periods are related in the usual way, f=1/Π. The f modes are short period, corresponding to a few kHz and the p modes are higher in frequency still. The core g modes are in the range of a few Hz to few tens of Hz. This is as analytically predicted by Cox [24] and the trend with increasingly relativistic models is as found by Andersson & Kokkotas [25] and McDermott et al [22], although their exact values are somewhat different due to different choices of neutron star model. For neutron stars, the f mode frequency increases as the square root of the mean density as expected, and hence also stellar mass.

Table 1. The f, p, g oscillation mode frequencies for a single-component neutron star are listed below. n refers to the polytropic index and azimuthal mode number $l = 2$. The mass is fixed at 1.4 $M_\odot$ and radius at 10 km in all cases.

$l = 2$	$f-$mode [kHz]	p_1-mode [kHz]	g_1-mode [Hz]
$n = 1.0$	3.59	8.23	4.63
$n = 1.5$	4.28	9.85	13.4
$n = 2.0$	5.11	11.5	24.4

Next we compute the f and p modes for a homogeneous quark star having an EoS defined as in Eq. 1, and study their dependence on the stellar mass. The mass of the strange quark, the value of the critical quark chemical potential, and the interaction parameters a_4 are tunable features in the equation of state. Although the oscillation spectra are normally shown as a function of stellar mass for fixed B (by changing central density), one should note that the Bag "constant" is really an effective parameter of the model, and not an actual physical quantity. We therefore keep the value of quark mass fixed at 100 MeV and hold the central density constant, but vary the value of the Bag constant and $\mu_{\rm crit}$ self-consistently from the equation of state to generate different mass configurations. In Fig.3, we show how the stellar mass varies with the Bag constant. As seen in Fig.5, the trend of the f modes is that as the bag constant is decreased (i.e, stellar mass increased), the frequencies gradually decrease, suggesting that the f modes scale with the Bag constant rather than stellar mass. The f mode lies in the 4-5 kHz range while the p1-mode lies in the 25-35 kHz range for $M \sim (1.2 - 2)M_\odot$. This range of frequencies for the f and p mode of strange quark stars is higher than found in the work of [9] who used general relativistic fluid equations, so the differences could be attributed to our Newtonian approximations to the fluid equations, since their results for hadronic stars also show a similar difference from Newtonian models. In addition, each pair of data corresponding to a fixed mass in Fig.5 represents a different EoS within the same family parametrized by Eq. 1, whereas the results of [9] were plotted against mass keeping the EoS the same. The Newtonian approximation to the oscillation equations appears to yield mode frequencies for strange stars similar to the n=2 polytropic neutron star (see Table 1), reflecting the fact that strange stars are more compact in comparison to the average neutron star. Furthermore, the longitudinal sound speed in quark stars is higher. These properties of the quark star are reflected in the slightly higher f and p-mode frequencies compared to, say, the n=1 polytrope for neutron stars.

The oscillation spectra for the two-component quark star (core+crust) was computed taking into account the shear modulus of the strangelet crust, which is estimated in [26]. The effect of the crust on the f modes is only a few Hz, since the mean density is hardly affected by adding the crust on top of the bare strange star. The p modes are modified to a slightly larger extent than the f mode, due to the shear modulus of the crust, but since the shear speed is numerically small compared to the longitudinal sound speed, the effect is not large (about 2% change). An important EoS parameter for the strangelet crust's properties is the strange quark

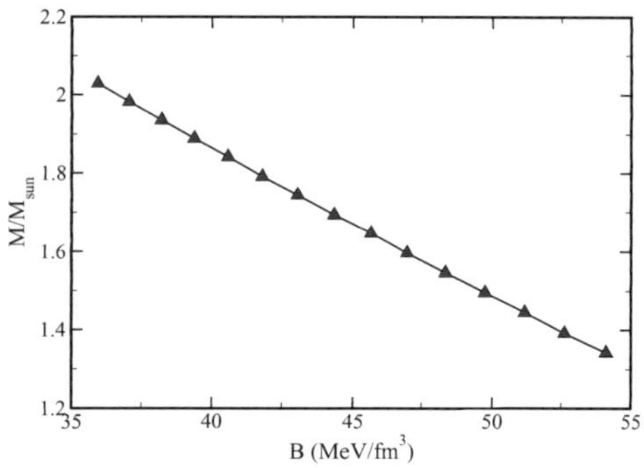

Figure 3. Stellar mass vs Bag constant B from Eq.1 for fixed central density of 3 x nuclear saturation density. $\mu_{\rm crit}$ is varied between 280 MeV to 320 MeV consistent with B, though values above $\mu_{\rm crit}$=310 MeV are excluded due to the requirement of self-bound quark matter.

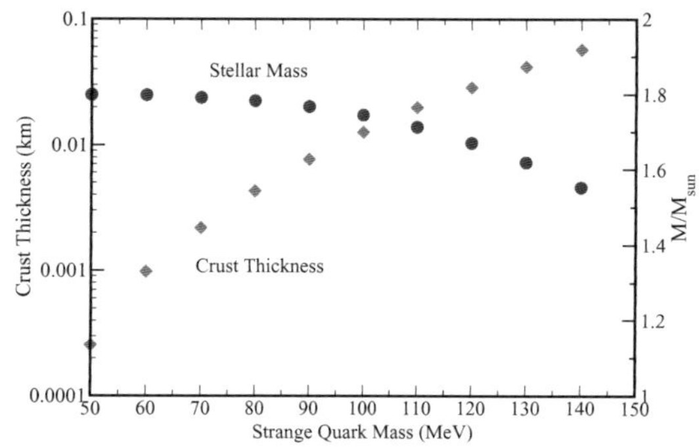

Figure 4. Crust thickness and stellar mass as a function of strange quark mass. The strangelet crust is constructed as a globally charge neutral phase atop the homogeneous phase which has zero pressure at $\mu_{\rm crit}$=310 MeV.

mass. In Fig. 4, we show how the crust thickness depends on the strange quark mass for a fixed $\mu_{\rm crit}$, along with the corresponding stellar mass. In Fig. 6, we show the f and p-mode of the 2-component strange star with varying stellar mass keeping $\mu_{\rm crit}$ (and hence also the Bag constant) fixed, but changing the central density. The mode frequencies now have a similar trend to [9], although the values differ due to reasons mentioned previously.

5. Conclusions

In this proceedings article, we have presented some preliminary work on the spheroidal non-radial oscillation modes of strange quark stars that are composed of a homogeneous core and a crust of strangelets in a globally neutral phase (Gibbs construction). In the Newtonian approximation to the fluid equations, and assuming the fluid perturbations obey the same equation of state as the background fluid, the f modes are in the range of 4-5 kHz while the p-modes are in the range of 25-35 kHz. Due to the small size of the crust and its low shear modulus, the impact on the modes from the case of a bare strange star is minimal. However, an extreme choice of parameters in the quark model can lead to a much thicker crust (hundreds of meters), and this affects the mode frequencies at almost the 10% level. Including a distinct EoS for the perturbations, considering a crystalline superconducting crust, rotation effects, or high temperature can all change the results presented here. In this work, we have not presented the results for the shear modes (s modes) or the interfacial modes (i modes) that arise as a result of the mixed phase crust, and which would have much smaller frequencies than the f, p modes. To resolve these modes, one needs a qualitatively new set of boundary conditions at the crust-core interface, since the energy density, and consequently the EoS varies sharply in that region. These modes are typically expected to lie between the f, p modes and the core g modes (few Hz to tens of Hz). Those findings and their comparison to the case of a neutron star with a crust or strange star with a nuclear crust, will be reported separately. The Newtonian approximation also appears to overestimate the mode frequencies when compared to including the perturbations in the gravitational potential [27]. A full general relativistic treatment along the lines in [27, 28] is currently underway, which

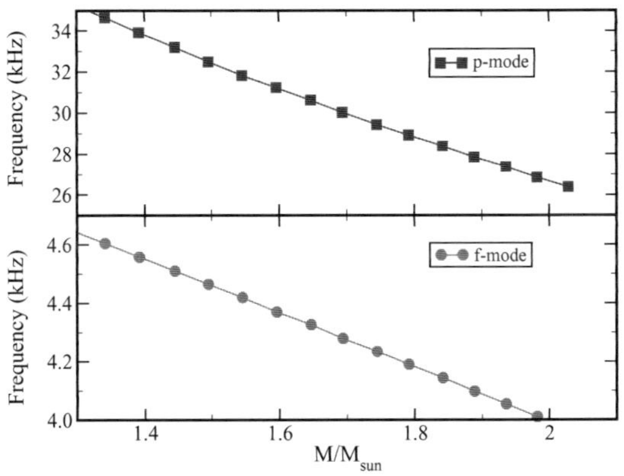

Figure 5. The f mode and first p mode frequencies for homogeneous strange quark stars with m_s=100 MeV and $\mu_{\rm crit}$ varied self-consistently with B. Each pair of f, p modes corresponding to a fixed mass represents a different EoS within the same family parametrized by Eq. 1.

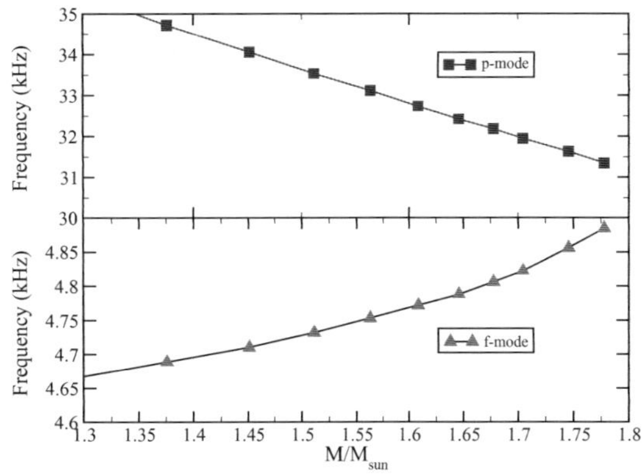

Figure 6. The f mode and first p mode frequencies for 2-component strange quark stars (core+strangelet crust) with m_s=100 MeV and $\mu_{\rm crit}$=310 MeV. Bag constant is held fixed as the central density is varied.

will yield the damping time and the amplitude of the modes for generating templates of the gravitational wave signal from pulsating strange stars.

Acknowledgments

P J thanks the organizers of the CSQCD V conference (Compact Stars in the QCD Phase Diagram V) held at the Gran Sasso Science Institute in L'Aquila, Italy, for their kind hospitality, and would also like to acknowledge fruitful discussions with G Rupak (Mississippi State University, USA) and C Vásquez-Flores (Universidade Federal do ABC, Brazil).

P J is supported by the U.S. National Science Foundation under Grant No. PHY 1608959.
J. A. acknowledges financial support from the Office of Research & Sponsored Programs at the California State University Long Beach.

References

[1] Abbott B P *et al.* (Virgo, LIGO Scientific) 2016 *Phys. Rev. Lett.* **116** 061102 (*Preprint* 1602.03837)
[2] Abbott B P *et al.* (Virgo, LIGO Scientific) 2016 *Phys. Rev. Lett.* **116** 241103 (*Preprint* 1606.04855)
[3] Creighton T 2003 *Class. Quant. Grav.* **20** S853–S869
[4] Bauswein A, Janka H T, Hebeler K and Schwenk A 2012 *Phys. Rev.* **D86** 063001 (*Preprint* 1204.1888)
[5] Kokkotas K D and Schmidt B G 1999 *Living Rev. Rel.* **2** 2 (*Preprint* gr-qc/9909058)
[6] Kokkotas K D, Apostolatos T A and Andersson N 2001 *Mon. Not. Roy. Astron. Soc.* **320** 307–315 (*Preprint* gr-qc/9901072)
[7] Pena-Garay C and Serenelli A 2008 (*Preprint* 0811.2424)
[8] Sotani H, Nakazato K, Iida K and Oyamatsu K 2012 *Phys. Rev. Lett.* **108** 201101 (*Preprint* 1202.6242)
[9] Flores C V and Lugones G 2014 *Class. Quant. Grav.* **31** 155002 (*Preprint* 1310.0554)
[10] Kojima Y and Sakata K 2002 *Prog. Theor. Phys.* **108** 801–808 (*Preprint* astro-ph/0209320)
[11] Moraes P H R S and Miranda O D 2014 *Mon. Not. Roy. Astron. Soc.* **445** L11–L15 (*Preprint* 1408.0929)
[12] Chatziioannou K, Yagi K, Klein A, Cornish N and Yunes N 2015 *Phys. Rev.* **D92** 104008 (*Preprint* 1508.02062)

[13] Fu W, Bai Z and Liu Y 2017 (*Preprint* 1701.00418)
[14] Jaikumar P, Reddy S and Steiner A W 2006 *Phys. Rev. Lett.* **96** 041101 (*Preprint* nucl-th/0507055)
[15] Alford M G, Rajagopal K, Reddy S and Steiner A W 2006 *Phys. Rev.* **D73** 114016 (*Preprint* hep-ph/0604134)
[16] Sotani H, Maruyama T and Tatsumi T 2013 *Nucl. Phys.* **A906** 37–49 (*Preprint* 1207.4055)
[17] Witten E 1984 *Phys. Rev. D* **30** 272–285
[18] Alford M, Braby M, Paris M W and Reddy S 2005 *Astrophys. J.* **629** 969–978 (*Preprint* nucl-th/0411016)
[19] Fraga E S, Pisarski R D and Schaffner-Bielich J 2001 *Phys. Rev.* **D63** 121702 (*Preprint* hep-ph/0101143)
[20] Rupak G and Jaikumar P 2013 *Phys. Rev.* **C88** 065801 (*Preprint* 1209.4343)
[21] Kastaun W 2008 *Phys. Rev.* **D77** 124019 (*Preprint* 0804.1151)
[22] McDermott P N, van Horn H M and Hansen C J 1988 *Astrophys. J.* **325** 725–748
[23] Yoshida S and Lee U 2002 *Astron. Astrophys.* **395** 201–208 (*Preprint* astro-ph/0210591)
[24] Cox J P 1980 **ed The Princeton University Press, New Jersey**
[25] Andersson N and Kokkotas K D 1998 *Mon. Not. Roy. Astron. Soc.* **299** 1059–1068 (*Preprint* gr-qc/9711088)
[26] Watts A L and Reddy S 2007 *Mon. Not. Roy. Astron. Soc.* **379** L63 (*Preprint* astro-ph/0609364)
[27] Sotani H and Harada T 2003 *Phys. Rev.* **D68** 024019 (*Preprint* gr-qc/0307035)
[28] Benhar O, Ferrari V and Gualtieri L 2004 *Phys. Rev.* **D70** 124015 (*Preprint* astro-ph/0407529)

AUTHOR INDEX